CHILTON'S
REPAIR & TUNE-UP GUIDE

MW01469147

GM C-BODY
1985

All U.S. and Canadian models of front-wheel drive
BUICK Electra Limited, Electra T-Type, Park Avenue •
OLDSMOBILE 98 Regency, 98 Regency Brougham •
CADILLAC Fleetwood, DeVille

President LAWRENCE A. FORNASIERI
Vice President and General Manager JOHN P. KUSHNERICK
Executive Editor KERRY A. FREEMAN, S.A.E.
Senior Editor RICHARD J. RIVELE, S.A.E.
Editor RICHARD J. RIVELE, S.A.E.

CHILTON BOOK COMPANY
Radnor, Pennsylvania
19089

SAFETY NOTICE

Proper service and repair procedures are vital to the safe, reliable operation of all motor vehicles, as well as the personal safety of those performing repairs. This book outlines procedures for servicing and repairing vehicles using safe, effective methods. The procedures contain many NOTES, CAUTIONS and WARNINGS which should be followed along with standard safety procedures to eliminate the possibility of personal injury or improper service which could damage the vehicle or compromise its safety.

It is important to note that repair procedures and techniques, tools and parts for servicing motor vehicles,a s well as the skill and experience of the individual performing the work vary widely. It is not possible to anticipate all of the conceivable ways or conditions under which vehicles may be serviced, or to provide cautions as to all of the possible hazards that may result. Standard and accepted safety precautions and equipment should be used when handling toxic or flammable fluids, and safety goggles or other protection should be used during cutting.

Some procedures require the use of tools specially designed for a specific purpose. Before substituting another tool or procedure, you must be completely satisfied that neither your personal safety, nor the performance of the vehicle will be endangered.

Although information in this guide is based on industry sources and is as complete as possible at the time of publication, the possibility exists that the manufacturer made later changes which could not be included here. While striving for total accuracy, Chilton Book Company cannot assume responsibility for any errors, changes, or omissions that may occur in the compilation of this data.

PART NUMBERS

Part numbers listed in this reference are not recommendations by Chilton for any product by brand name. They are references that can be used with interchange manuals and aftermarket supplier catalogs to locate each brand supplier's discrete part number.

SPECIAL TOOLS

Special tools are recommended by the vehicle manufacturer to perform their specific job. Use has been kept to a minimum, but where absolutely necessary, they are referred to in the text by the part number of the tool manufacturer. These tools can be purchased, under the appropriate part number, from the Service Tool Division, Kent-Moore Corporation, 29784 Little Mack, Roseville, MI. 48066-2290, or an equivalent tool can be purchased locally from a tool supplier or parts outlet. Before substituting any tool for the one recommended, read the SAFETY NOTICE at the top of this page.

ACKNOWLEDGMENTS

The Chilton Book Company expresses its appreciation to the General Motors Corporation, Detroit, Michigan for their generous assistance.

Information ahs been selected from shop manuals, owners manuals, service bulletins and technical training manuals.

Copyright © 1985 by Chilton Book Company
All Rights Reserved
Published in Radnor, Pennsylvania 19089, by Chilton Book Company

Manufactured in the United States of America
234566789   432109876

Chilton's Repair & Tune-Up Guide: GM C-Body 1985
ISBN 0-8019-7587-5 pbk
Library of Congress Catalog Card No. 84-45468

CONTENTS

Quick Reference Specifications For Your Vehicle

Fill in this chart with the most commonly used specifications for your vehicle. Specifications can be found in Chapters 1 through 3 or on the tune-up decal under the hood of the vehicle.

 **Tune-Up**

Firing Order_____

Spark Plugs:

    Type_____

    Gap (in.)_____

Torque (ft. lbs.)_____

Idle Speed (rpm)_____

Ignition Timing (°)_____

    Vacuum or Electronic Advance (Connected/Disconnected)_____

Valve Clearance (in.)

    Intake_____   Exhaust_____

 **Capacities**

Engine Oil Type (API Rating)_____

    With Filter Change (qts)_____

    Without Filter Change (qts)_____

Cooling System (qts)_____

Manual Transmission (pts)_____

    Type_____

Automatic Transmission (pts)_____

    Type_____

Front Differential (pts)_____

    Type_____

Rear Differential (pts)_____

    Type_____

Transfer Case (pts)_____

    Type_____

FREQUENTLY REPLACED PARTS

Use these spaces to record the part numbers of frequently replaced parts.

PCV VALVE	OIL FILTER	AIR FILTER	FUEL FILTER
Type_____	Type_____	Type_____	Type_____
Part No._____	Part No._____	Part No._____	Part No._____

General Information and Maintenance

HOW TO USE THIS BOOK

Chilton's Repair & Tune-Up Guide for General Motors C-Body cars is intended to help you learn more about the inner workings of your car and save you money on its upkeep and operation.

The first two chapters will be the most used, since they contain maintenance and tune-up information and procedures. Studies have shown that a properly tuned and maintained car can get at least 10% better gas mileage (which translates into lower operating costs) and periodic maintenance will catch minor problems before they turn into major repair bills. The other chapters deal with the more complex systems of your car. Operating systems from engine through brakes are covered to the extent that the average do-it-yourselfer becomes mechanically involved. This book will not explain such things as rebuilding the differential for the simple reason that the expertise required and the investment in special tools make this task uneconomical. It will give you the detailed instructions to help you change your own brake pads and shoes, tune-up the engine, replace spark plugs and filters, and do many more jobs that will save you money, give you personal satisfaction and help you avoid expensive problems.

A secondary purpose of this book is a reference guide for owners who want to understand their car and/or their mechanics better. In this case, no tools at all are required. Knowing just what a particular repair job requires in parts and labor time will allow you to evaluate whether or not you're getting a fair price quote and help decipher itemized bills from a repair shop.

Before attempting any repairs or service on your car, read through the entire procedure outlined in the appropriate chapter. This will give you the overall view of what tools and supplies will be required. There is nothing more frustrating than having to walk to the bus stop on Monday morning because you were short one gasket on Sunday afternoon. So read ahead and plan ahead. Each operation should be approached logically and all procedures thoroughly understood before attempting any work. Some special tools that may be required can often be rented from local automotive jobbers or places specializing in renting tools and equipment. Check the yellow pages of your phone book.

All chapters contain adjustments, maintenance, removal and installation procedures, and overhaul procedures. When overhaul is not considered practical, we tell you how to remove the failed part and then how to install the new or rebuilt replacement. In this way, you at least save the labor costs. Backyard overhaul of some components (such as the alternator or water pump) is just not practical, but the removal and installation procedure is often simple and well within the capabilities of the average car owner.

Two basic mechanic's rules should be mentioned here. First, whenever the LEFT side of the car or engine is referred to, it is meant to specify the DRIVER side of the car. Conversely, the RIGHT side of the car means the PASSENGER side. Second, all screws and bolts are removed by turning counterclockwise, and tightened by turning clockwise.

Safety is always the most important rule. Constantly be aware of the dangers involved in working on or around an automobile and take proper precautions to avoid the risk of personal injury or damage to the vehicle. See the section in this chapter "Servicing Your Vehicle Safely" and the SAFETY NOTICE on the acknowledgment page before attempting any service procedures and pay attention to the instructions provided. There are 3 common mistakes in mechanical work:

1. Incorrect order of assembly, disassembly or adjustment. When taking something apart or putting it together, doing things in the wrong order usually just costs you extra time; however it CAN break something. Read the entire procedure before beginning disassembly. Do everything in the order in which the instructions say you should do it, even if you can't immediately see a reason for it. When you're taking apart something that is very intricate (for example a carburetor), you might want to draw a picture of how it looks when assembled at one point in order to make sure you get everything back in its proper position. We will supply exploded views whenever possible, but sometimes the job requires more attention to detail than an illustration provides. When making adjustments (especially tune-up adjustments), do them in order. One adjustment often affects another and you cannot expect satisfactory results unless each adjustment is made only when it cannot be changed by any other.

2. Overtorquing (or undertorquing) nuts and bolts. While it is more common for overtorquing to cause damage, undertorquing can cause a fastener to vibrate loose and cause serious damage, especially when dealing with aluminum parts. Pay attention to torque specifications and utilize a torque wrench in assembly. If a torque figure is not available remember that, if you are using the right tool to do the job, you will probably not have to strain yourself to get a fastener tight enough. The pitch of most threads is so slight that the tension you put on the wrench will be multiplied many times in actual force on what you are tightening. A good example of how critical torque is can be seen in the case of spark plug installation, especially where you are putting the plug into an aluminum cylinder head. Too little torque can fail to crush the gasket, causing leakage of combustion gases and consequent overheating of the plug and engine parts. Too much torque can damage the threads or distort the plug, which changes the spark gap at the electrode. Since more and more manufacturers are using aluminum in their engine and chassis parts to save weight, a torque wrench should be in any serious do-it-yourselfer's tool box.

There are many commercial chemical products available for ensuring that fasteners won't come loose, even if they are not torqued just right (a very common brand is Loctite®). If you're worried about getting something together tight enough to hold, but loose enough to avoid mechanical damage during assembly, one of these products might offer substantial insurance. Read the label on the package and make sure the product is compatible with the materials, fluids, etc. involved before choosing one.

3. Crossthreading. This occurs when a part such as a bolt is screwed into a nut or casting at the wrong angle and forced, causing the threads to become damaged. Crossthreading is more likely to occur if access is difficult. It helps to clean and lubricate fasteners, and to start threading with the part to be installed going straight in, using your fingers. If you encounter resistance, unscrew the part and start over again at a different angle until it can be inserted and turned several times without much effort. Keep in mind that many parts, especially spark plugs, use tapered threads so that gentle turning will automatically bring the part you're threading to the proper angle if you don't force it or resist a change in angle. Don't put a wrench on the part until it's been turned in a couple of times by hand. If you suddenly encounter resistance and the part has not seated fully, don't force it. Pull it back out and make sure it's clean and threading properly.

Always take your time and be patient; once you have some experience, working on your car will become an enjoyable hobby.

TOOLS AND EQUIPMENT

Naturally, without the proper tools and equipment it is impossible to properly service your vehicle. It would be impossible to catalog each tool that you would need to perform each or every operation in this book. It would also be unwise for the amateur to rush out and buy an expensive set of tools an the theory that he may need one or more of them at sometime.

The best approach is to proceed slowly, gathering together a good quality set of those tools that are used most frequently. Don't be misled by the low cost of bargain tools. It is far better to spend a little more for better quality. Forged wrenches, 10 or 12 point sockets and fine tooth ratchets are by far preferable to their less expensive counterparts. As any good mechanic can tell you, there are few worse experiences than trying to work on a car or truck with bad tools. Your monetary savings will be far outweighed by frustration and mangled knuckles.

Begin accumulating those tools that are used most frequently; those associated with routine maintenance and tune-up.

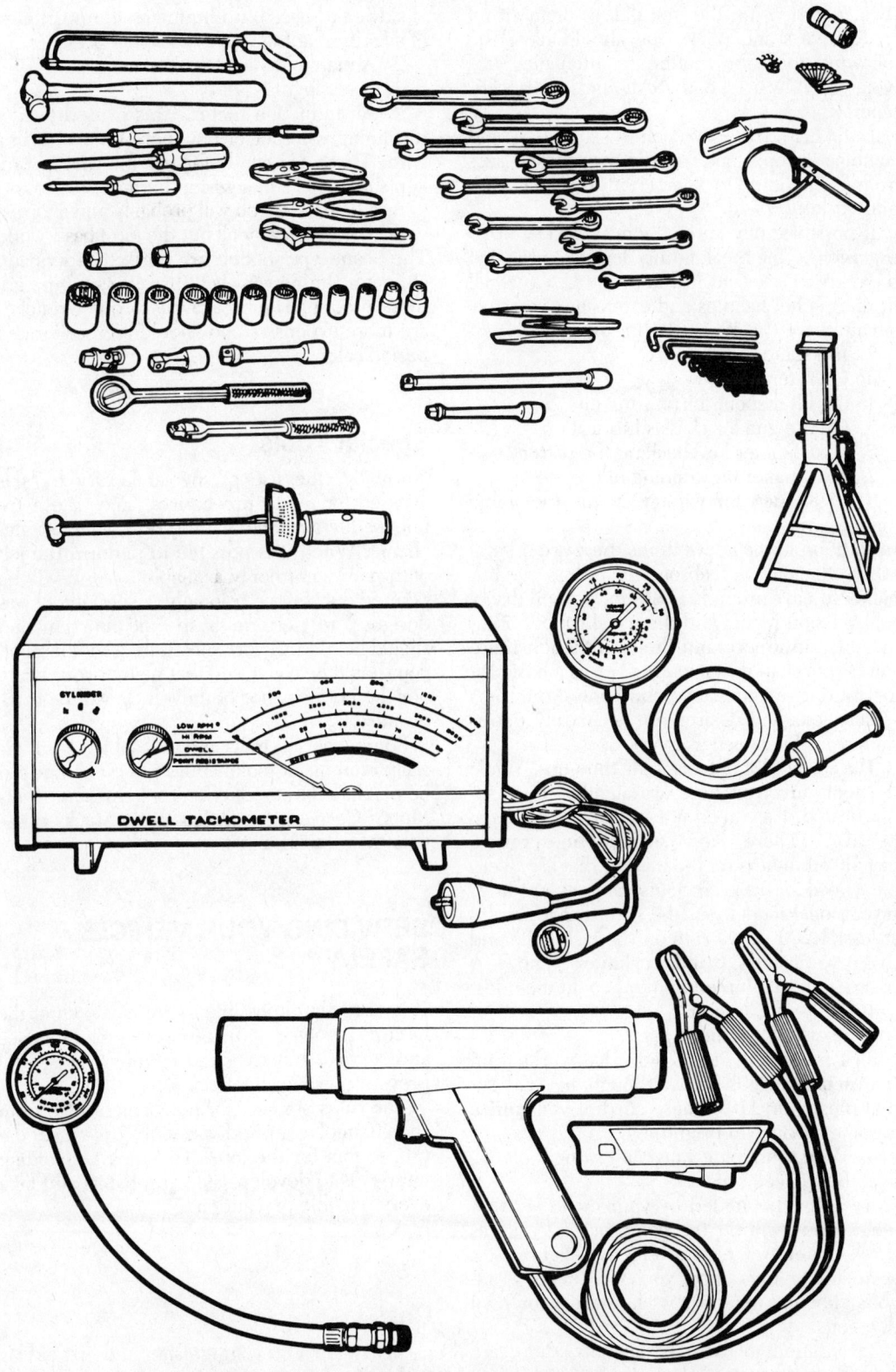

This basic collection of hand tools will handle most service needs

In addition to the normal assortment of screwdrivers and pliers you should have the following tools for routine maintenance jobs (your C-body uses both SAE and metric fasteners):

1. SAE/Metric wrenches—sockets and combination open end/box end wrenches in sizes from ⅛ in. (3mm) to ¾ in. (19mm); and a spark plug socket (¹³⁄₁₆ in.)

If possible, buy various length socket drive extensions. One break in this department is that the metric sockets available in the U.S. will all fit the ratchet handles and extensions you may already have (¼, ⅜, and ½ in. drive).

2. Jackstands for support
3. Oil filter wrench
4. Oil filter spout for pouring oil
5. Grease gun for chassis lubrication
6. Hydrometer for checking the battery
7. A container for draining oil
8. Many rags for wiping up the inevitable mess.

In addition to the above items there are several others that are not absolutely necessary, but handy to have around. These include oil dry, a transmission funnel and the usual supply of lubricants, antifreeze and fluids, although these can be purchased as needed. This is a basic list for routine maintenance, but only your personal needs and desires can accurately determine your list of necessary tools.

The second list of tools is for tune-ups. While the tools involved here are slightly more sophisticated, they need not be outrageously expensive. There are several inexpensive tach/dwell meters on the market that are every bit as good for the average mechanic as a $100.00 professional model. Just be sure that it goes to at least 1,200–1,500 rpm on the tach scale and that it works on 4, 6 and 8 cylinder engines. A basic list of tune-up equipment could include:

1. Tach-dwell meter
2. Spark plug wrench
3. Timing light (a DC light that works from the car's battery is best, although an AC light that plugs into 110V house current will suffice at some sacrifice in brightness)
4. Wire spark plug gauge/adjusting tools
5. Set of feeler blades.

Here again, be guided by your own needs. A feeler blade will set the point gap as easily as dwell meter will read dwell, but slightly less accurately. And since you will need a tachometer anyway . . . well, make your own decision.

In addition to these basic tools, there are several other tools and gauges you may find useful. These include:

1. A compression gauge. The screw-in type is slower to use, but eliminates the possibility of a faulty reading due to escaping pressure

2. A manifold vacuum gauge
3. A test light
4. An induction meter. This is used for determining whether or not there is current in a wire. These are handy for use if a wire is broken somewhere in a wiring harness.

As a final note, you will probably find a torque wrench necessary for all but the most basic work. The beam type models are perfectly adequate, although the newer click (breakaway) type are more precise. The breakaway torque wrenches are more expensive and should be recalibrated periodically.

Special Tools

Normally, the use of special factory tools is avoided for repair procedures, since these are not readily available for the do-it-yourself mechanic. When it is possible to perform the job with more commonly available tools, it will be pointed out, but occasionally, a special tool was designed to perform a specific function and should be used. Before substituting another tool, you should be convinced that neither your safety nor the performance of thevehicle will be compromised.

Some special tools are available commercially from major tool manufacturers. Others can be purchased from: Service Tool Division, Kent-Moore Corporation, 29784 Little Mack, Roseville, MI, 48066–2290.

SERVICING YOUR VEHICLE SAFELY

It is virtually impossible to anticipate all of the hazards involved with automotive maintenance and service, but care and common sense will prevent most accidents.

The rules of safety for mechanics range from "don't smoke around gasoline," to "use the proper tool for the job." The trick to avoiding injuries is to develop safe work habits and take every possible precaution.

Dos

• Do keep a fire extinguisher and first aid kit within easy reach.
• Do wear safety glasses or goggles when cutting, drilling or prying, even if you have 20–

20 vision. If you wear glasses for the sake of vision, they should be made of hardened glass that can also serve as safety glasses, or wear safety goggles over your regular glasses.

• Do shield your eyes whenever you work around the battery. Batteries contain sulphuric acid; in case of contact with the eyes or skin, flush the area with water or a mixture of water and baking soda and get medical attention immediately.

• Do use safety stands for any undercar service. Jacks are for raising vehicles; safety stands are for making sure the vehicle stays raised until you want it to come down. Whenever the vehicle is raised, block the wheels remaining on the ground and set the parking brake.

• Do use adequate ventilation when working with any chemicals. Like carbon monoxide, the asbestos dust resulting from brake lining wear can be poisonous in sufficient quantities.

• Do disconnect the negative battery cable when working on the electrical system. The primary ignition system can contain up to 40,000 volts.

• Do follow manufacturer's directions whenever working with potentially hazardous materials. Both brake fluid and antifreeze are poisonous if taken internally.

• Do properly maintain your tools. Loose hammerheads, mushroomed punches and chisels, frayed or poorly grounded electrical cords, excessively worn screwdrivers, spread wrenches (open end), cracked sockets, slipping ratchets, or faulty droplight sockets can cause accidents.

• Do use the proper size and type of tool for the job being done.

• Do when possible, pull on a wrench handle rather than push on it, and adjust your stance to prevent a fall.

• Do be sure that adjustable wrenches are tightly adjusted on the nut or bolt and pulled so that the face is on the side of the fixed jaw.

• Do select a wrench or socket that fits the nut or bolt. The wrench or socket should sit straight, not cocked.

• Do strike squarely with a hammer—avoid glancing blows.

• Do set the parking brake and block the drive wheels if the work requires that the engine be running.

Don'ts

• Don't run an engine in a garage or anywhere else without proper ventilation—EVER! Carbon monoxide is poisonous; it takes a long time to leave the human body and you can build up a deadly supply of it in your system by simply breathing in a little every day. You may not realize you are slowly poisoning yourself. Always use power vents, windows, fans or open the garage doors.

• Don't work around moving parts while wearing a necktie or other loose clothing. Short sleeves are much safer than long, loose sleeves and hard-toed shoes with neoprene soles protect your toes and give a better grip on slippery surfaces. Jewelry such as watches, fancy belt buckles, beads or body adornment of any kind is not safe working around a car. Long hair should be hidden under a hat or cap.

• Don't use pockets for toolboxes. A fall or bump can drive a screwdriver deep into your body. Even a wiping cloth hanging from the back pocket can wrap around a spinning shaft or fan.

• Don't smoke when working around gasoline, cleaning solvent or other flammable material.

• Don't smoke when working around the battery. When the battery is being charged, it gives off explosive hydrogen gas.

• Don't use gasoline to wash your hands; there are excellent soaps available. Gasoline may contain lead, and lead can enter the body through a cut, accumulating in the body until you are very ill. Gasoline also removes all the natural oils from the skin so that bone dry hands will suck up oil and grease.

• Don't service the air conditioning system unless you are equipped with the necessary tools and training. The refrigerant, R-12, is extremely cold and when exposed to the air, will instantly freeze any surface it comes in contact with, including your eyes. Although the refrigerant is normally non-toxic, R-12 becomes a deadly poisonous gas in the presence of an open flame. One good whiff of the vapors from burning refrigerant can be fatal.

SERIAL NUMBER IDENTIFICATION

Vehicle

The Vehicle Identification Number (V.I.N.) is a seventeen digit alphanumeric sequence stamped on a plate which is located at the top left-hand side of the instrument panel. As far as the car owner is concerned, many of the digits of the V.I.N. are of little or no value. However, it may occasionally be necessary to refer to the V.I.N. to interpret certain information

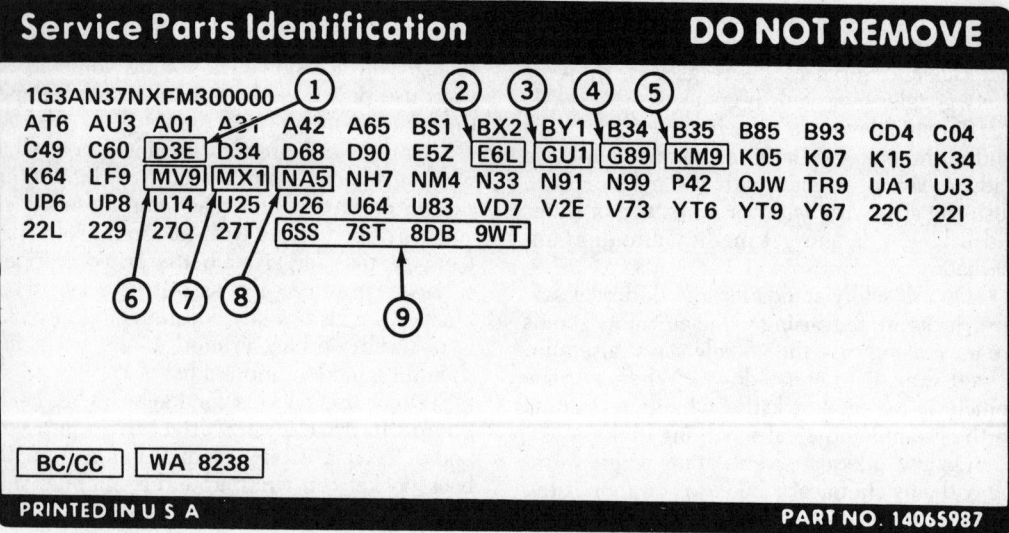

1. SPEEDOMETER DRIVEN GEAR
2. SPEEDOMETER DRIVEN GEAR ADAPTER USAGE
3. SPEEDOMETER DRIVEN GEAR SLEEVE
4. FINAL DRIVE OR REAR AXLE RATIO
5. REAR AXLE RING GEAR SIZE
6. TRANSMISSION TYPE
7. TRANSMISSION TYPE
8. EMISSION SYSTEM
9. SPRING CODES; 6=LF, 7=RF, 8=LR, 9=RR.
   TWO ALPHAS ARE THE CODE

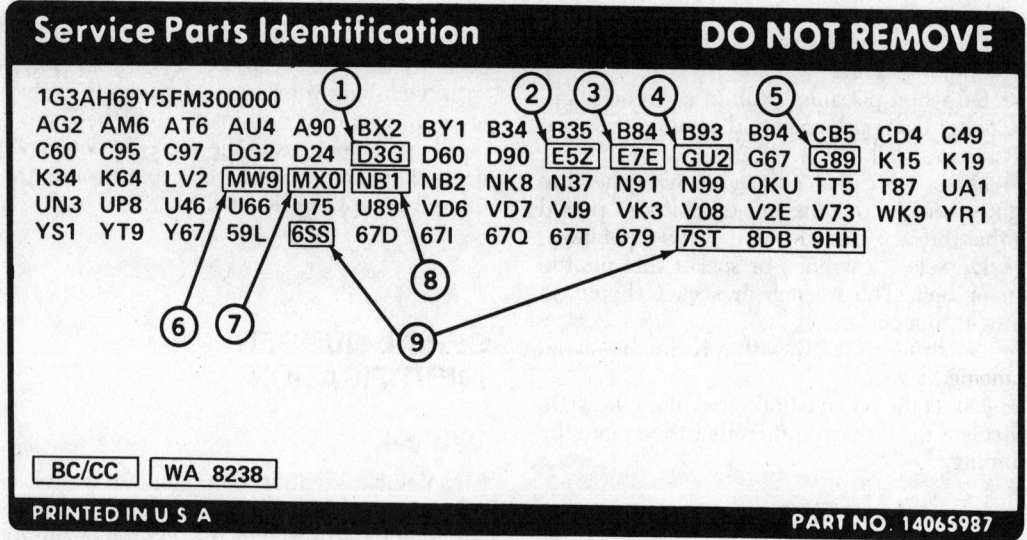

Reading the Service Parts Identification Label

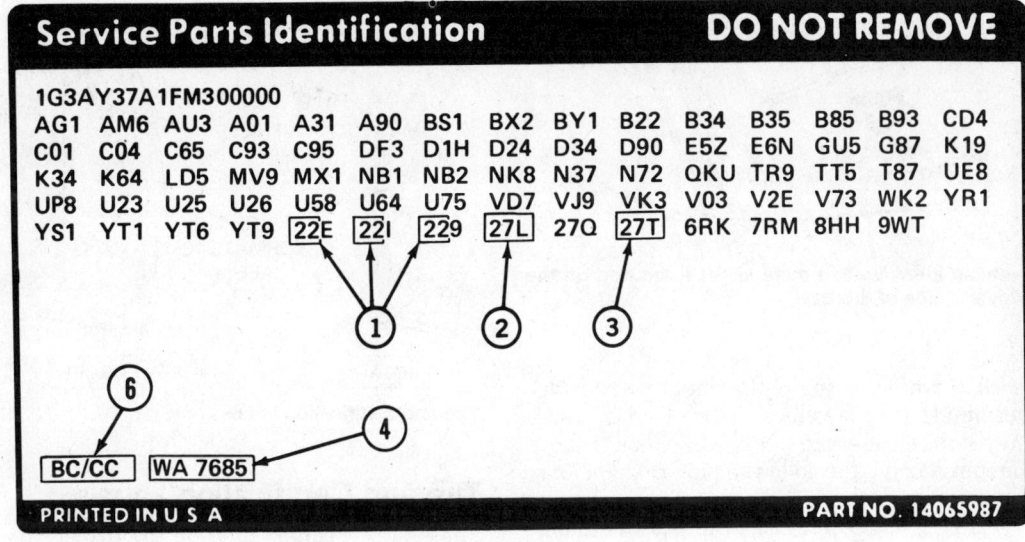

1. INTERIOR COLOR
2. EXTERIOR COLOR, LOWER
3. EXTERIOR COLOR, UPPER
4. FISHER BODY PAINT NUMBER
5. SPECIAL CAR ORDER INFORMATION
6. PAINT TECHNOLOGY

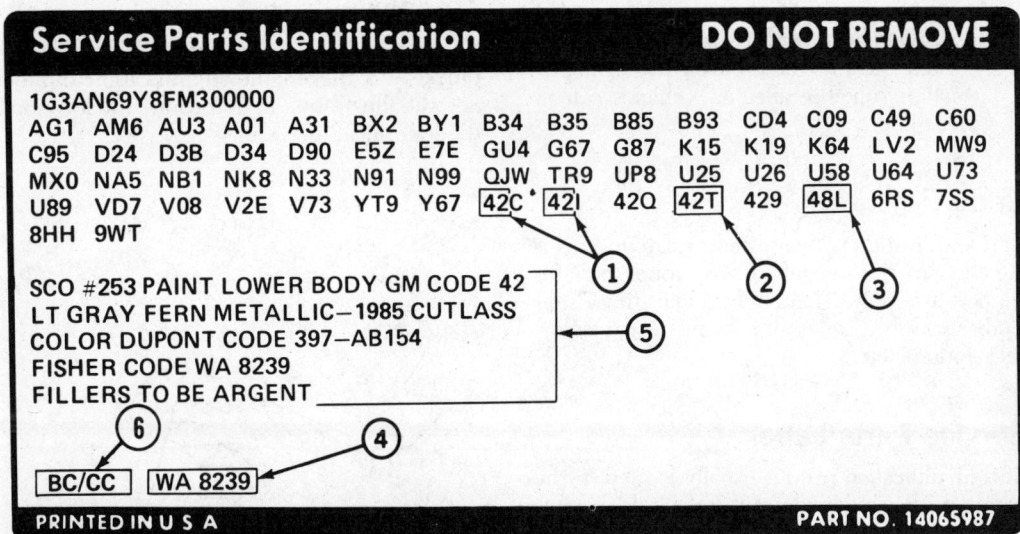

Reading the Service Parts Identification Label

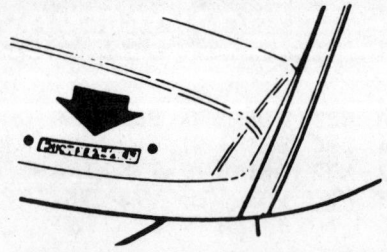

Vehicle identification number (VIN) located on the driver's side of the dash

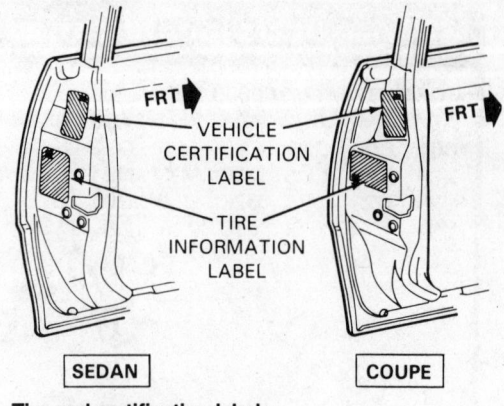

SEDAN   COUPE

Tire and certification labels

such as whe ordering replacement parts or determining if your vehicle is involved in a factory service campaign (recall). In either of these circumstances, the following information may be helpful:

- 1ST DIGIT—indicates the place of manufacture. A "1" indicates the car was built in the USA; a "2" indicates it was built in Canada
- 8TH DIGIT—indicates the type and the manufacturer of the original engine which was installed in the vehicle. See "Engine" for more details
- 10TH DIGIT—indicates the model year of the vehicle; a "D" stands for 1983, an "E" for 1984, and so on.
- 11TH DIGIT—indicates the specific plant where the vehicle was assembled.
- 12TH–17TH DIGITS—indicate the plant sequential number, which identifies the specific number of each vehicle within a production run. In the event of an engineering change or a recall involving only a certain quantity of vehicles within a production run, the affected vehicles can be identified.

Body

An identification plate for body-related items is attached to the top left of the upper radiator support assembly. This label contains the Fisher Body assembly codes pertaining to car color, seat options, etc.

Service Parts Label

This identification plate is usually located in the trunk on the underside of the deck lid, or under the spare tire cover. This label contains information such as the vehicle identification number, special car order information (if special ordered) and regular product options (RPO) in alphanumeric sequence. This information can assist in servicing and determining the correct replacement parts, if necessary.

Tire and Certification Labels

These are usually located on the driver's door jamb and contain such information as vehicle gross weight, tire pressures, and capacities.

Engine

The engine identification code will sometimes be necessary to order replacement engine parts. The code is stamped in different locations, depending on the size of the engine. Refer to the accompanying illustrations to determine the location of the engine identification number location for your engine.

Transaxle

The transaxle code number serves the same purpose as the engine identification number. See the illustration to determine the location of the identification code.

3.0 L / 3.8 L

3.0L & 3.8L V6 engine identification label

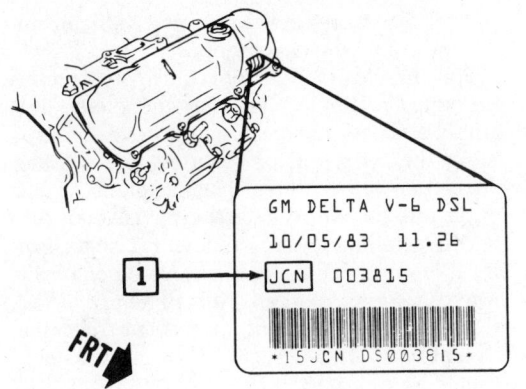

4.3 L

4.3L diesel engine identification label

111 = JULIAN DATE
5BA = 4 — MODEL YEAR
         BA — MODEL
1427 = SERIAL NO.

Transaxle code identification plates

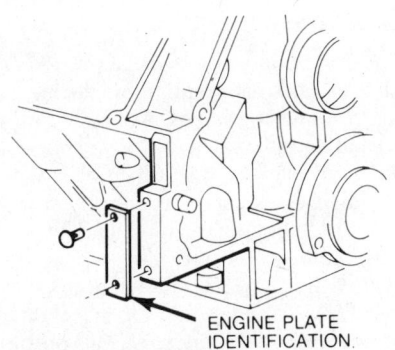

ENGINE PLATE
IDENTIFICATION

4.1L V8 engine identification label

ROUTINE MAINTENANCE

Proper maintenance of any vehicle is the key to long and trouble-free vehicle life. As a conscientious car owner, you should set aside a Saturday morning (about once a month) to check or replace normal maintenance items that can cause major problems later. Keep your own personal log to jot down which services you performed, how much parts cost, the date and the exact odometer reading at the time. Recording this information on the sales receipt for the parts you buy is a system many use, although a separate log book kept in the glove box is probably the most common method. Keep all receipts for such items as engine oil and filter changes, so they may be referred to in case of related problems or to determine operating expenses. As a do-it-yourselfer, these receipts are the only proof you have that the required maintenance was performed and in the event of a warranty problem, these receipts can be invaluable.

The owner's literature provided with your car when it was originally delivered includes the factory-recommended maintenance schedule. If you do not have the factory schedule, we have provided a general example of the GM schedule pertaining to the C-body. No matter which schedule is used, it should be followed to the letter. Even if your C-body was previously owned, it is important that the maintenance be performed, as the effects of poor maintenance

Engine Identification Chart

Cylinder Configuration	Displacement			Fuel System	VIN Code	Application	Mfg. by
	cu. in.	cc	Liters				
V6	181	2966.76	3.0	2-bbl	E	O, B	Buick
V6	231	3792.10	3.8	MFI	3	O, B	Buick
V6	263	4303.30	4.3	Diesel	T	O, B, C	Oldsmobile
V8	250	4088.96	4.1	DFI	8	C	Cadillac

2-bbl: E2MC two barrel carburetor
MFI: Multi-port fuel injection
DFI: Digital fuel injection
O: Oldsmobile
B: Buick
C: Cadillac

can at least be halted by initiating a regular maintenance program.

Air Cleaner

Regular air cleaner element replacement is a must, since a partially clogged element will cause a performance loss, decreased fuel mileage and engine damage if enough dirt gets into the cylinders and contaminates the engine oil. The air cleaner element is a dry paper type and should be replaced every 30,000 miles (every 15,000 miles if the car is driven in heavy traffic or under extremely dusty conditions.

The air cleaner element must be checked periodically. Replacement of the element is simply a matter of removing the cover by loosening the wing nut or releasing the retainer clips, then lifting the air cleaner element out. Wipe out the interior of the air cleaner case before installing the new element. Use a clean, lint-free rag to wipe any accumulated oil or dirt from the sides or bottom. Never attempt to soak the air cleaner element in gasoline, cleaning solvent or oil. It is designed to be a throw-away item.

NOTE: *The car should not be operated for any length of time without an air cleaner element in place. Small particles of dirt can get into the engine and damage components, especially on fuel injected models.*

Unscrew the wing nut and remove the cover

Remove and discard the old filter

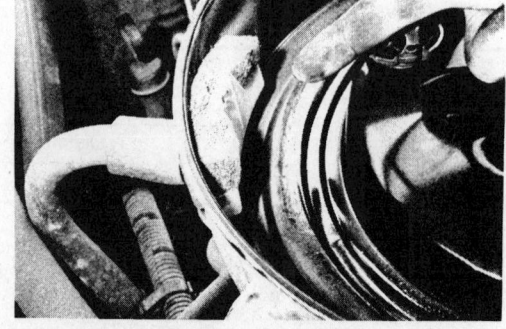

Check the small crankcase breather

PCV Valve

The positive crankcase ventilation (PCV) valve regulates crankcase ventilation during various engine running conditions. At high vacuum (idle speed and partial load range), it will open slightly and at low vacuum (full throttle) it will open fully. This causes vapors to be drawn from the crankcase by engine vacuum and then pulled into the combustion chamber where they are burned with the regular fuel charge.

The PCV valve should be checked or replaced every 30,000 miles. Details on the PCV system, including system tests are given in

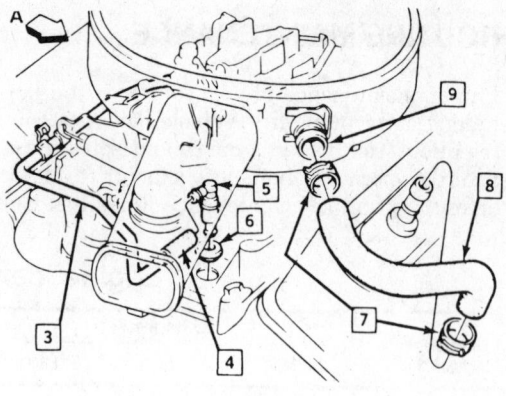

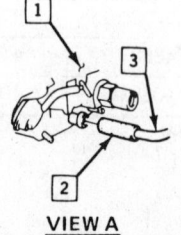

1. Carburetor assembly
2. Hose—carb. to pipe
3. PCV pipe
4. Hose—PCV valve to pipe
5. PCV valve
6. Grommet—PCV valve
7. Clamp
8. Crankcase vent tube
9. Filter

VIEW A
Carbureted V6 PCV system

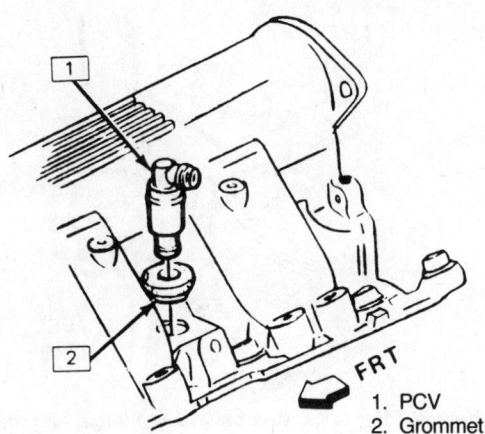

1. PCV
2. Grommet

3.8L V6 PCV system

Chapter 4. The valve is located in a rubber grommet in the valve cover, connected to the air cleaner housing by a large diameter rubber hose. To replace the valve:

1. Pull the valve (with the hose attached) from the rubber grommet in the valve cover.
2. Remove the valve from the hose.
3. Install the new valve into the hose.
4. Install the valve into the grommet.

PCV Filter

Carbureted engines have a PCV filter located in the air cleaner housing which must be re-placed periodically, usually along with the PCV valve. To replace the PCV filter:

1. Remove the air cleaner housing cover.
2. Slide back the filter retaining clip.
3. Pull the old filter from the hose.
4. Install the new filter, replace the clip (usually included with the new filter) and re-place the air cleaner cover.

Evaporation Control System Canister Filter

All models have a charcoal canister located in the engine compartment as part of the evapo-rative control system. Details on the system can be found in Chapter 4. Every 30,000 miles or 24 months, the filter on the bottom of the char-coal canister must be changed. Cut the re-placement interval in half if the car is operated in heavy traffic or under dusty conditions. To remove the filter:

1. Locate the canister in the right front of the engine compartment. It is held at the base by two bolts. Unbolt and lift the canister with-out disconnecting any of the hoses.

NOTE: *If there is not enough slack in the hoses to allow you to get at the filter, label them before removal. It's very important to connect them properly.*

2. Turn the canister over. The filter is in-stalled in the bottom and can simply be pulled out.

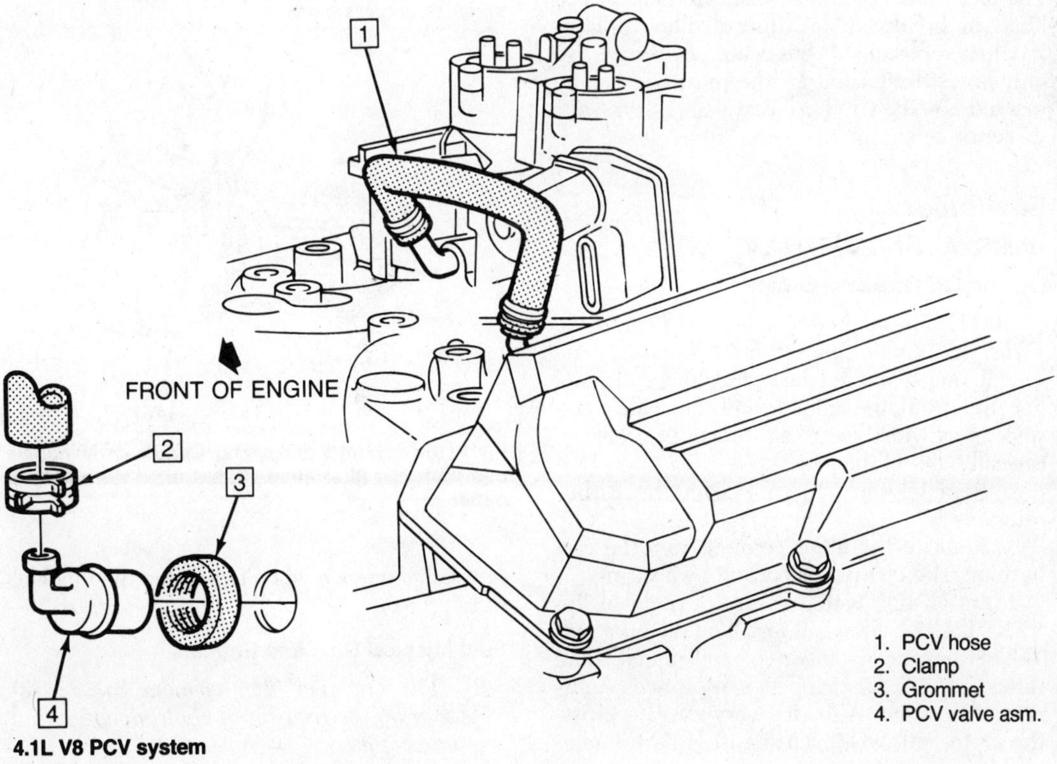

FRONT OF ENGINE

1. PCV hose
2. Clamp
3. Grommet
4. PCV valve asm.

4.1L V8 PCV system

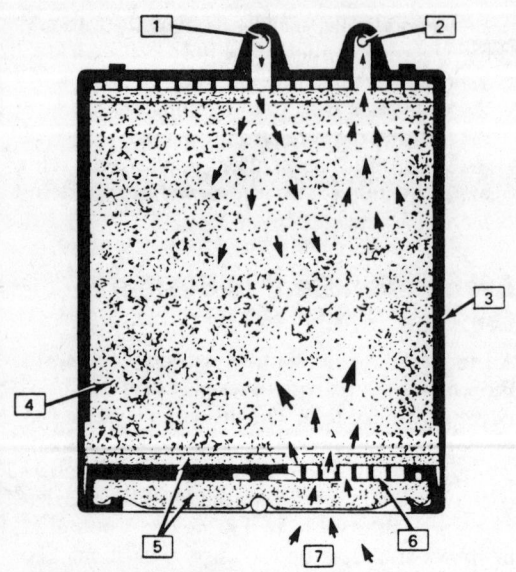

1. Vapor from fuel tank
2. Canister purge vacuum
3. Canister body
4. Carbon
5. Filter
6. Grid
7. Air flow during purge

Typical evaporative canister

3. Install a new filter into the bottom of the canister.

4. Install the canister and its hold down bolts. The ECS hoses should be inspected for cracks, kinks or breaks at the time of filter replacement. If replacement hoses are necessary, use only hoses designated for the purpose, usually marked EVAP. This hose is available from your dealer or an automotive parts store.

Fuel Filter

REMOVAL AND INSTALLATION

Carbureted Gasoline Engines

NOTE: *The fuel filter is located behind the fuel inlet connection on the carburetor.*

1. Using two wrenches, disconnect the fuel line nut from the inlet fitting and catch any spilled fuel with a clean rag. Move the fuel line, carefully, out of the way.

2. Remove the inlet nut from the carburetor.

3. Remove the filter element from the carburetor. The element is backed by a spring.

4. Installation is the reverse of removal. Be VERY CAREFUL when installing the inlet nut! The carburetor is made of soft metal. The threads are easily stripped if the nut is tightened too much! Also, it's very easy to cross-thread the nut when starting it. Hold the inlet

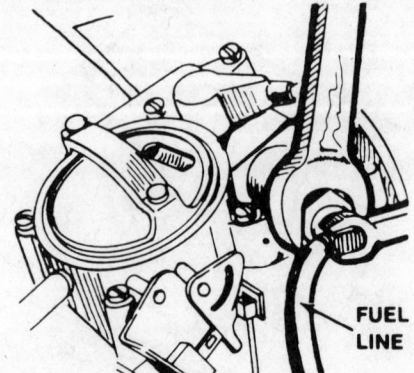

The fuel filter is located behind the large fuel line inlet nut on the carburetor

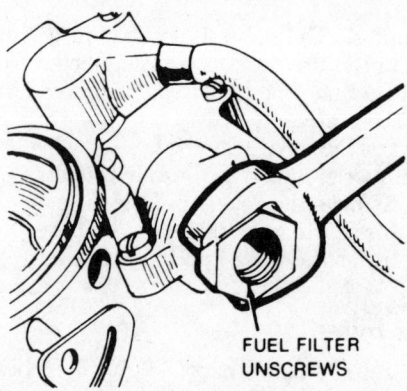

Remove the retaining nut and the filter will pop out under spring pressure

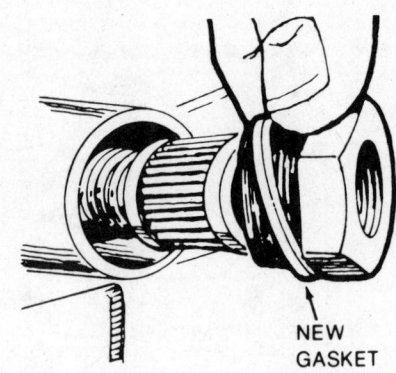

Install the new filter and spring. Certain early models use a bronze filter element, but most are made of paper

nut with a wrench when tightening the fuel line nut.

Fuel Injected Gasoline Engines

NOTE: *The fuel filter element can be replaced by unscrewing the bottom cover and removing it.*

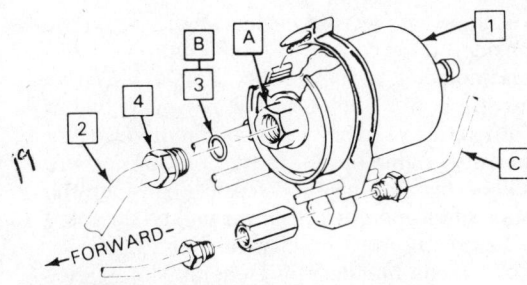

1. Fuel filter
2. Fuel feed line
3. O ring
4. 30 N m (22 ft. lbs.)
A. Use back-up wrench at this location
B. Also make sure O ring is on pipe prior to installation
C. Brake pipe

3.8L V6 fuel filter

1. Bleed the pressure from the fuel delivery system as outlined in Steps 1–4 of the "Chassis-Mounted Fuel Pump" removal procedure and remove the fuel inlet and outlet hoses from the fuel filter.

2. Remove the two screws retaining the fuel filter to the bracket and remove the filter from the engine or frame.

3. Remove the inlet and outlet fittings from the filter assembly if they are needed for the new filter.

4. Install the fittings to the new filter, using a sealer on the threads.

5. Attach the filter to the bracket and tighten the retaining screws to 12 ft. lbs.

6. Connect the inlet and outlet line, using new clamps.

NOTE: *It may require considerable cranking before the engine starts due to the drained fuel lines.*

Diesel Engines

The fuel filter is a square assembly located at the back of the engine above the intake manifold. Disconnect the fuel lines and remove the filter. Install the lines to the new filter. Start the engine and check for leaks.

FLUIDS AND LUBRICANTS

Fuel Recommendations

C-Body cars with gasoline engines are designed to use only unleaded gasoline, with an octane rating of at least 87.

Using leaded gasoline can damage the emission control system by decreasing the effectiveness of the catalyst in the catalytic converter and by damaging the oxygen sensor which is part of the "Computer Command Control System".

Do not use gasolines containing more than 5 percent methanol even if they contain cosolvents and corrosion inhibitors.

Although gasolines containing 5 percent or less methanol and appropriate cosolvents and inhibitors for methanol may be suitable for use in your car, General Motors does not endorse their use, at this time.

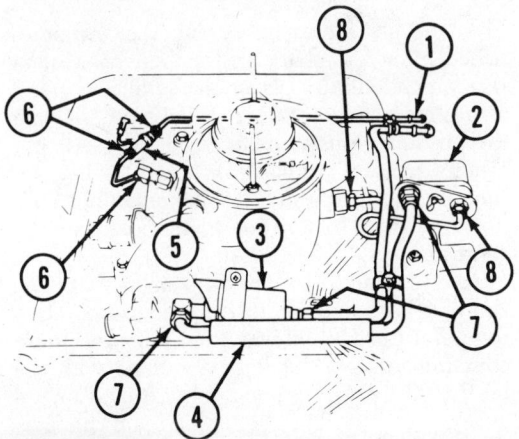

1. Return line
2. Fuel filter
3. Fuel pump
4. Fuel line heater (optional)
5. Housing pressure altitude advance
6. 13 N·m (10 lbs. ft.)
7. 26 N·m (19 lbs. ft.)
8. 15 N·m (11 lbs. ft.)

Diesel fuel filter on systems without a water separator

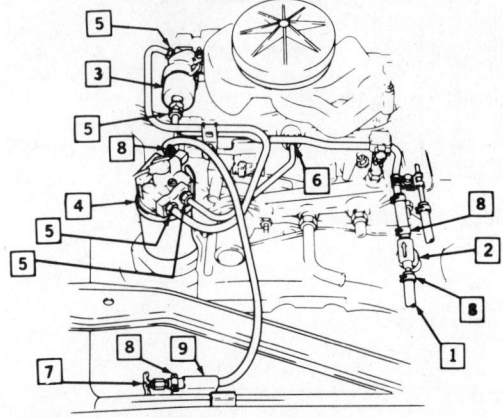

1. From fuel tank
2. In-line filter assembly
3. Fuel pump assembly
4. Fuel filter assembly
5. 25 N·m (18 lbs. ft.)
6. 30 N·m (22 lbs. ft.)
7. Drain valve
8. Clamps
9. Bracket

Diesel fuel filter on systems with a water separator

Diesel engined cars are designed to run on No. 2 diesel fuel whenever the temperature is consistantly above 20°F (−7°C). Whenever temperatures fall consistantly below this point, the use of No. 1 diesel fuel is recommended. In some areas, a blend of Nos. 1 & 2 can be found. This blend should be used throughout the colder months. Commercially available cold weather additives can be used when No. 1 fuel is not available.

Engine

OIL RECOMMENDATION

Engine oils are labeled on the containers with various API (American Petroleum Institute) designations of quality. For gasoline engines, make sure that the oil you use has the API designation "SF", either alone or shown with other designations such as SF/CC or SF/CD. Oils with a label on which the designation "SF" does not appear should not be used.

For diesel engines, use only oils designated SF/CC or SF/CD.

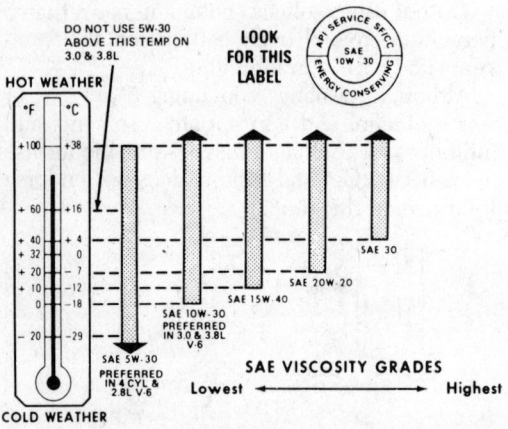

Gasoline engine oil viscosity chart

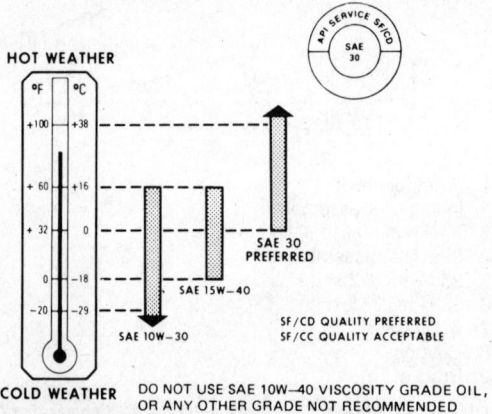

Diesel engine oil viscosity chart

Engine oil viscosity (thickness) should be considered according to temperature weather conditions. Lower viscosity engine oils can provide better fuel economy; however, higher temperature weather conditions require higher viscosity engine oils for satisfactory lubrication. When selecting an oil viscosity, consider the range of temperature your car will be operated in before the next oil change. For the oil viscosity recommended by General Motors, see the accompanying illustrations.

OIL LEVEL CHECK

At every fuel stop, check the engine oil in the following manner:

1. Park the car on a level area.

2. The engine oil may be either hot or cold when checking the oil level. However, if it is hot, wait a few minutes after the engine has been shut off to allow the oil to drain back into the oil pan. If the engine is cold, do not start it before checking the oil level.

3. Open the engine compartment and locate the dipstick. Pull the dipstick from its tube, wipe it clean and reinsert it.

NOTE: *Make sure the dipstick is fully seated when checking the oil level to assure accurate readings.*

4. Pull the dipstick out again and while holding it horizontally, read the oil level. The oil level should be above the "ADD" line but not above the "FULL" line. Do not overfill.

OIL AND FILTER CHANGE

The milage figure given in the "Maintenance Intervals" chart are the General Motors recommended intervals for oil and filter changes assuming average driving. If your car is being driven under dusty, polluted, or off road conditions, or used regularly for trailer towing, cut the milage intervals in half. The same thing goes for cars driven in stop-and-go traffic or for only short distances.

Always drain the oil after the engine has been running long enough to bring it to operating temperature. Hot oil will flow easier and more contaminants will be removed along with the oil than if it were drained cold. You will need a large capacity drain pan which you can purchase at any auto store. Another necessity is containers for used oil. You will find that plastic bottles such as those used for detergents, bleaches etc., make excellent storage jugs. One ecologically desirable solution to the used oil disposal problem is to find a cooperative gas station owner who will allow you to dump your used oil into his tank.

General Motors recommends changing both the oil and filter during the first oil change and the filter every other oil change thereafter. For the small price of an oil filter, its cheap insurance to replace the filter at every oil change. One of the larger filter manufacturers points out in its advertisements that not changing the filter leaves a quanity of dirty oil in the engine, which could be as much as a quart on some models. This claim is true and should be kept inmined when changing your oil. Change your oil as follows:

1. Run the engine until it reaches normal operating temperature.

2. Jack up the car and support it safely with jackstands.

3. Slide a drain pan under the oil pan.

4. Loosen the drain plug. Turn the plug out by hand. By keeping an inward pressure on the plug as you unscrew it, oil won't escape past the threads and you can remove it without being burnt with hot oil.

5. Allow the oil to drain completely and then install the drain plug. Be careful not to overtighten the plug and strip the threads in the oil pan.

6. Using a strap wrench, remove the oil filter. Keep in mind that it's holding dirty, hot oil.

7. Empty the old filter into the drain pan and dispose of the filter.

8. Using a clean rag, wipe off the filter

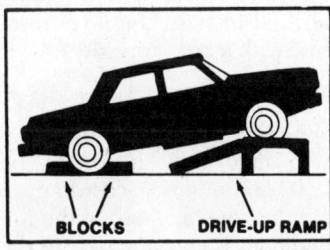

1. Warm the car up before changing your oil. Raise the front end of the car and support it on drive-on ramps or jackstands.

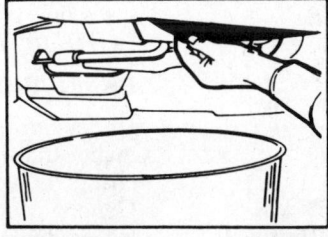

2. Locate the drain plug on the bottom of the oil pan and slide a low flat pan of sufficient capacity under the engine to catch the oil. Loosen the plug with a wrench and turn it out the last few turns by hand. Keep a steady inward pressure on the plug to avoid hot oil from running down your arm.

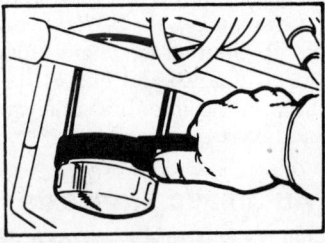

3. Remove the oil filter with a filter wrench. The filter can hold more than a quart of oil, which will be hot. Be sure the gasket comes off with the filter and clean the mounting base on the engine.

4. Lubricate the gasket on the new filter with clean engine oil. A dry gasket may not make a good seal and will allow the filter to leak.

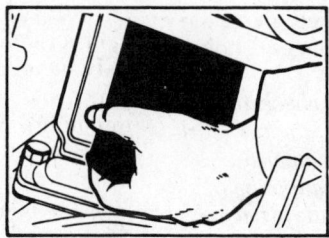

5. Position a new filter on the mounting base and spin it on by hand. Do not use a wrench. When the gasket contacts the engine, tighten it another ½–1 turn by hand.

6. Using a rag, clean the drain plug and the area around the drain hole in the oil pan.

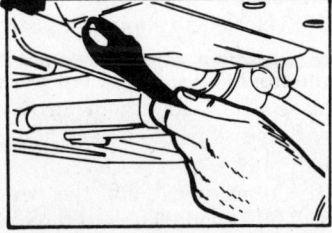

7. Install the drain plug and tighten it finger-tight. If you feel resistance, stop and be sure you are not cross-threading the plug. Finally, tighten the plug with a wrench.

8. Locate the oil cap on the valve cover. An oil spout is the easiest way to add oil, but a funnel will do just as well.

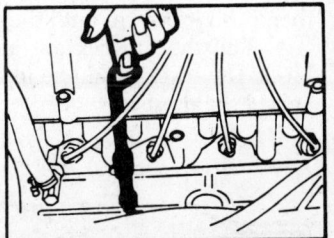

9. Start the engine and check for leaks. The oil pressure warning light will remain on for a few seconds; when it goes out, stop the engine and check the level on the dipstick.

adapter on the engine block. Be sure that the rag doesn't leave any lint which could clog an oil passage.

9. Coat the rubber gasket on the new filter with fresh oil. Spin it onto the engine by hand; when the gasket touches the adapter surface give it another ½–¾ turn. Do not overtighten or you may squash the gasket and cause it to leak.

10. Refill the engine with a correct amount of fresh oil.

11. Check the oil level on the dipstick. It is normal for the level to be a bit above the full mark. Start the engine and allow it to idle a few minutes.

CAUTION: *Do not run the engine above idle speed until it has built up oil pressure, indicated when the oil light goes out.*

12. Shut off the engine, allow the oil to drain for a minute, and check the oil level. Check around the filter and drain plug for any leaks, and correct as necessary.

Automatic Transaxle

FLUID RECOMMENDATION AND LEVEL CHECK

The automatic transaxle fluid level should be checked at each engine oil change. When adding or changing the automatic transaxle fluid use only fluid labeled Dexron®II.

1. Set the parking brake and start the engine with the transaxle in "P" (Park).

2. With the service brakes applied, move the shift lever through all the gear ranges, ending in "P" (Park).

NOTE: *The fluid level must be checked with the engine running at slow idle, the car level and the fluid at least at room temperature.*

NOTE: *The correct fluid level cannot be read if you have just driven the car for a long time at high speed, city traffic in hot weather or if the car has been pulling a trailer. In these cases, wait at least 30 minutes for the fluid to cool down.*

3. Remove the dipstick located at the rear end of the engine compartment, wipe it clean, then push it back in until the cap seats.

4. Pull the dipstick out and read the fluid level. The level should be in the cross-hatched area of the dipstick.

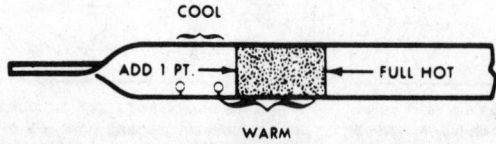

Automatic transmission dipstick marks; the proper level is within the shaded area

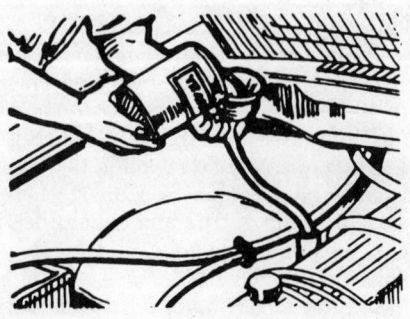

Add automatic transmission fluid through the dipstick tube

5. Add fluid using a long plastic funnel in the dipstick tube. Keep in mind that it only takes one pint of fluid to raise the level from "ADD" to "FULL" with a hot transaxle.

DRAIN AND REFILL

Under normal operating conditions, the automatic transmission fluid only needs to be changed every 100,000 miles, according to General Motors, unless one or more of the following driving conditions is encountered. In the following cases the fluid and filter should be changed every 15,000 miles:

a. Driving in heavy traffic when the outside temperature reaches 90°F.

b. Driving regularly in hilly or mountainous areas.

c. Towing a trailer.

d. Using a vehicle as a taxi or police car or for delivery purposes.

Remember, these are factory recommendations, and in this case are considered to be minimum. You must determine a change interval which fits your driving habits. If your vehicle is never subjected to these conditions, a 100,000 mile change interval is adequate. If you are a normal driver, a two-year/30,000 mile interval will be more than sufficient to maintain the long life for which your automatic transaxle was designed.

NOTE: *Use only fluid labeled Dexron®II. Use of other fluids could cause erratic shifting and transmission damage.*

1. Jack up your vehicle and support it safely with jackstands.

2. Remove the front and side pan bolts.

3. Loosen the rear bolts about four turns.

4. Carefully pry the oil pan loose and allow the fluid to drain.

5. Remove the remaining bolts, the pan, and the gasket or RTV sealant. Discard the old gasket.

6. Clean the pan with solvent and dry it thoroughly, with compressed air.

7. Remove the strainer and O-ring seal.

8. Install a new strainer and O-ring seal, locating the strainer against the dipstick stop.

NOTE: *Always replace the filter with a new one. Do not attempt to clean the old one.*

9. Install a new gasket or RTV sealant then tighten the pan bolts to 12 ft. lbs.

10. Lower the car and add about 4 quarts of Dexron®II transmission fluid.

11. Start the engine; let it idle. Block the wheels and apply the parking brake.

12. Move the shift lever through the ranges. With the lever in Park, check the fluid level and add as necessary.

NOTE: *The transmission fluid currently being used may appear to be darker and have a strong odor. This is normal and not a sign of required maintenance or transmission failure.*

Battery

The C-body uses a maintenance free battery as standard equipment, eliminating the need for fluid level checks and the possibility of specific gravity tests. Nevertheless, the battery does require some attention. The single battery used in models equipped with gasoline engines is located at the left front corner of the engine compartment. Because of the added cranking loads, diesel models use two batteries; one in each front corner of the engine compartment.

The major cause of slow engine cranking or a "no-start" condition is battery terminals which are loose, corroded, or dirty. Every three months or so, disconnect the battery and clean both the terminals and the cable connectors. Cleaning tools for this purpose are available at any automotive parts store, but a small wire brush or steel wool will also work to remove the light corrosion that is considered normal. When you buy a cleaning tool, be sure to specify that you have a side terminal battery, as the type of tool for top terminal batteries is different. To use a terminal cleaning tool, it may be necessary to remove the windshield washer bottle or the coolant recovery tank to gain access to the terminals with the tool.

CAUTION: *When loosening or tightening the positive battery cable screw(s) at the battery, DO NOT touch the wrench to any metal surface. Personal injury or other component damage can result.*

Check the battery cables for signs of wear or chafing. If corrosion is present on the cable or if the cable is visible through the cable jacket, the cable assembly should be replaced. If cable replacement is necessary, it is best to purchase a high quality cable that has the cable jacket sealed to the terminal ends. Batteries themselves can be cleaned using a solution of baking soda and water. Surface coatings on battery cases can actually conduct electricity which will cause a slight voltage drain, so make sure the battery case is clean. To remove the battery:

1. Raise the hood and remove the front end diagonal brace from above the battery.

2. Disconnect the battery cables from the battery. It may be necessary to use a small box end wrench or a ¼ in. drive ratchet to get in between the battery and windshield washer (or coolant recovery) tank. Avoid using an open-end wrench for the cable bolts.

3. Loosen and remove the battery hold-down bolt and block. The use of a long extension which places the ratchet above the battery makes it very easy to get at the hold-down bolt.

4. Carefully lift the battery from the engine compartment. It may be necessary to remove the air cleaner intake duct for clearance.

5. Clean the battery and the battery tray thoroughly with baking soda/water solution, but don't allow the solution to get into the small vent holes on the battery.

6. Rinse the battery with clear water and wipe it dry with a couple of clean paper towels, then dispose of the towels. Don't use the towels for anything else, as they will probably contain traces of sulfuric acid.

7. Thoroughly flush the battery tray and the surrounding area with clear water. Using a wire brush, remove any rust which may be on the tray. Clear away the rust and allow the tray to dry.

8. Coat the battery tray liberally with anti-rust paint, then thoroughly clean the battery

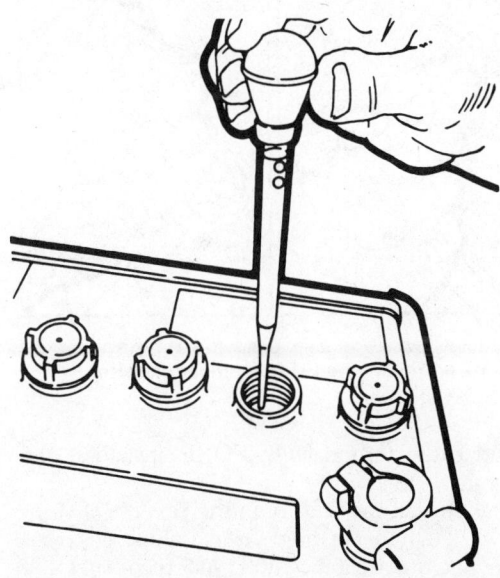

Specific gravity can be checked with an hydrometer

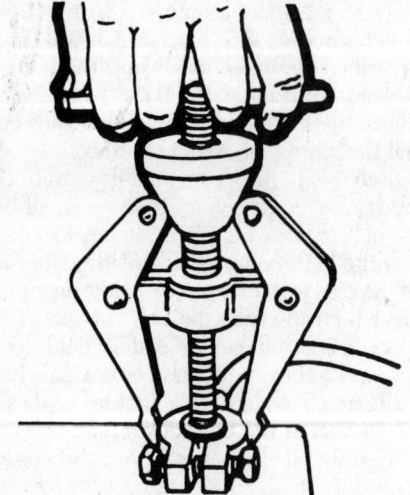

Pullers make clamp removal easier

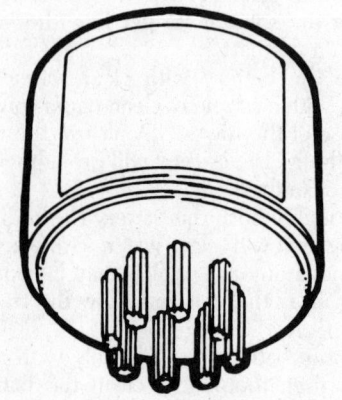

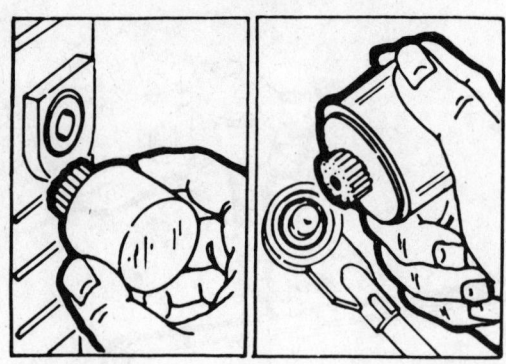

Special tools are also available for cleaning the posts and clamps on side terminal batteries

not meant specifically for this purpose. Do not apply the corrosion inhibitor to the mating surfaces of the terminals. Any time the engine won't crank, check the color of the battery condition indicator (which is actually a built-in hydrometer). If the indicator color is green, the battery is sufficiently charged and in good condition, and a complete check of the starter and related wiring should be performed. If the indicator is darkened, the battery is discharged. In this case, the reason for the discharge should be determined (low alternator output, voltage draw, etc.) then the battery itself should be tested and recharged. If the indicator light is without a green dot or is yellow in color, the battery should be replaced. DO NOT attempt to test or recharge a battery with this indicator condition. Test the electrical system after the battery has been replaced.

Brake Master Cylinder

NOTE: *The brake fluid reservoir is part of the brake master cylinder, and is located under the front engine compartment lid, on the driver's side of the car. Check the fluid level each time your engine oil is changed.*

FLUID RECOMMENDATION AND LEVEL CHECK

The master cylinder reservoir cover is retained by a bail wire. To remove the wire, pry it off with a screwdriver or other suitable tool. The levels in both the front and rear chambers must be maintained at a point about ¼ inch below the top of the chamber. Delco Supreme No. 11 or other DOT 3 specification brake fluid is used in most cars; add as necessary. Note, however, that some cars may come filled with silicone based brake fluid. In these cases, only

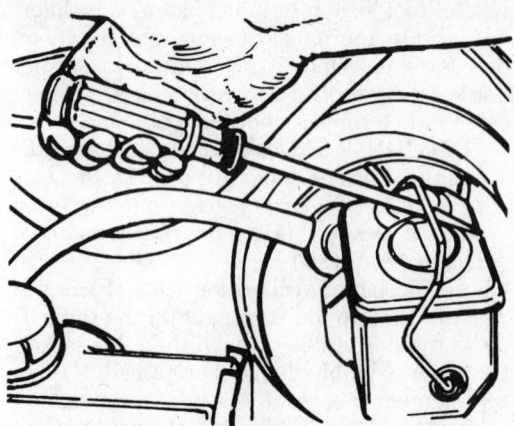

Pry the retaining bail from the master cylinder reservoir cap to check the fluid level

and cable terminals BEFORE installing the battery.

9. Install the battery in the reverse of steps 1–4. Tighten the hold-down bolt snugly, but do not overtighten. After you reconnect the battery, apply a corrosion inhibitor to the terminals. Stay away from any substance which is

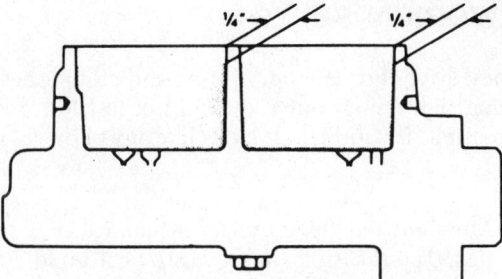

The fluid level in the master cylinder reservoir should be within ¼ in. of the top edge

silicone based fluid may be added to the system. Never mix DOT 3 and silicone fluids!

CAUTION: *Do not allow anyone to depress the brake pedal while the brake fluid reservoir cover is removed. Also, brake fluid will remove paint. If spilled, flush the area immediately with water.*

Drive Belts

BELT TENSION ADJUSTMENT

Every 12 months or 15,000 miles (every 5000 miles on diesel engines), check the drive belts for proper tension. Also look for signs of wear, fraying, separation or glazing and replace the belts as required. Belt tension should be checked with a gauge made for the purpose. If a gauge is not available, tension can be checked with moderate thumb pressure applied to the belt at its longest span between pulleys. If the belt has a free span less than 12 inches, it should deflect approximately ⅛–¼ inch. If the span is

longer than 12 inches, deflection can range between ⅛–⅜ inch.

NOTE: *Models with diesel engines use a serpentine belt which is automatically adjusted*

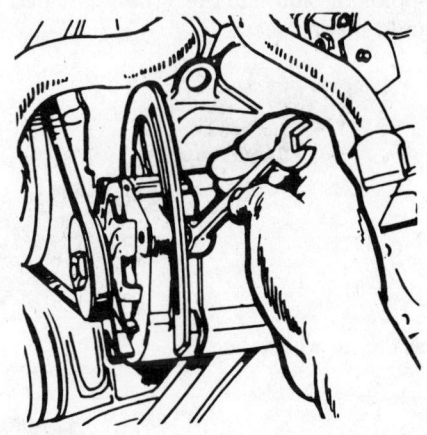

To adjust belt tension or to replace belts, first loosen the component's mounting and adjusting bolts slightly

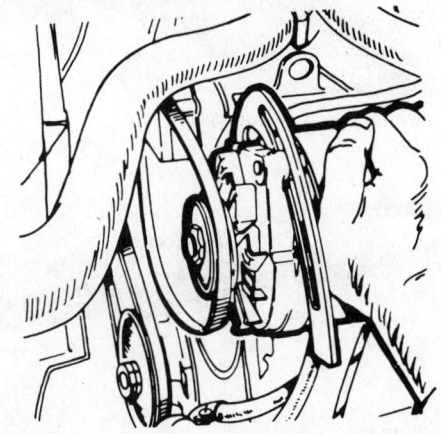

Push the component toward the engine and slip off the belt

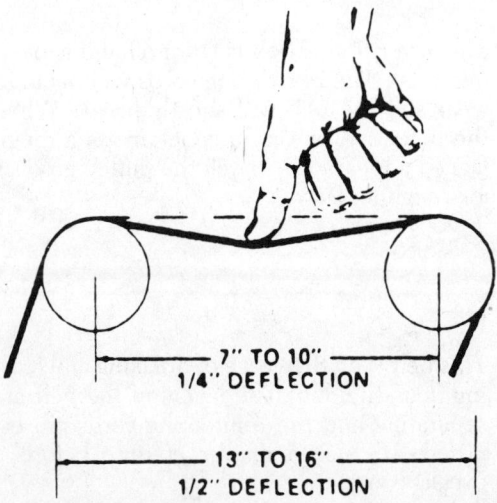

7″ TO 10″
1/4″ DEFLECTION

13″ TO 16″
1/2″ DEFLECTION

A gauge is recommended, but you can check belt tension with thumb pressure

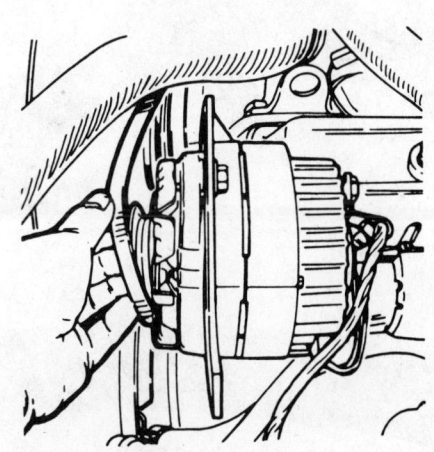

Slip the new belt over the pulley

HOW TO SPOT WORN V-BELTS

V-Belts are vital to efficient engine operation—they drive the fan, water pump and other accessories. They require little maintenance (occasional tightening) but they will not last forever. Slipping or failure of the V-belt will lead to overheating. If your V-belt looks like any of these, it should be replaced.

This belt has deep cracks, which cause it to flex. Too much flexing leads to heat build-up and premature failure. These cracks can be caused by using the belt on a pulley that is too small. Notched belts are available for small diameter pulleys.

Cracking or weathering

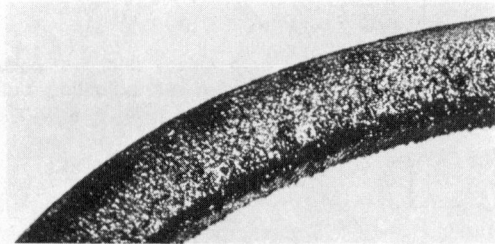

Oil and grease on a belt can cause the belt's rubber compounds to soften and separate from the reinforcing cords that hold the belt together. The belt will first slip, then finally fail altogether.

Softening (grease and oil)

Glazing is caused by a belt that is slipping. A slipping belt can cause a run-down battery, erratic power steering, overheating or poor accessory performance. The more the belt slips, the more glazing will be built up on the surface of the belt. The more the belt is glazed, the more it will slip. If the glazing is light, tighten the belt.

Glazing

The cover of this belt is worn off and is peeling away. The reinforcing cords will begin to wear and the belt will shortly break. When the belt cover wears in spots or has a rough jagged appearance, check the pulley grooves for roughness.

Worn cover

This belt is on the verge of breaking and leaving you stranded. The layers of the belt are separating and the reinforcing cords are exposed. It's just a matter of time before it breaks completely.

Separation

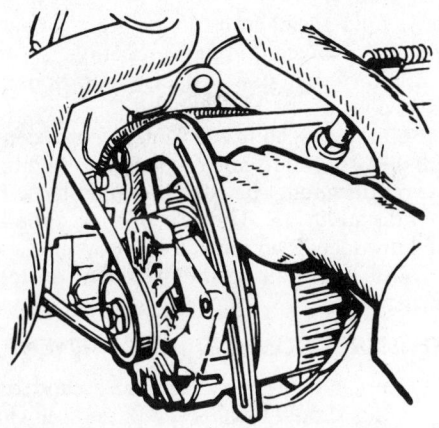

Pull outward on the component and tighten the mounting bolts

by a spring loaded tensioner. Adjustments are not normally required.

1. Loosen the driven accessory's pivot and mounting bolts.

2. Move the accessory toward or away from the engine until the tension is correct. You may use a wooden broomstick as a lever, but don't use anything metallic.

3. Tighten the bolts and recheck the tension. If new belts have been installed, run the engine for a few minutes, then recheck and readjust as necessary. It is better to have belts too loose than too tight, because overtight belts will lead to bearing failure, particularly in the water pump and alternator. However, loose belts place an extremely high impact load on the driven component due to the whipping action of the belt.

Cooling System

Every 12 months or 15,000 miles the following services should be performed:

1. Wash and inspect the radiator cap and the filler neck.

2. Check the coolant level and the degree of antifreeze protection.

3. If a pressure tester is available, pressure test the system and the radiator cap.

4. Inspect the hoses of the cooling system for cracks, leaks or deterioration (expect to replace the hoses at 24 months/30,000 miles).

5. Check the fins of the radiator (or air conditioning condenser) for blockage or damage.

RADIATOR CAP INSPECTION

Allow the engine to cool sufficiently before attempting to remove the radiator cap. Use a rag to cover the cap, then remove by pressing down and turning counterclockwise to the first stop. If any hissing is noted (indicating the release of pressure), wait until the hissing stops completely, then press down again and turn counterclockwise until the cap can be removed.

CAUTION: *DO NOT attempt to remove the radiator cap while the engine is hot. Severe personal injury from steam burns can result.*

Check the condition of the radiator cap gasket and seal inside of the cap. The radiator cap is designed to seal the cooling system under normal operating conditions which allows the build up of a certain amount of pressure (this pressure rating is stamped or printed on the cap). The pressure in the system raises the boiling point of the coolant to help prevent overheating. If the radiator cap does not seal, the boiling point of the coolant is lowered and overheating will occur. If the cap must be replaced, purchase the new cap according to the pressure rating which is specified for your vehicle.

Prior to installing the radiator cap, inspect and clean the filler neck. If you are reusing the old cap, clean it thoroughly with clear water. After turning the cap on, make sure the arrows align with the overflow hose.

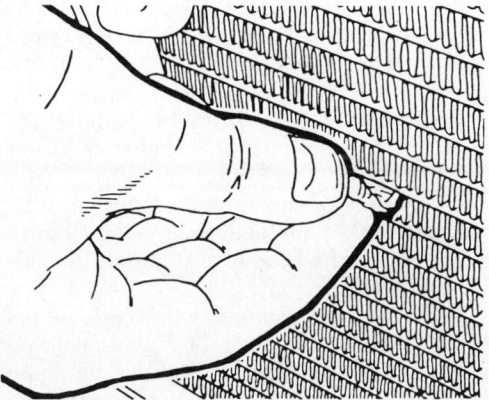

Clean the radiator fins of debris

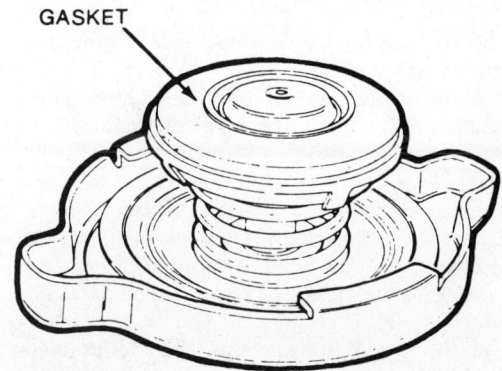

GASKET

Check the radiator cap gasket for cuts or cracks

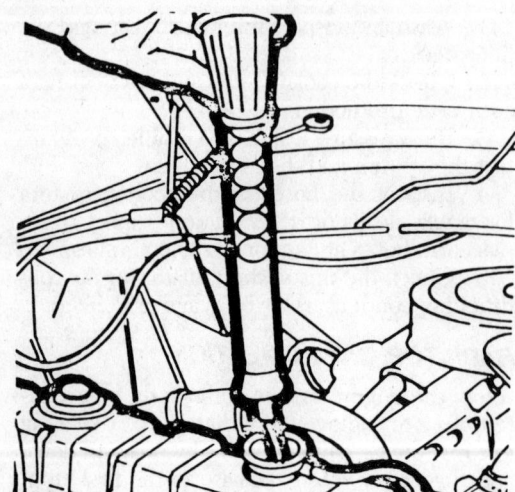
Testing coolant protection with an antifreeze tester

CHECKING THE COOLANT

Any time the hood is raised, check the level of the coolant in the "see-through" plastic coolant recovery tank. With the engine cold, the coolant level should be near the "ADD" mark on the recovery tank. At normal operating temperature, the level should be near the "FULL" mark. If coolant must be added to the tank, use a 50/50 mix of coolant and water to adjust the fluid level on models with gasoline engines. On models with diesel engines, use straight, undiluted coolant to adjust the level in the tank. See "Coolant Requirements" for additional information.

An inexpensive tester may be purchased to test the freezing protection of the coolant. Follow the instructions provided with the tester. The coolant used in models with gasoline engines should protect to –37 degrees F on gasoline engines. Diesel engines must have protection to –75 degrees F.

COOLANT REQUIREMENTS

The coolant used in any General Motors engine must:

a. Be a high quality ethylene glycol-based solution. Do not use alcohol or methanol-based solutions at any time.

b. Have built-in rust inhibitors.

c. Be designed for year-round use.

d. Offer complete protection for a minimum of 2 years/30,000 miles without replacement, as long as the proper concentration is maintained.

e. Meet GM specification 1825-M (as specified on the container). This point is critical for diesel engines; coolant meeting other specifications could result in cooling system damage and engine damage due to overheating.

f. Be mixed in the proper proportions of water to coolant.

The use of "self-sealing" coolants is not recommended. Also, the use of a coolant meeting the above requirements negates the need for supplemental additives. The use of such supplemental products is an unnecessary expense and may cause less than optimum cooling system performance.

HOSE INSPECTION AND REPLACEMENT

The upper and lower radiator hoses and the heater hoses should be checked periodically for deterioration, leaks and loose clamps. GM recommends this be done every 12 months or 15,000 miles. For your own peace of mind, it may be wise to check these items at least every spring and fall, since the summer and winter months wreak the most havoc with your cooling system. Expect to replace the hoses about every 24 months or 30,000 miles. To replace the hoses:

1. Allow the engine to cool and drain the cooling system.

2. Loosen the hose clamps at each end of the hose to be removed. If the clamps are of the type which have a screw positioned vertically in relation to the hose, loosen the screw and gently tap the head of the screw toward the hose. Repeat this until the clamp is loose enough. If corrosion on the clamp prevents loosening in this manner, carefully cut the clamp off with cutters and replace the clamp with a new one.

3. Once the clamps are out of the way, grasp the hose and twist it off at the hose connection using only moderate force. If the hose won't break loose, DON'T use excessive force as doing so can easily damage the heater core and/or radiator tubes. Using a razor blade, carefully slit the portion of the hose which covers the connection point, peel the hose off of the connection and then disconnect the hose.

4. If so equipped, disconnect the hose routing clamps from the hose.

5. Remove the hose and clean the hose connection points. Use steel wool to remove any corrosion from around the radiator or heater core tubes.

6. Slip the (loosened) hose clamps onto the hose ends and install the new hose, being careful to position the hose so that no interference is encountered.

7. Position the clamps at the ends of the hoses, beyond the sealing bead and centered on the clamping surface. Tighten the hose clamps with a screwdriver—don't use a wrench on the screw hex heads as overtightening can

HOW TO SPOT BAD HOSES

Both the upper and lower radiator hoses are called upon to perform difficult jobs in an inhospitable environment. They are subject to nearly 18 psi at under hood temperatures often over 280°F., and must circulate nearly 7500 gallons of coolant an hour—3 good reasons to have good hoses.

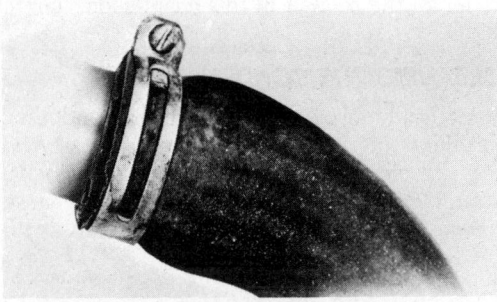

Swollen hose

A good test for any hose is to feel it for soft or spongy spots. Frequently these will appear as swollen areas of the hose. The most likely cause is oil soaking. This hose could burst at any time, when hot or under pressure.

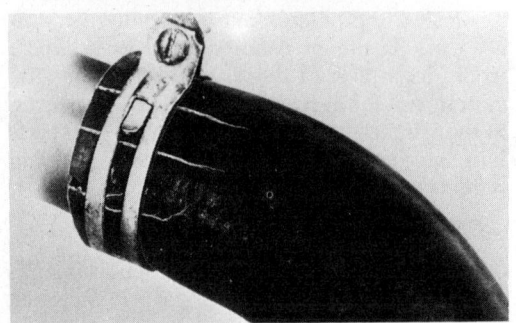

Cracked hose

Cracked hoses can usually be seen but feel the hoses to be sure they have not hardened; a prime cause of cracking. This hose has cracked down to the reinforcing cords and could split at any of the cracks.

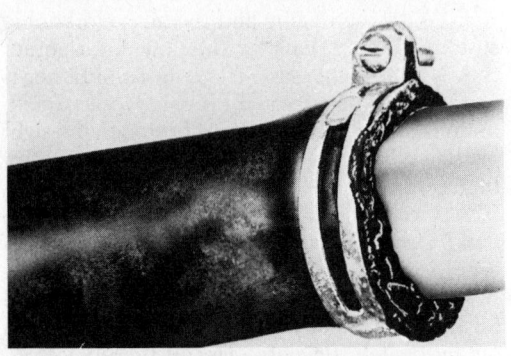

Frayed hose end (due to weak clamp)

Weakened clamps frequently are the cause of hose and cooling system failure. The connection between the pipe and hose has deteriorated enough to allow coolant to escape when the engine is hot.

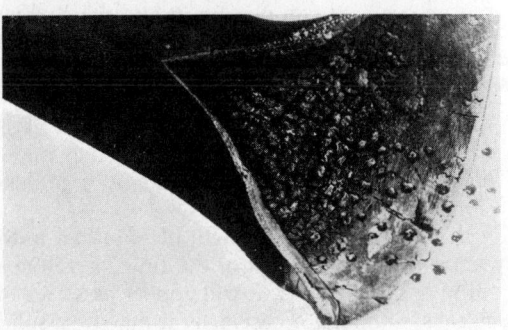

Debris in cooling system

Debris, rust and scale in the cooling system can cause the inside of a hose to weaken. This can usually be felt on the outside of the hose as soft or thinner areas.

damage both the hose and the connection points.

8. Refill the cooling system as detailed below and check for leaks.

DRAINING AND REFILLING

At least every two years or 30,000 miles (whichever comes first), the cooling system should be completely drained and refilled with the proper mixture of coolant and water. Many mechanics recommend that this be done once a year for extra protection against corrosion and subsequent overheating.

Though most coolants are labeled "permanent," this only means that the coolant will retain its anti-freezing characteristics. The required rust inhibitors and other chemicals which were added to the coolant during its manufacture will become less effective over a period of time. The following procedure covers the factory-recommended procedure for draining and refilling the system.

1. Allow the engine to cool, the remove the radiator cap.

2. Raise the front of the vehicle and support it safely with jackstands.

3. Open the radiator drain fitting (located at the bottom of the radiator)

Air Conditioning
SAFETY PRECAUTIONS

There are two particular hazards associated with air conditioning systems and they both relate to refrigerant gas.

First, the refrigerant gas is an extremely cold substance. When exposed to air, it will instantly freeze any surface it comes in contact with, including your eyes.

Second: fire. Although normally non-toxic, refrigerant gas becomes highly poisonous in the presence of an open flame. One good whiff of the vapor formed by burning refrigerant can be fatal. Keep all forms of fire (including cigarettes) well clear of the air conditioning system.

Any repair work to an air conditioning system should be left to a professional. DO NOT, under any circumstances, attempt to loosen or tighten any fittings or perform any work other than that outlined here.

CHECKING FOR OIL LEAKS

Refrigerant leaks show up only as oily areas on the various components because the compressor oil is transported around the entire system along with the refrigerant. Look for oily spots on all the hoses and lines, and especially on the hose and tube connections. If there are oily deposits, the system may have a leak, and you

should have it checked by a qualified repairman.

NOTE: *A small area of oil on the front of the compressor is normal and no cause for alarm.*

CHECK THE COMPRESSOR BELT

Refer to the section in this chapter on "Drive Belts".

CLEANING THE CONDENSER

Periodically inspect the front of the condenser for bent fins or foreign material (dirt, bugs, leaves, etc.) If any cooling fins are bent, straighten them carefully with needle nose pliers. You can remove any debris with a stiff bristle brush or hose.

OPERATE THE A/C SYSTEM PERIODICALLY

A lot of A/C problems can be avoided by simply running the air conditioner at least once a week, regardless of the season. Simply let the system run for at least 5 minutes a week (even in the winter), and you'll keep the internal parts lubricated as well as preventing the hoses from hardening.

REFRIGERANT LEVEL CHECK

The first order of business when checking the sight glass is to find the sight glass. It is located in the head of the receiver/drier. Once you've found it, wipe it clean and proceed as follows:

1. With the engine and the air conditioning system running, look for the flow of refrigerant through the sight glass. If the air conditioner is working properly, you'll be able to see a continuous flow of clear refrigerant through the sight glass, with perhaps an occasional bubble at very high temperatures.

2. Cycle the air conditioner on and off to make sure what you are seeing is clear refrigerant. Since the refrigerant is clear, it is possible to mistake a completely discharged system for one that is fully charged. Turn the system off and watch the sight glass. If there is refrigerant in the system, you'll see bubbles during the off cycle. If you observe no bubbles when the system is running, and the air flow from the unit in the car is delivering cold air, everything is OK.

3. If you observe bubbles in the sight glass while the system is operating, the system is low on refrigerant. Have it checked by a professional.

4. Oil streaks in the sight glass are an indication of trouble. Most of the time, if you see oil in the sight glass, it will appear as series of streaks, although occasionally it may be a solid stream of oil. In either case, it means that part of the charge has been lost.

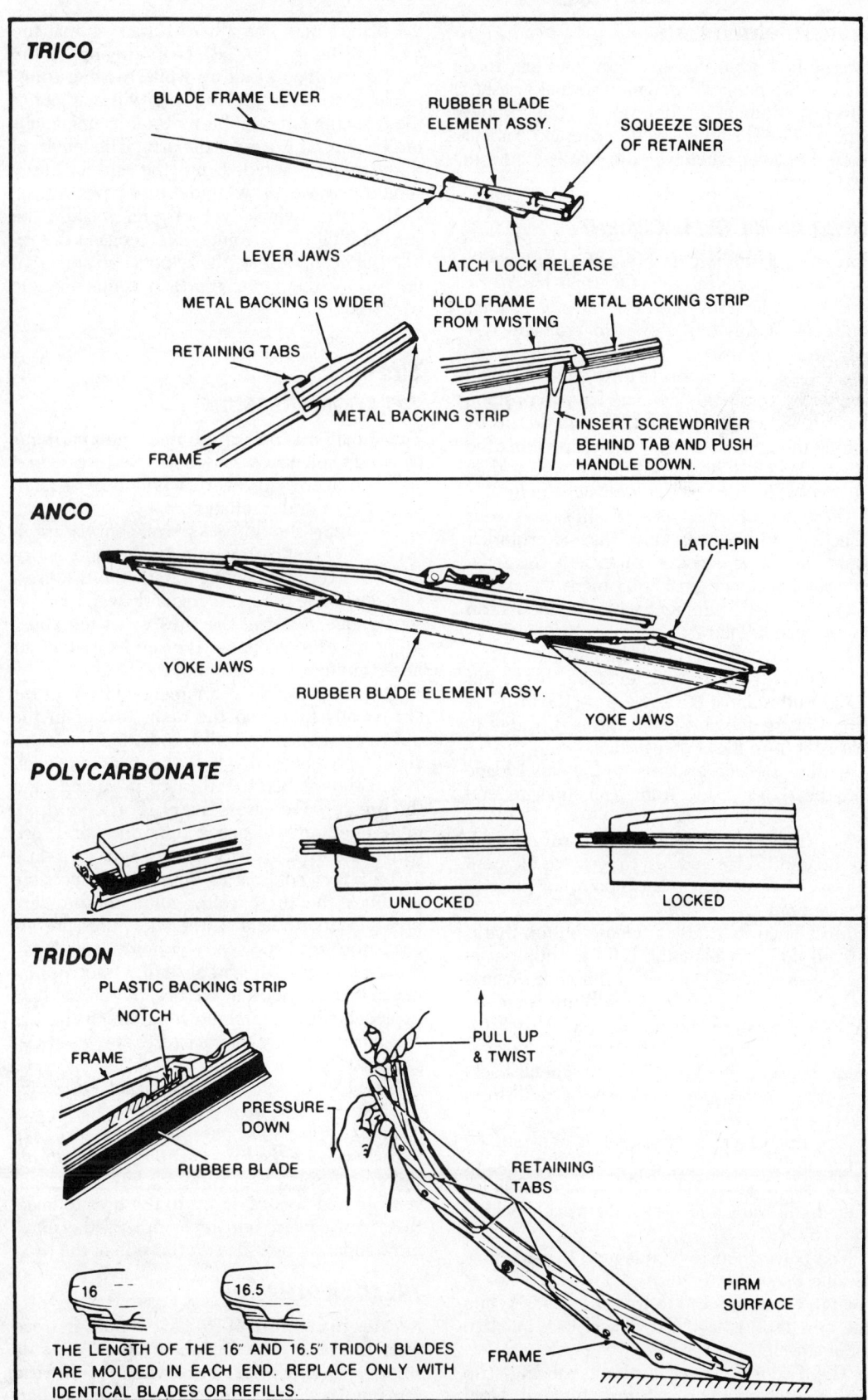

TRICO

BLADE FRAME LEVER

RUBBER BLADE ELEMENT ASSY.

SQUEEZE SIDES OF RETAINER

LEVER JAWS

LATCH LOCK RELEASE

METAL BACKING IS WIDER

HOLD FRAME FROM TWISTING

METAL BACKING STRIP

RETAINING TABS

METAL BACKING STRIP

FRAME

INSERT SCREWDRIVER BEHIND TAB AND PUSH HANDLE DOWN.

ANCO

LATCH-PIN

YOKE JAWS

RUBBER BLADE ELEMENT ASSY.

YOKE JAWS

POLYCARBONATE

UNLOCKED

LOCKED

TRIDON

PLASTIC BACKING STRIP

NOTCH

FRAME

PULL UP & TWIST

PRESSURE DOWN

RUBBER BLADE

RETAINING TABS

16

16.5

THE LENGTH OF THE 16" AND 16.5" TRIDON BLADES ARE MOLDED IN EACH END. REPLACE ONLY WITH IDENTICAL BLADES OR REFILLS.

FRAME

FIRM SURFACE

Wiper insert replacement

Windshield Wipers

Intense heat from the sun, snow and ice, road oils and the chemicals used in windshield washer solvents combine to deteriorate the rubber wiper refills. The refills should be replaced about twice a year or whenever the blades begin to streak or chatter.

WIPER REFILL REPLACEMENT

Normally, if the wipers are not cleaning the windshield properly, only the refill has to be replaced. The blade and arm usually require replacement only in the event of damage. It is only necessary (except on Tridon refills) to remove the arm or the blade to replace the refill (rubber part), though you may have to position the arm higher on the glass. You can do this by turning the ignition switch on and operating the wipers. When they are positioned where they are accessible, turn the ignition switch off.

There are several types of refills and your vehicle could have any kind, since aftermarket blades and arms may not use exactly the same type refill as the original equipment.

The original equipment wiper elements can be replaced as follows:

1. Lift the wiper arm off the glass.
2. Depress the release lever on the center bridge and remove the blade from the arm.
3. Lift the tab and pinch the end bridge to release it from the center bridge.
4. Slide the end bridge from the wiper blade and the wiper blade from the opposite end bridge.
5. Install a new element and be sure the tab on the end bridge is down to lock the element in place. Check each release point for positive engagement.

Most Trico styles use a release button that is pushed down to allow the refill to slide out of the release jaws. The new refill slide in and locks in place. Some Trico refills are removed by locating where the metal backing strip or the refill is wider. Insert a small screwdriver blade between the frame and the metal backing strip. Press down to release the refill from the retaining tab.

The Anco style is unlocked at one end by squeezing 2 metal tabs, and the refill is slid out of the frame jaws. When the new refill is installed, the tabs will click into place, locking the refill.

The polycarbonate type is held in place by a locking lever that is pushed downward out of the groove in the arm to free the refill. When the new refill is installed, it will lock in place automatically.

The Tridon refill has a plastic backing strip with a notch about an inch from the end. Hold the blade (frame) on a hard surface so that the frame is tightly bowed. Grip the tip of the backing strip and pull up while twisting counterclockwise. The backing strip will snap out of the retaining tab. Do this for the remaining tabs until the refill is free of the arm. The length of these refills is molded into the end and they should be replaced with identical types.

No matter which type of refill you use, be sure that all of the frame claws engage the refill. Before operating the wipers, be sure that no part of the metal frame is contacting the windshield.

Tires

INFLATION PRESSURE

Tire inflation is one of the most ignored items of auto maintenance. Gasoline mileage can drop as much as .8% for every 1 pound per square inch (psi) of under inflation.

Two items should be a permanent fixture in every glove compartment: a tire pressure gauge and a tread depth gauge. Check the tire pressure (including the spare) regularly with a pocket type gauge. Kicking the tires won't tell you a thing, and the gauge on the service station air hose is notoriously inaccurate.

The tire pressures recommended for your car are usually found on the door post or in the owner's manual. Ideally, inflation pressure should be checked when the tires are cool. When the air becomes heated it expands and the pressure increases. Every 10° rise (or drop) in temperature means a difference of 1 psi, which also explains why the tire appears to lose air on a very cold night. When it is impossible to check the tires "cold," allow for pressure build-up due to heat. If the "hot" pressure exceeds the "cold" pressure by more than 15 psi, reduce your speed, load or both. Otherwise internal heat is crated in the tire. When the heat approaches the temperature at which the tire was cured, during manufacture, the tread can separate from the body.

CAUTION: *Never counteract excessive pressure build-up by bleeding off air pressure (letting some air out). This will only further raise the tire operating temperature.* Before starting a long trip with lots of luggage, you can add about 2–4 psi to the tires to make them run cooler, but never exceed the maximum inflation pressure on the side of the tire.

TREAD DEPTH

All tires made since 1968, have 8 built-in tread wear indicator bars that show up as ½" wide smooth bands across the tire when ⅟₁₆" of tread remains. The appearance of tread wear indica-

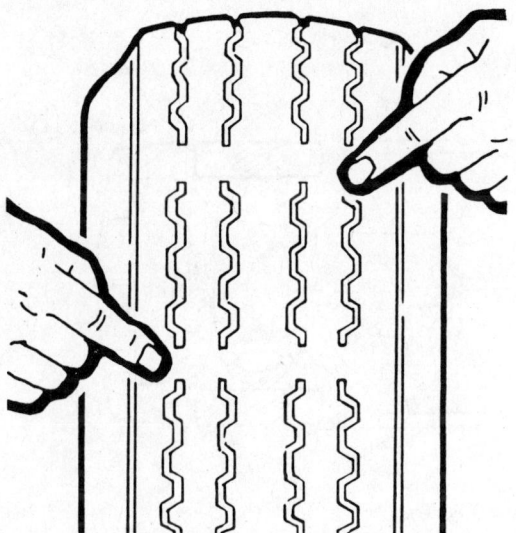

Replace a tire that shows the built-in "bump strip"

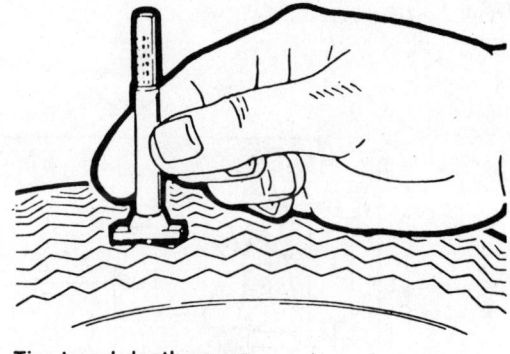

Tire tread depth gauge

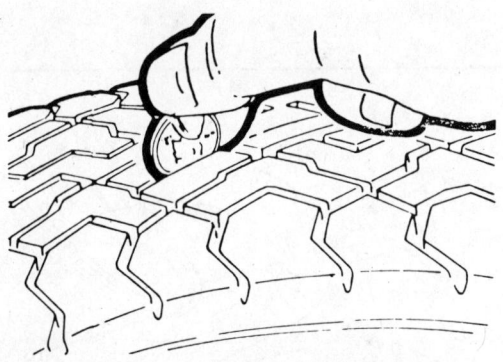

A penny used to determine tread depth

tors means that the tires should be replaced. In fact, many states have laws prohibiting the use of tires with less than $1/16$" tread.

You can check your own tread depth with an inexpensive gauge or by using a Lincoln head penny. Slip the Lincoln penny, upside down, into several tread grooves. If you can see the top of Lincoln's head in 2 adjacent grooves, the tires have less than $1/16$" tread left and should be replaced. You can measure snow tires in the same manner by using the "tails" side of the Lincoln penny. If you can see the top of the Lincoln memorial, it's time to replace the snow tires.

TIRE ROTATION

Tire wear can be equalized by switching the position of the tires about every 6000 miles. Including a conventional spare in the rotation pattern can give up to 20% more tire life.

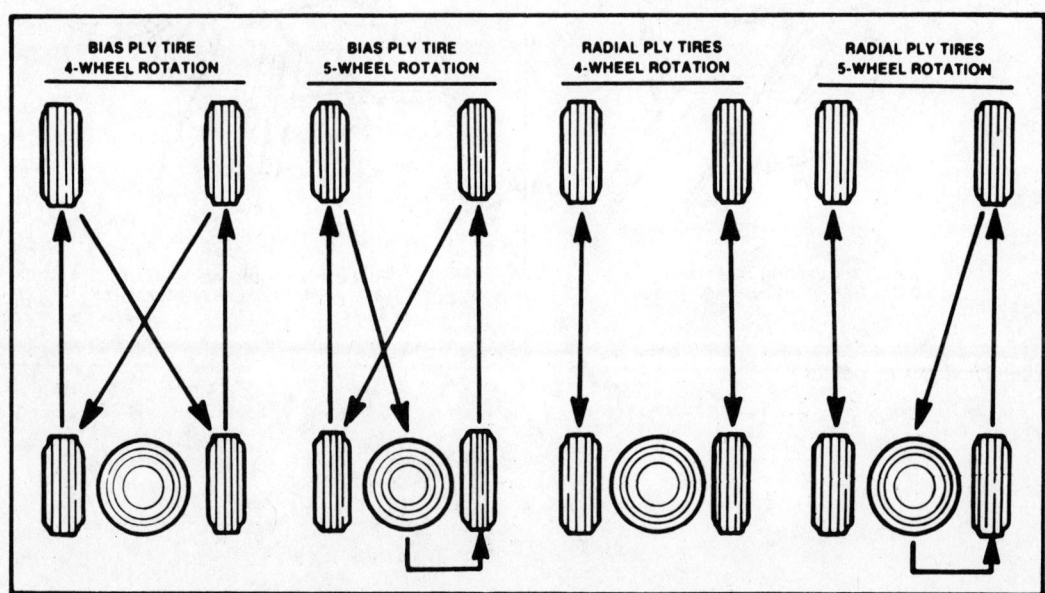

Tire rotation patterns

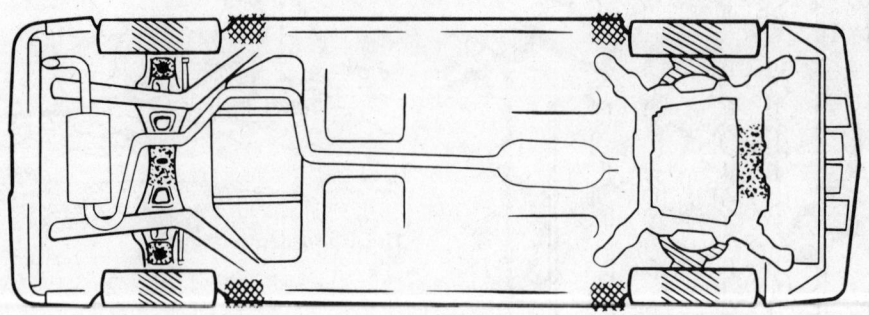

FRT ▶

▨▨▨ FRAME CONTACT HOIST

░░░ FLOOR JACK

▨▨▨ SUSPENSION CONTACT HOIST

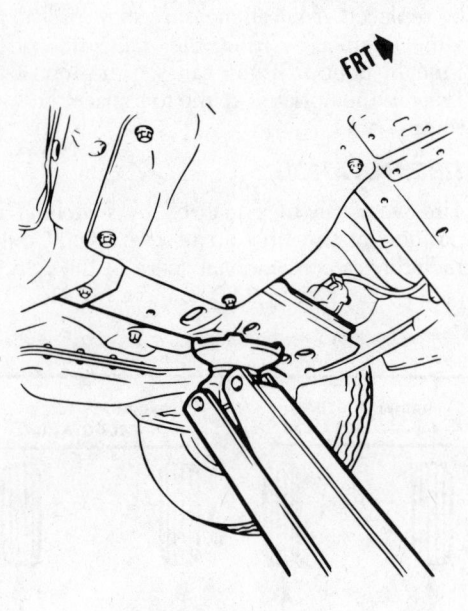

WHEN USING FLOOR JACK, LIFT
ON CENTER OF FRONT CROSSMEMBER

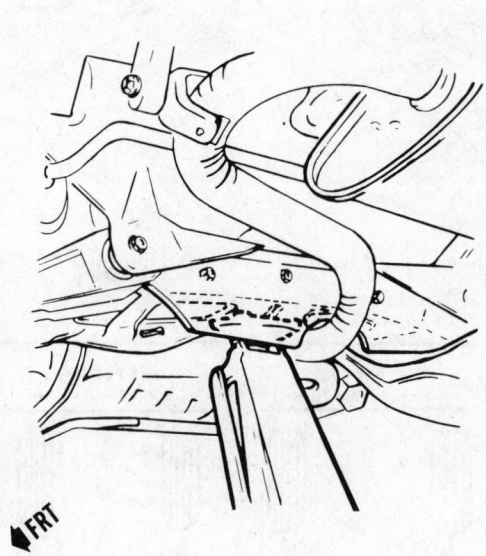

WHEN USING FLOOR JACK, LIFT ON
REAR SUSPENSION CENTER SUPPORT

Lifting and jacking points

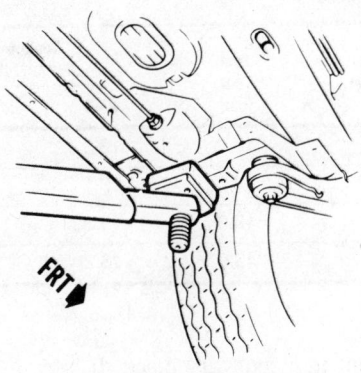

FRAME CONTACT HOIST
(REARWARD OF FRONT TIRE)

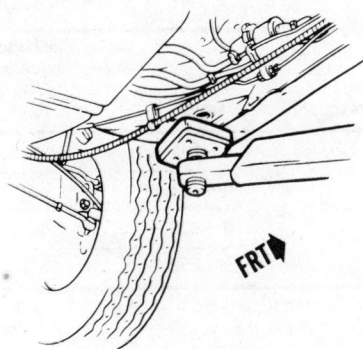

FRAME CONTACT HOIST
(FORWARD OF REAR TIRE)

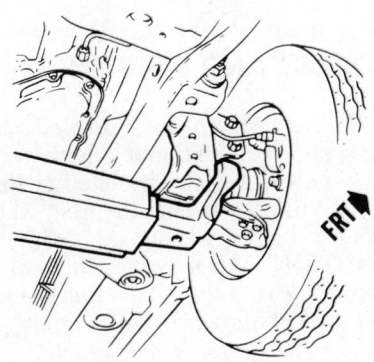

SUSPENSION CONTACT HOIST
(UNDER FRONT LOWER CONTROL ARM)

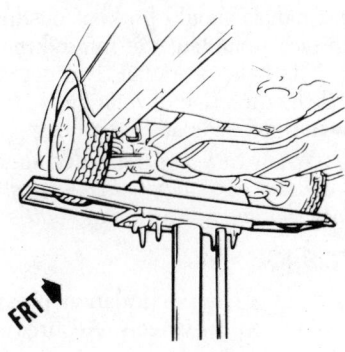

SUSPENSION CONTACT HOIST
(LIFTING ON REAR TIRES)

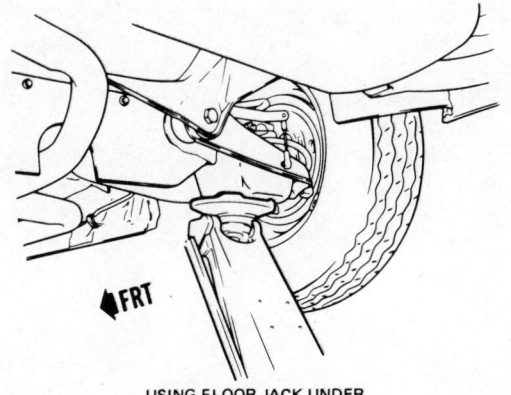

USING FLOOR JACK UNDER
REAR CONTROL ARM

Lifting and jacking points

Capacities Chart

Years	Engine	Crankcase Includes Filter (qt)	Transaxle (pts) *	Fuel Tank (gal)	Cooling System (qt)	
					w/AC	woAC
1985–86	6—181	5.0	12.0	18.0	13.3	13.6
	6—231	5.0	12.0	18.0	13.1	13.2
	6—263	6.5	11.6	18.0	13.3	13.3
	8—250	5.0	12.0	18.0	13.2	13.2

*Does not include torque converter

CAUTION: *Do not include the new "Space-saver" rotation pattern.*

There are certain exceptions to tire rotation, however. Studded snow tires should not be rotated, and radials should be kept on the same side of the car (maintain the same direction of rotation). The belts on radial tires get set in a pattern. If the direction of rotation is reversed, it can cause rough ride and vibration.

NOTE: *When radials or studded snows are taken off the car, mark them, so you can maintain the same direction of rotation.*

TIRE STORAGE

Store the tires at proper inflation pressure if they are mounted on wheels. All tires should be kept in a cool, dry place. If they are stored in the garage or basement, do not let them stand on a concrete floor, set them on strips of wood.

JACKING AND HOISTING

Jack the car at the engine cradle crossbar. The car can also be raised under the lower control arms. The car can also be lifted at the frame side rails, directly behind the front wheels and in front of the rear wheels.

CAUTION: *When raising the car by the frame side rails, be certain that the jack does not contact the catalytic converter.*

Tune-Up and Performance Maintenance

2

TUNE-UP PROCEDURES

The procedures listed here are intended as specific procedure. More general procedures are given in Chapter 9, Troubleshooting.

Neither tune-up nor troubleshooting can be considered independently, since each has a direct bearing on the other.

An engine tune-up is a service designed to restore the maximum capability of power, performance, economy and reliability in an engine, and, at the same time, assure the owner of a complete check and more lasting results in efficiency and trouble-free performance. Engine tune-up becomes increasingly important each year, to ensure that pollutant levels are in compliance with federal emissions standards.

It is advisable to follow a definite and thorough tune-up procedure. Tune-up consists of three separate steps: Analysis, the process of determining whether normal wear is responsible for performance loss, and whether parts require replacement or service; Parts Replacement or Service; and Adjustment, where engine adjustments are returned to the original factory specifications.

The extent of an engine tune-up is usually determined by the length of time since the previous service, although the type of driving and general mechanical conditioning of the engine must be considered. Specific maintenance should also be performed at regular intervals, depending on operating conditions.

Troubleshooting is a logical sequence of procedures designed to lead the owner or service man to the particular cause of trouble. The troubleshooting chapter of this manual is general in nature, yet specific enough to locate the problem. Service usually comprises two areas; diagnosis and repair. While the apparent cause of trouble, in many cases, is worn or damaged parts, performance problems are less obvious. The first job is to locate the problem and cause.

Once the problem has been isolated, refer to the appropriate section for repair, removal or adjustment procedures.

It is advisable to read the entire chapter before beginning a tune-up, although those who are more familiar with tune-up procedures may wish to go directly to the instructions.

Spark Plugs

A typical spark plug consists of a metal shell surrounding a ceramic insulator. A metal electrode extends downward through the center of the insulator and protrudes a small distance. Located at the end of the plug and attached to the side of the outer metal shell is the side electrode. The side electrode bends in at a 90° angle so that its tip is even with, and parallel to, the tip of the center electrode. The distance between these two electrodes (measured in thousandths of an inch) is called the spark plug

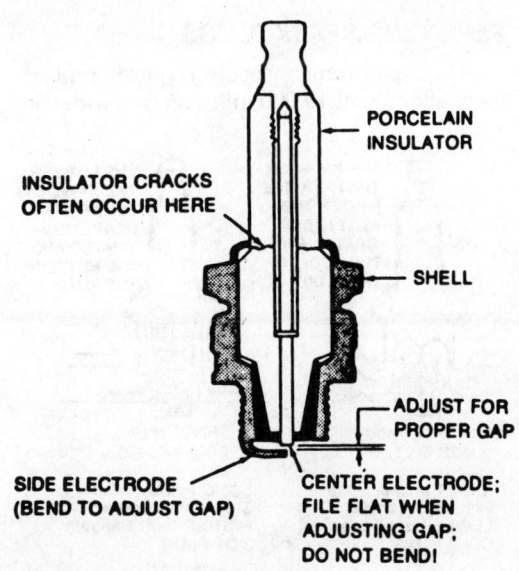

PORCELAIN INSULATOR

INSULATOR CRACKS OFTEN OCCUR HERE

SHELL

ADJUST FOR PROPER GAP

SIDE ELECTRODE (BEND TO ADJUST GAP)

CENTER ELECTRODE; FILE FLAT WHEN ADJUSTING GAP; DO NOT BEND!

Cross-section of a spark plug

gap. The spark plug in no way produces a spark but merely provides a gap across which the current can arc. The coil produces anywhere from 20,000 to 40,000 volts which travels to the distributor where it is distributed through the spark plug wires to the spark plugs. The current passes along the center electrode and jumps the gap to the side electrode, and, in so doing, ignited the air/fuel mixture in the combustion chamber.

SPARK PLUG HEAT RANGE

Spark plug heat range is the ability of the plug to dissipate heat. The longer the insulator (or the farther it extends into the engine), the hotter the plug will operate; the shorter the insulator the cooler it will operate. A plug that absorbs little heat and remains too cool will quickly accumulate deposits of oil and carbon since it is not hot enough to burn them off. This leads to plug fouling and consequently to misfiring. A plug that absorbs too much heat will have no deposits, but, due to the excessive heat, the electrodes will burn away quickly and in some instances, preignition may result. Preignition takes place when plug tips get so hot that they glow sufficiently to ignite the fuel/air mixture before the actual spark occurs. This early ignition will usually cause a pinging during low speeds and heavy loads.

The general rule of thumb for choosing the correct heat range when picking a spark plug is: if most of your driving is long distance, high speed travel, use a colder plug; if most of your driving is stop and go, use a hotter plug. Original equipment plugs are compromise plugs, but most people never have occasion to change their plugs from the factory-recommended heat range.

REPLACING SPARK PLUGS

A set of spark plugs usually requires replacement after about 10,000 miles on cars with con-

ventional ignition systems and after about 20,000 to 30,000 miles on cars with electronic ignition, depending on your style of driving. In normal operation; plug gap increased about 0.001 in. for every 1,000–2,500 miles. As the gap increased, the plug's voltage requirement also increases. It requires a greater voltage to jump the wider gap and about two or three times as much voltage to fire a plug at high speeds than at idle.

When you're removing spark plugs, you should work on one at a time. Don't start by removing the plug wires all at once, because unless you number them, they may become mixed up. Take a minute before you begin and number the wires with tape. The best location for numbering is near where the wires come out of the cap.

1. Twist the spark plug boot and remove the boot and wire from the plug. Do not pull on the wire itself as this will ruin the wire.

2. If possible, use a brush or rag to clean the area around the spark plug. Make sure that all the dirt is removed so that none will enter the cylinder after the plug is removed.

3. Remove the spark plug using the proper size socket. Turn the socket counterclockwise to remove the plug. Be sure to hold the socket

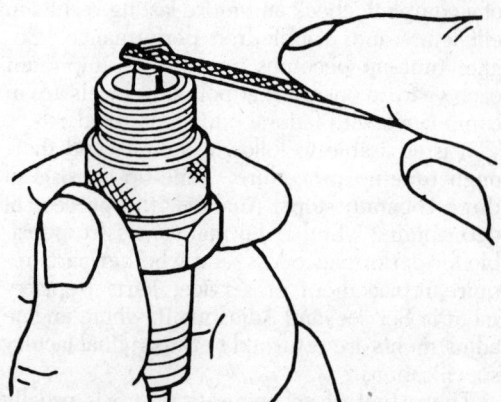

Plugs that are in good condition can be cleaned with a file and reused

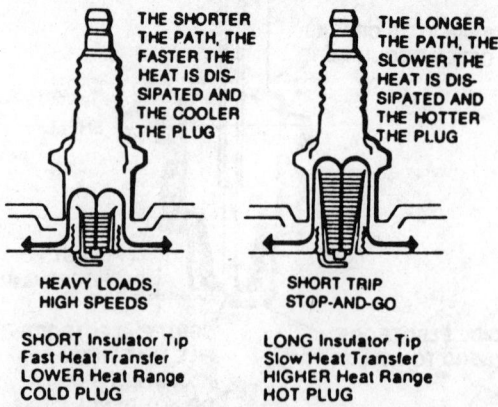

THE SHORTER THE PATH, THE FASTER THE HEAT IS DISSIPATED AND THE COOLER THE PLUG

THE LONGER THE PATH, THE SLOWER THE HEAT IS DISSIPATED AND THE HOTTER THE PLUG

HEAVY LOADS,
HIGH SPEEDS

SHORT TRIP
STOP-AND-GO

SHORT Insulator Tip
Fast Heat Transfer
LOWER Heat Range
COLD PLUG

LONG Insulator Tip
Slow Heat Transfer
HIGHER Heat Range
HOT PLUG

Spark plug heat range

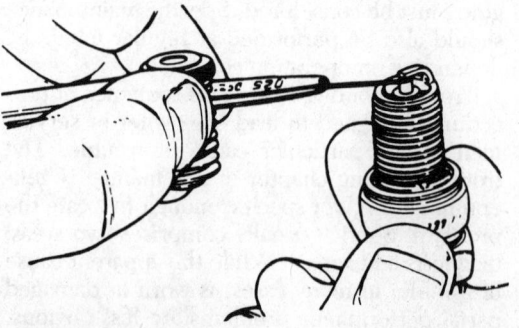

Checking spark plug gap

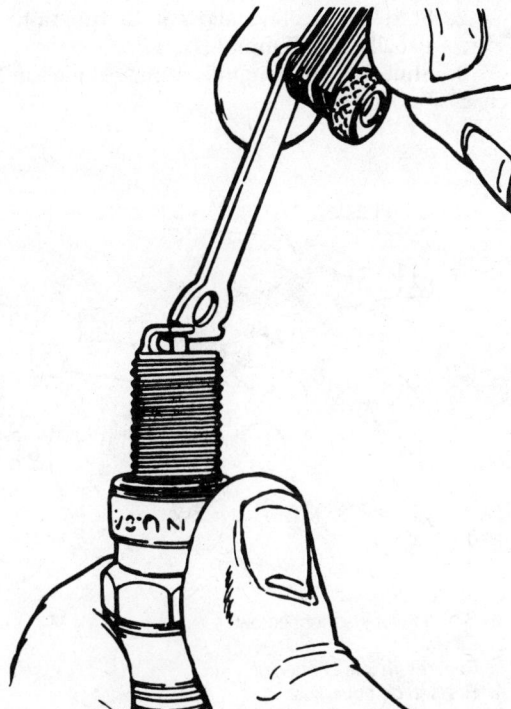

Bending the side electrode to adjust spark plug gap

straight on the plug to avoid breaking the plug, or rounding off the hex on the plug.

4. Once the plug is out, check it against the plugs shown in the color section to determine engine condition. This is crucial since plug readings are vital signs of engine condition.

5. Use a round wire feeler gauge to check the plug gap. The correct size gauge should pass through the electrode gap with a slight drag. If you're in doubt, try one sizesmaller and one larger. The smaller gauge should go through easily while the larger one shouldn't go through at all. If the gap is incorrect, use the electrode bending tool on the end of the gauge to adjust the gap. When adjusting the gap, always bend the side electrode. The center electrode is non-adjustable.

6. Squirt a drop of penetrating oil on the threads of the new plug and install it. Don't oil the threads too heavily. Turn the plug in clockwise by hand until it is snug.

7. When the plug is finger tight, tighten it with a wrench. If you don't have a torque wrench, tighten the plug as shown.

8. Install the plug boot firmly over the plug. Proceed to the next plug.

CHECKING AND REPLACING SPARK PLUG CABLES

Visually inspect the spark plug cables for burns, cuts, or breaks in the insulation. Check the spark plug boots and the nipples on the distributor

cap and coil. Replace any damaged wiring. If no physical damage is obvious, the wires can be checked with an ohmmeter for excessive resistance. (See the tune-up and troubleshooting section).

When installing a new set of spark plug cables, replace the cables one at a time so there will be no mixup. Start by replacing the longest cable first. Install the boot firmly over the spark plug. Route the wire exactly the same as the original. Insert the nipple firmly into the tower on the distributor cap. Repeat the process for each cable.

Ignition Timing
ADJUSTMENT

NOTE: *The 8–250 engine incorporates a magnetic timing probe hole for use with special electronic timing equipment. Consult manufacturer's instructions before using this system.*

1. Connect a timing light to the No. 1 spark plug wire according to the light manufacturer's instructions. DO NOT PIERCE THE SPARK PLUG WIRE TO CONNECT THE TIMING LIGHT.

2. Follow the instructions on the tune up label located in the engine compartment.

3. On models with Electronic Spark Timing (EST) distributor, disconnect the 4 terminal plug at the distributor. Some models may require grounding the diagnostic connector (ALCL) located under the left side of the dash.

4. Start the engine and run it at idle speed.

5. Aim the timing light at the degree scale just over the harmonic balancer.

6. Adjust the timing by loosening the securing clamp and rotating the distributor until the desired ignition advance is achieved. When the correct timing is set, tighten the clamp.

NOTE: *On the 8–250 cu., a special tool No. J-29791 is used to loosen the hold down nut.*

7. Adjust the timing, then replace and tighten the clamp. To advance the timing, rotate the

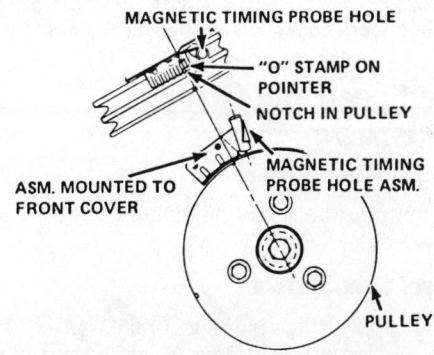

The 4.1L V8 uses a magnetic timing probe hole

distributor opposite the normal direction of rotor rotation. Retard the timing by rotating the distributor in the normal direction of rotor rotation.

NOTE: *On DFI systems (Digital Fuel Injection), the malfunction trouble codes must be cleared after removal or adjustment of the distributor. This is accomplished by removing battery voltage to terminal "R" for 10 seconds.*

HEI SYSTEM TACHOMETER HOOKUP

On all models, there is a terminal on the distributor cap marked TACH (usually next to the BAT terminal). Connect one tachometer lead to this terminal and the other lead to a suitable ground. On some tachometers, the leads must be connected to the TACH terminal and then to the positive battery terminal.

CAUTION: *Never ground the TACH terminal; serious module and ignition coil damage will result. If there is any doubt as to the correct tachometer hookup, check with the tachometer manufacturer.*

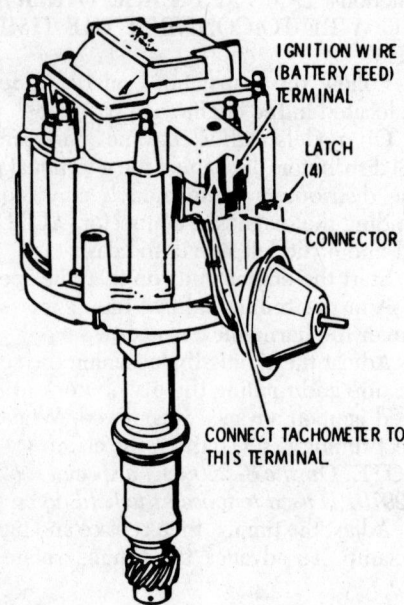

Tachometer connection for the HEI system

Idle Speed and Mixture

ADJUSTMENT

Gasoline Fuel System

No routine idle speed or mixture adjustments are possible.

Diesel Fuel System

1. Apply the parking brake, place the transmission selector lever in "park" and block the drive wheels.

2. Start engine and allow it to run until warm, usually 10–15 minutes.

3. Shut off the engine, remove the air cleaner assembly.

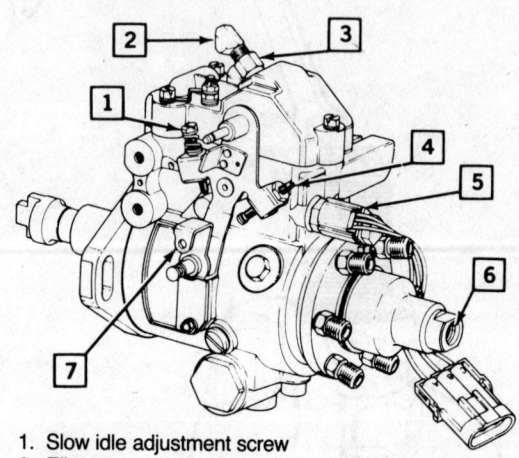

1. Slow idle adjustment screw
2. Elbow
3. Fuel return line connector
4. Pre-set, do not adjust
5. Metering valve sensor
6. Inlet
7. Throttle lever

Diesel injection pump adjustment points

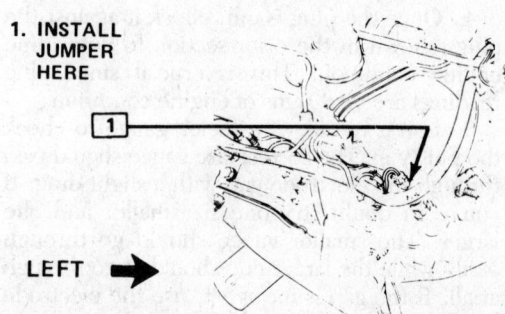

1. INSTALL JUMPER HERE

LEFT →

Jumper wire installation on diesel injection pump

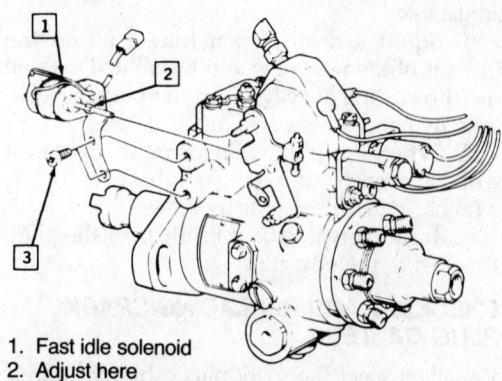

1. Fast idle solenoid
2. Adjust here
3. 8 N·m (69 lbs. in.)

Diesel injection pump fast idle solenoid

Engine Tune-Up Specifications

Years	Engine	Spark Plugs Type	Gap	Ignition Timing (deg.)	Idle Speed (rpm)
1985–86	6—181	R44TSX	.060	15B	computer controlled
	6—231	R44TS8	.080	15B	computer controlled
	8—250	R42CLTS6	.060	see underhood specifications sticker	
	6—263	Diesel		①	②

① Exc. Calif.: 5½ @ 1300 rpm in PARK
  Calif.: 6½ @ 1300 rpm in PARK
② Slow idle: 675
  Fast idle: 750

4. Clean the front cover RPM counter (probe holder) and the crankshaft balancer trim.

5. Install the magnetic pick-up probe of tool J-26925 fully into the RPM counter. Connect the battery leads; red to positive and black to negative.

6. Disconnect the two-lead connector at the generator.

7. Turn off all electrical accessories.

8. Allow no one to touch either the steering wheel or service brake pedal.

9. Start the engine and place the transmission selector lever in "Drive".

10. Check the slow idle speed reading against the one given on the "Vehicle Emission Information Label". Reset if required.

11 Unplug the connector from the fast idle cold advance (engine temp.) switch and install a jumper between the connector terminals. Do not allow the jumper to touch ground.

12. Check the fast idle solenoid speed against the one given on the "Vehicle Emission Information Label." Reset if required.

13. Remove the jumper and reconnect the connector to the temperature switch.

14. Recheck and reset the slow idle speed if necessary.

15. Shut off the engine.

16. Reconnect the lead at the generator.

17. Disconnect and remove the tachometer.

18. If equipped with cruise control adjust the servo throttle cable to minimum slack then install the clip on the servo stud.

Engine and Engine Rebuilding

3

IGNITION SYSTEM

The Delco-Remy High Energy Ignition (HEI) System is a breakerless, pulse triggered, transistor controlled, inductive discharge ignition system used on all GM passenger car engines as standard equipment. The ignition coil is located in the top of the distributor cap on all V6 and V8 engines.

The magnetic pick-up assembly located inside the distributor contains a permanent magnet, a pole pieces with internal teeth, and a pick-up coil. When the teeth of the rotating timer core and pole piece align, an induced voltage in the pick-up coil signals the electronic module to open the coil primary circuit. As the primary current decreases, a high voltage is induced in the secondary windings of the ignition coil, directing a spark through the rotor and high voltage leads to fire the spark plugs. The dwell period is automatically controlled by the electronic module and is increased with increasing engine rpm. The condenser (capacitor) located within the HEI distributor is provided for noise (static) suppression purposes only and is not a regularly replaced ignition system component.

All models are equipped with an Electronic Spark Timing (EST) distributor, which is part of the CCC System. Ignition timing is determined by the CCC Electronic Control Module (ECM). The EST module has seven terminals. Some models are equipped with Electronic Module Retard (EMR). This system uses a five terminal module which retards ignition timing a calibrated number of crankshaft degrees. Distributors with this system are equipped with vacuum advance. When replacing modules on these systems, be certain to obtain the correct part as the modules are not interchangeable.

Internal Ignition Coil
REMOVAL AND INSTALLATION

1. Disconnect the feed and module wire terminal connectors from the distributor cap.
2. Remove the ignition wire set retainer.
3. Remove the 4 coil cover-to-distributor cap screws and the coil cover.
4. Remove the 4 coil-to-distributor cap screws.
5. Using a blunt drift, press the coil wire spade terminals up out of distributor cap.
6. Lift the coil up out of the distributor cap.
7. Remove and clean the coil spring, rubber seal washer and coil cavity of the distributor cap.
8. Reverse the above procedures to install.

Distributor Cap
REMOVAL AND INSTALLATION

1. Remove the feed and module wire terminal connectors from the distributor cap.
2. Remove the retainer and spark plug wires from the cap.
3. Depress and release the 4 distributor cap-to-housing retainers and lift off the cap assembly.
4. Remove the coil from the old cap and install into the new cap.
5. Using a new distributor cap, reverse the above procedures to assemble.

ROTOR REPLACEMENT—ALL ENGINES

1. Disconnect the feed and module wire connectors from the distributor.
2. Depress and release the 4 distributor cap to housing retainers and lift off the cap assembly.

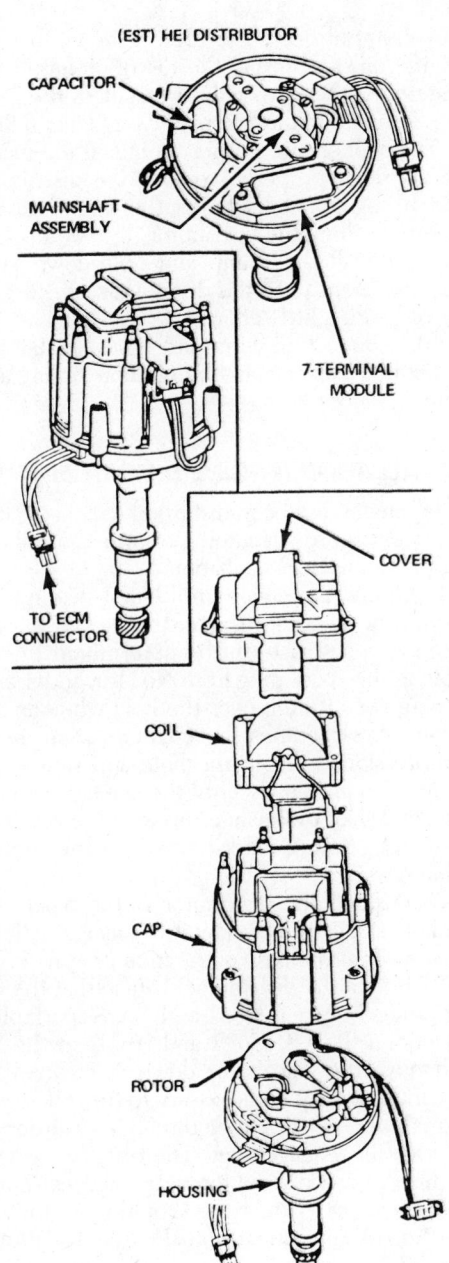

Typical HEI distributor

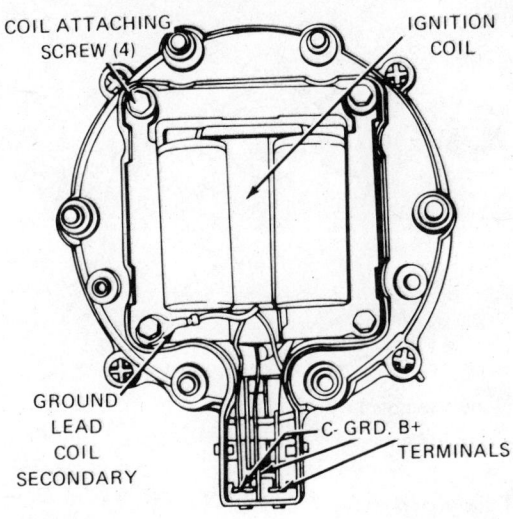

Internal coil removal

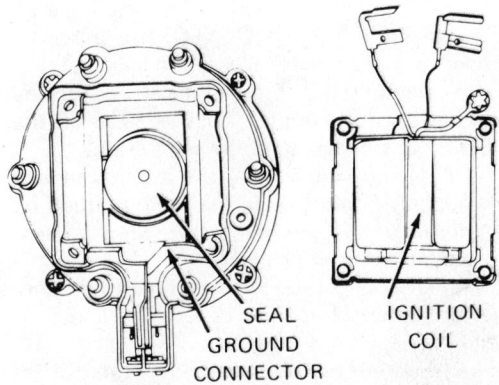

Coil removed from cap

3. Remove the two rotor attaching screws and rotor.

4. Reverse the above procedure to install.

Vacuum Advance Unit
REMOVAL AND INSTALLATION

1. Remove the distributor cap and rotor as previously described.

2. Disconnect the vacuum hose from the vacuum advance unit. Remove the module.

3. Remove the two vacuum advance retainings crews, pull the advance unit outward, rotate and disengage the operating rod from its tang.

4. Reverse the above procedure to install.

Control Module
REMOVAL AND INSTALLATION

1. Remove the distributor cap and rotor as previously described.

2. Disconnect the harness connector and pick-up coil spade connectors from the module (note their positions).

3. Remove the two screws and module from the distributor housing.

4. Coat the bottom of the new module with silicone dielectric compound.

NOTE: *If a five terminal or seven terminal module is replaced, the ignition timing must be checked and reset as necessary.*

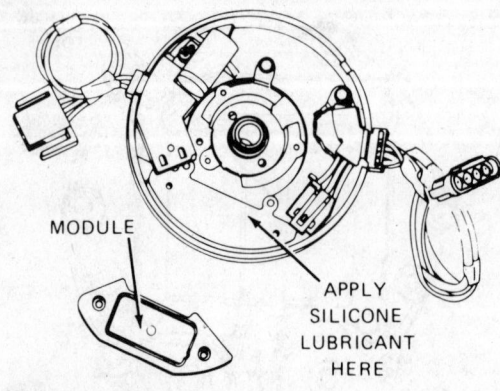

MODULE

APPLY
SILICONE
LUBRICANT
HERE

Module removal

Distributor
REMOVAL AND INSTALLATION

1. Disconnect the negative battery cable.

2. Tag and disconnect all wires leading from the distributor cap. DO NOT use a screwdriver or other tool to release the locking tabs.

3. Remove the distributor cap by turning the four latches counterclockwise. Lift off the distributor cap and carefully set it aside.

NOTE: *The location of the distributor cap "doghouse" must be in the same position on reinstallation in order to provide sufficient clearance for adjustment.*

4. Disconnect the four terminal ECM connector harness from the distributor if not already done.

5. Loosen, but do not remove, the distributor hold-down clamp.

NOTE: *On the 8–250 engine, a special tool (No. J-29791) will be required to loosen the hold-down clamp.*

6. Scribe a mark on the distributor body to note the initial position of the rotor. Pull the distributor upward until the rotor just stops turning (counterclockwise); note the position of the rotor once again. Remove the distributor.

NOTE: *Do not crank the engine with the distributor removed.*

7. On certain models, a thrust washer is used between the distributor drive gear and the crankcase. This washer may stick to the bottom of the distributor as it is removed. Always make sure that this washer is at the bottom of the distributor bore before installation. On DFI systems (Digital Fuel Injection), the malfunction trouble codes must be cleared after removal or adjustment of the distributor. This is accomplished by removing battery voltage to terminal "R" for 10 seconds.

8. To install the distributor, rotate the distributor shaft until the rotor aligns with the second mark you made (when the shaft stopped moving). Lubricate the drive gear with clean engine oil, and install the distributor into the engine. As the distributor is installed, the rotor should rotate to the first mark you made in Step 6. This will ensure proper timing. If the marks do not align properly, remove the distributor and try again. Don't forget the thrust washer when installing the distributor, if so equipped.

9. Install the clamp and hold-down bolt. Tighten them until the distributor can just be moved with a little effort.

10. Connect all wires and hoses. Install the distributor cap. Check the ignition timing and adjust if necessary.

INSTALLATION (ENGINE DISTURBED)

If the engine has been disturbed (cranked) after removing the distributor, perform the following procedure for installation:

1. Crank the engine until No. 1 piston is at the top of its compression stroke (TDC). The compression stroke can be determined by removing the spark plug from No. 1 cylinder and placing your thumb over the hole while an assistant slowly cranks the engine. Crank until compression is felt at the hole and then continue cranking slowly until the timing mark on the crankshaft pulley lines up with the zero degrees (O°) timing mark located on the timing chain cover.

2. Position the distributor in the block but do not, at this time, allow it to engage with its drive gear at the base of the mounting hole.

3. Rotate the distributor shaft so that the rotor points between No. 1 and No. 8 spark plug towers on the V8, No. 1 and No. 6 on the V6 and push the distributor down to engage the camshaft. It may be necessary to turn the rotor a small amount in either direction in order to achieve this engagement. The rotor will rotate slightly as the distributor gear engages. If installed correctly, the rotor should point toward the No. 1 spark plug terminal in the distributor cap.

4. Press down firmly on the distributor housing. This will ensure that the distributor shaft engages the oil pump shaft, thereby allowing the distributor to fully contact the engine block.

5. Install the hold-down clamp and tighten the bolt until it is snug.

6. Install the distributor cap, making sure that the rotor points to No. 1 terminal in the cap.

7. Attach all wires and hoses.

8. Start the engine. If it fails to start, or runs roughly, the distributor may be 180° out of time. Lift up on the distributor, turn the rotor one-half revolution, and install the distributor. Re-

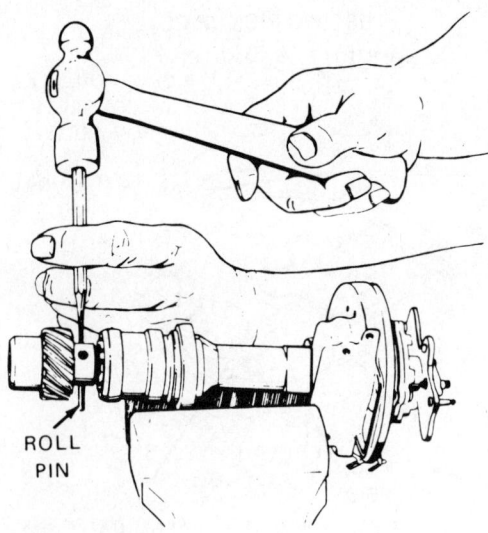

ROLL
PIN

Roll pin removal

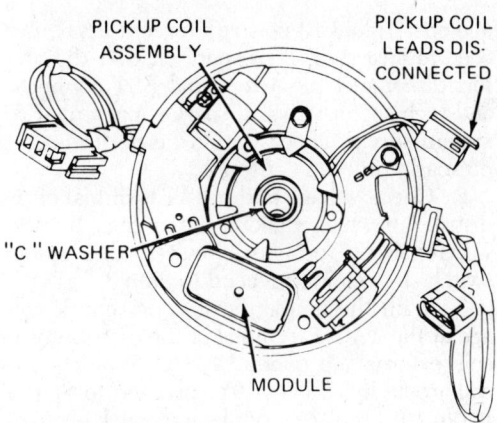

PICKUP COIL
ASSEMBLY

PICKUP COIL
LEADS DIS-
CONNECTED

"C" WASHER

MODULE

Magnetic shield removal

peat Steps 1–8 if the engine continues to run poorly.

9. Check the timing and change it as necessary.

SERVICE PROCEDURES (DISTRIBUTOR REMOVED)

Driven Gear Replacement

1. Mark the distributor shaft and gear so they can be reassembled in the same position. With the distributor removed, use a ⅛ in. pin punch and tap out the driven gear roll pin.
2. Hold the rotor end of shaft and rotate the driven gear to shear any burrs in the roll pin hole.
3. Remove the driven gear from the shaft.
4. Reverse the above procedure to install.

Mainshaft Replacement

1. With the driven gear and rotor removed, gently pull the mainshaft out of the housing.
2. Remove the advance springs, weights and slide the weight base plate off the mainshaft.
3. Reverse the above procedure to install.

Pole Piece, Magnet Or Pick-Up Coil Replacement

The pole piece, magnet, and pickup coil are serviced as an assembly. With the mainshaft out of its housing, remove the thee screws and the magnetic shield, remove the thin "C" washer on top of the pickup coil assembly, remove the pickup coil leads from the module, and remove the pickup coil as an assembly. Note the alignment marks when the drive gear is reinstalled.

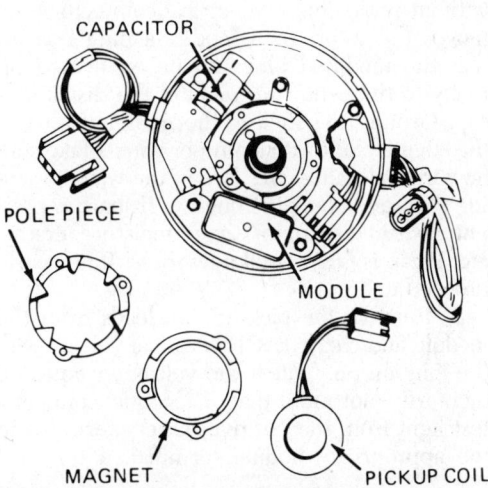

CAPACITOR

POLE PIECE

MODULE

MAGNET

PICKUP COIL

Pick-up coil removed and disassembled

TROUBLESHOOTING THE HEI SYSTEM

An accurate diagnosis is the first step to problem solution and repair. For several of the following steps, a modified spark plug (sides electrode removed) is needed. GM makes a modified plug (tool ST 125) which also has a spring clip to attach it to ground. Use of this tool is recommended, as there is less chance of being shocked. If a tachometer is connected to the TACH terminal on the distributor, disconnect it before proceeding with this test.

Engine Cranks But Will Not Run

1. Check for spark at the spark plugs by attaching the modified spark plug to one of the plug wires, grounding the modified plug shell on the engine and cranking the starter. Wear heavy gloves, use insulated pliers and make sure the ground is good. If no spark on one wire, check a second. If spark is present, HEI system is good. Check fuel system, plug wires,

and spark plugs. If no spark (except EST), proceed to next step. If no spark on EST distributor, disconnect the 4 terminal EST connector and recheck for spark. If spark is present, EST system service check should be performed. If no spark, proceed to Step 2.

2. Check voltage at the BAT terminal of the distributor while cranking the engine. If under 7V, repair the primary circuit to the ignition switch. If over 7V, proceed to Step 3.

3. With the ignition switch on, check voltage at the TACH terminal of the distributor or coil (external). If under 1V, coil connection or coil are faulty. If over 10V, proceed to Step 4. If 1 to 10V, replace module and check for spark from coil. See Step 4.

4. Remove distributor cap from distributor without removing its electrical connectors, removes the rotor, then modify a plug boot so that the modified plug can be connected directly to the center terminal of the distributor cap. Ground the shell of the modified plug to the engine block with a jumper wire. Make sure no wires, clothing, etc., are in the way of moving parts and crank the engine. If spark is present, inspect distributor cap for moisture, cracks, etc. If cap is OK, install new rotor. If no spark, proceed to Step 5.

5. Remove the pick-up coil leads from the module and check TACH terminal voltage with the ignition on. Watch the voltmeter and momentarily (not more than 5 seconds) connect a test light from the positive battery terminal to the appropriate module terminal: 4 terminal

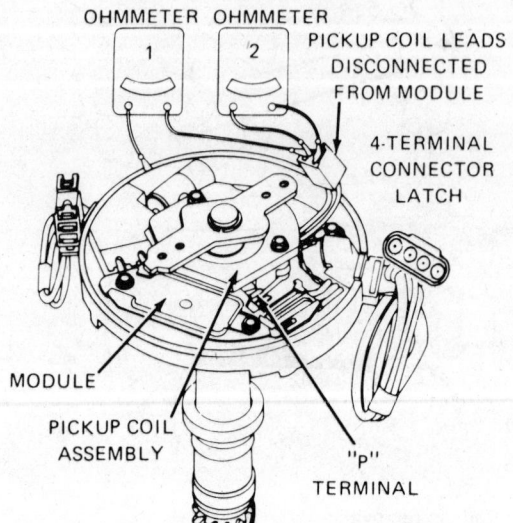

TESTING PICKUP COIL

Pick-up coil test

module, terminal "G" (small terminal); 5 terminal module (ESS or ESC), terminal "D"; 5 terminal module (EMR) terminal "H"; 7 terminal module, terminal "P". If no drop in voltage, test the module, check module ground, and check for open in wires from cap to distributor. If OK, replace module. If voltage drops, proceed to nextstep.

NOTE: *4 terminal modules may be tested with simple tools, according to the Module Test procedure which follows.*

6. Reconnect modified plug to ignition coil as instructed in Step 4, and check for spark as the test light is removed from the appropriate module terminal (see Step 5 for appropriate terminal). Do not connect test light for more than 5 seconds. If spark is present, problem is with pick-up coil or connections. Pick-up coil resistance should be 500–1500 ohms and not grounded. If no spark, proceed to next step.

7. Check the coil ground by attaching a test light from the BAT terminal of the cap to the coil ground wire. If the light lights when the ignition is on, replace the ignition coil and repeat Step 6. If the light does not light, repair the ground. If no spark is present, replace the module and reinstall the original coil. Repeat Step 6 again. If no spark is present, replace the original ignition coil with a good one.

MODULE TESTING

NOTE: *This procedure applies only to 4 terminal HEI modules.*

1. Remove the module from the distributor as previously outlined. Connect a 12VDC test lamp between the B and C module terminals.

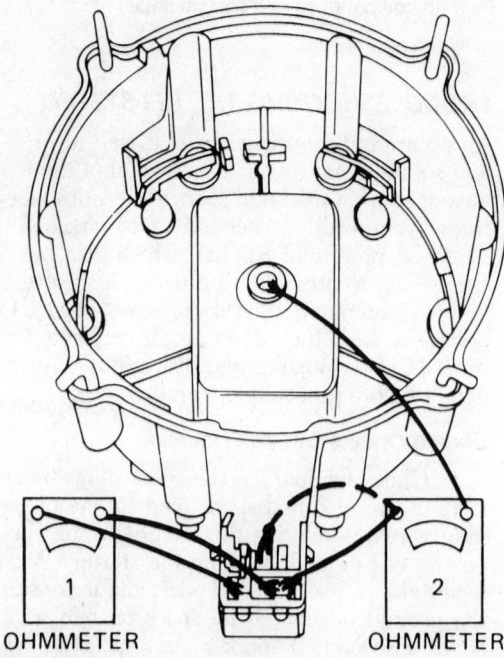

Ignition coil test

Alternator and Regulator Specifications

| Engine | Model | Alternator | | Regulator | |
		Field Current @ 12v (amps)	Output (amps)	Manufacturer	Volts @ 75[DE]F
6—181	Delco 15SI		85	Delco	<16
6—231	Delco 17SI		97	Delco	<16
6—263	Delco 12SI		94	Delco	<16
8—250	Delco 17SI	Ratings are stamped on the alternator case			

2. Connect a jumper wire from a 12VDC source to the B module terminal.

3. Connect the module ground terminal to a good ground. If the test lamp lights, the module is defective and must be replaced.

4. Connect a jumper wire between the B and G module terminals. The test lamp will light if the module is okay.

NOTE: *When using an auxiliary starter switch on AEI systems, the distributor BATT lead must be disconnected. Failure to do this may cause damage to the grounding circuit in the ignition switch.*

HEI SYSTEM TACHOMETER HOOKUP

On all models, there is a terminal on the distributor cap marked TACH (usually next to the BAT terminal). Connect one tachometer lead to this terminal and the other lead to a suitable ground. On some tachometers, the leads must be connected to the TACH terminal and then to the positive battery terminal.

CAUTION: *Never ground the TACH terminal; serious module and ignition coil damage will result. If there is any doubt as to the correct tachometer hookup, check with the tachometer manufacturer.*

CHARGING SYSTEM

Alternator

ALTERNATOR PRECAUTIONS

To prevent damage to the alternator and regulator, the following precautions should be taken when working with the electrical system:

1. Never reverse the battery connections.

2. Booster batteries for starting must be connected properly: positive-to-positive and negative-to-negative.

3. Disconnect the battery cables before using a fast charger; the charger has a tendency to force current through the diodes in the opposite direction for which they were designed. This burns out the diodes.

4. Never use a fast charger as a booster for starting the vehicle.

5. Never disconnect the voltage regulator while the engine is running.

6. Avoid long soldering times when replacing diodes or transistors. Prolonged heat is damaging to AC generators.

7. Do not use test lamps of more than 12 volts (V) for checking diode continuity.

8. Do not short across or ground any of the terminals on the AC generator.

9. The polarity of the battery, generator, and regulator must be matched and considered before making any electrical connections within the system.

10. Never operate the alternator on an open circuit. Make sure that all connections within the circuit are clean and tight.

11. Disconnect the battery terminals when performing any service on the electrical system. This will eliminate the possibility of accidental reversal of polarity.

12. Disconnect the battery ground cable if arc welding is to be done on any part of the car.

REMOVAL AND INSTALLATION

1. Disconnect the negative battery cable.

2. Tag and disconnect the battery charge wire, 3-prong connector and the ground wire at the back of the alternator.

3. Remove the brace at the back of the alternator (if so equipped).

4. Loosen the adjusting bolt, swivel the alternator in and remove the drive belt.

5. Loosen the power steering pump brace mounting bolts.

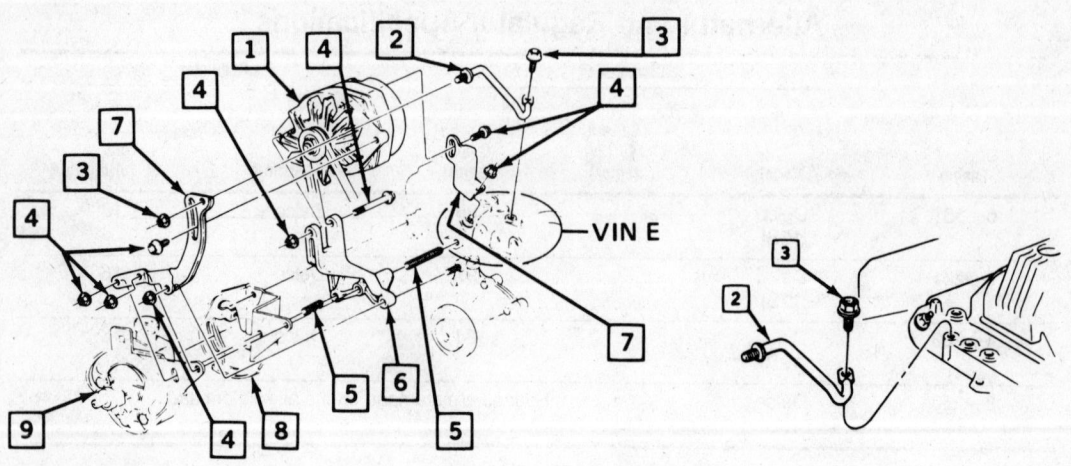

1. Generator
2. Brace
3. 17 N·m (20 lbs. ft.)
4. 50 N·m (37 lbs. ft.)
5. 15 N·m (11 lbs. ft.)

6. Support
7. Bracket
8. Power steering bracket
9. Power steering pump and bracket

Alternator mounting on the 3.0 & 3.8L V6

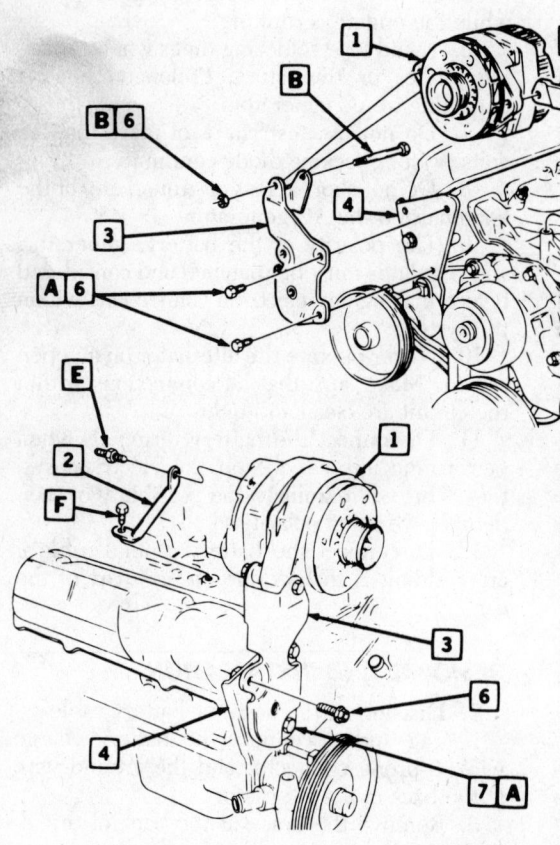

1. Generator
2. Brace
3. Generator support
4. Power steering support

5. Bracket
6. 55 N·m (41 lbs. ft.)
7. 28 N·m (21 lbs. ft.)
8. Air conditioning bracket

FASTENER TIGHTENING SEQUENCE

1. P/S pump support must first be assembled and tightened to engine
2. Bolts designated **A** must be tightened before other fasteners shown
3. Loose assemble bolts/studs **B**, **C**, and **D**, then tighten bolt and nut
4. With bracket **5** pushed so it contacts brace **8** and generator **1**, tighten bolt/stud **C**.
5. Tighten bolt **D**, brace must first be located by A/C compressor bolt
6. Loose assembly bolt/stud **E** and **F**. Tighten stud **E**, and then bolt **F**

Alternator mounting on the V6 diesel

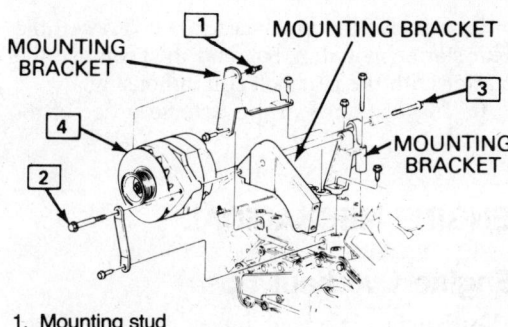

1. Mounting stud
2. 3. Mounting bolts
4. Generator

Alternator mounting on the 4.1L V8

6. Support the alternator, remove the mounting bolts and then remove the alternator.

7. Installation is in the reverse order of removal. Adjust the drive belt to have ¼–½ in. play midway along the longest freespan of the belt.

Voltage Regulator

An alternator with an integral voltage regulator is standard equipment. There are no adjustments possible with this unit.

STARTING SYSTEM

Starter

REMOVAL AND INSTALLATION

All Except Diesel

1. Disconnect the negative battery cable.
2. Raise and support the vehicle on jackstands.
3. Tag and disconnect all wires at the solenoid. Note color coding of wires for reinstallation.

NOTE: *On some models it may be necessary to remove the crossover pipe to complete this procedure.*

Starter Specifications

Engine	Model	Lock Test			No-Load Test			Brush* Spring Tension (oz.)
		Amps	Volts	Torque (ft. lb.)	Amps	Volts	RPM	
6—181	Delco 5MT	not recommended			50–75	10	6,000–11,900	35
6—231	Delco 5MT	not recommended			52–76	10	6,000–12,000	35
6—263	Delco ALU/GR	not recommended			125–170	10	3,200–** 4,100	35
8—250	Delco 5MT	not recommended			65–95	10	7,000–10,000	35

*Minimum tension
**Pinion speed

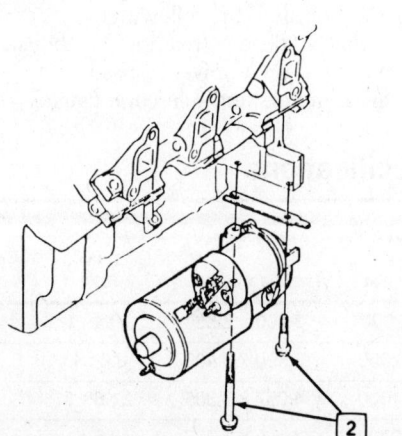

Starter mounting for the 3.0 & 3.8L V6; bolts marked #2 are torqued to 32 ft. lb.

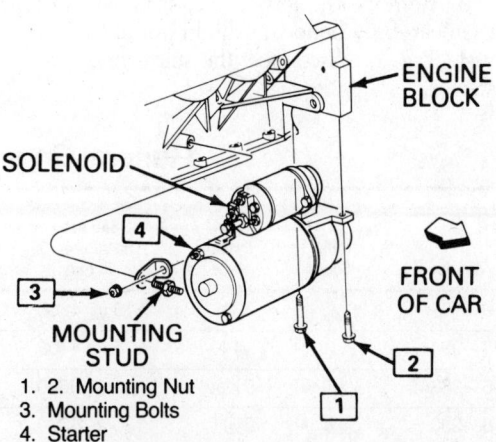

1. 2. Mounting Nut
3. Mounting Bolts
4. Starter

Starter mounting on the 4.1 L V8

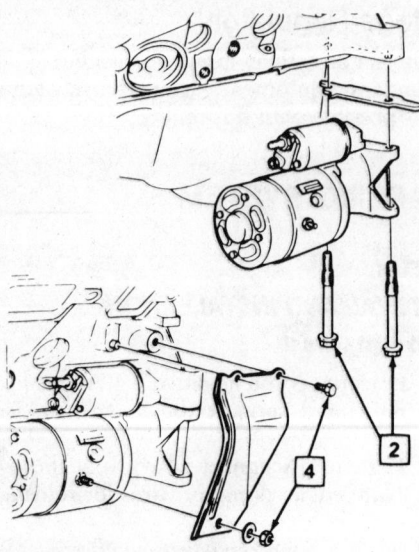

VIN T

Starter mounting on the V6 diesel; bolts marked #2 are torqued to 32 ft. lb. and those marked #4 are torqued to 18 ft. lb.

4. Remove starter support bracket mount bolts. On engines with solenoid heat shield, remove front bracket upper bolt and detach bracket from starter motor.

5. Loosen the front bracket bolt or nut and rotate bracket clear. Lower and remove starter. Note the location of any shims so that they may be replaced in the same positions upon installation.

6. Installation is in the reverse order of removal. Don't forget any shims.

Diesel

1. Disconnect the negative battery cable.

2. Raise and support the vehicle on jackstands.

3. Remove the lower starter shield nut and then carefully bend the shield out of the way.

4. Tag and disconnect the starter leads at the starter.

5. Remove the front starter bolt. Loosen the rear starter mounting bolt and then remove the starter with the rear bolt still in housing.

6. Installation is in the reverse order of removal.

ENGINE MECHANICAL

Engine Overhaul Tips

Most engine overhaul procedures are fairly standard. In addition to specific parts replacement procedures and complete specifications for your individual engine, this chapter also is a guide to accept rebuilding procedures. Examples of standard rebuilding practice are shown and should be used along with specific details concerning your particular engine.

Competent and accurate machine shop services will ensure maximum performance, reliability and engine life. Procedures marked with the symbol shown above should be performed by a competent machine shop, and are provided so that you will be familiar with the procedures necessary to a successful overhaul.

In most instances it is more profitable for the do-it-yourself mechanic to remove, clean and inspect the component, buy the necessary parts and deliver these to a shop for actual machine work.

On the other hand, much of the rebuilding work (crankshaft, block, bearings, piston rods, and other components) is well within the scope of the do-it-yourself mechanic.

TOOLS

The tools required for an engine overhaul or parts replacement will depend on the depth of your involvement. With a few exceptions, they will be the tools found in a mechanic's tool kit (see Chapter 1). More in-depth work will require any or all of the following:
- a dial indicator (reading in thousandths) mounted on a universal base
- micrometers and telescope gauges

General Engine Specifications

Engine	VIN Codes	Fuel System Type	SAE net Horsepower @ rpm	SAE net Torque ft. lb. @ rpm	Bore x Stroke	Comp Ratio	Oil Press. (psi.) @ 2000 rpm
6—181	E	2-bbl	110 @ 4800	145 @2600	3.800 x 2.660	8.4 : 1	40
6—231	3	MFI	125 @ 4400	195 @ 2000	3.800 x 3.400	8.0 : 1	40
6—263	T	Diesel	85 @ 3600	165 @ 1600	4.057 x 3.385	21.6 : 1	40
8—250	8	DFI	135 @ 4200	190 @ 2000	3.465 x 3.307	8.5 : 1	30

MIF: Multi-Point Fuel Injection
DFI: Digital Fuel Injection

- jaw and screw-type pullers
- scraper
- valve spring compressor
- ring groove cleaner
- piston ring expander and compressor
- ridge reamer
- cylinder hone or glaze breaker
- Plastigage®
- engine stand

The use of most of these tools is illustrated in this chapter. Many can be rented for a one-time use from a local parts jobber or tool supply house specializing in automotive work.

Occasionally, the use of special tools is called for. See the information on Special Tools and Safety Notice in the front of this book before substituting another tool.

Valve Specifications

Engine	Seat Angle (deg)	Face Angle (deg)	Spring Test Pressure (lbs. @ in.)	Spring Installed Height (in.)	Stem to Guide Clearance (in.)		Stem Diameter (in.)	
					Intake	Exhaust	Intake	Exhaust
6—181	45	45	220 @ 1.340	1.727	.0015–.0035	.0015–.0032	.3401–.3412	.3405–.3412
6—231	45	45	220 @ 1.340	1.727	.0015–.0035	.0015–.0032	.3401–.3412	.3405–.3412
6—263	①	②	210 @ 1.220	1.670	.0010–.0027	.0015–.0032	.3425–.3432	.3420–.3427
8—250	45	44	182 @ 1.280	—	.0010–.0030	.0010–.0030	.3413–.3420	.3411–.3418

① Intake: 45; Exhaust: 31
② Intake: 44; Exhaust: 30

Camshaft Specifications

Engine	Journal Diameter					Bearing Clearance	Lobe lift		End Play
	1	2	3	4	5		Int.	Exh.	
6—181	1.7680	1.7680	1.7680	1.7680	1.7680	①	.406	.406	—
6—231	1.7680	1.7680	1.7680	1.7680	1.7680	①	.406	.406	—
6—263	②	2.0155	1.9955	1.9755	—	.00395	.252	.279	.0018
8—250	—	—	—	—	—	.00275	.384	.396	—

① No. 1: .0005–.0025
  No. 2–5: .0005–.0035
② No. 1 bearing is not boreable and must be replaced as an assembly

Crankshaft and Connecting Rod Specifications

Engine	Crankshaft				Connecting Rod		
	Main Bearing Journal Dia.	Main Bearing Oil Clearance	Shaft End Play	Thrust on No.	Journal Dia.	Oil Clearance	Side Clearance
6—181	2.4990–2.5000	.0003–.0018	.0030–.0110	2	2.2487–2.2495	.0005–.0026	.006–.023
6—231	2.4990–2.5000	.0003–.0018	.0030–.0110	2	2.2487–2.2495	.0005–.0026	.006–.023
6—263	2.9993–3.0003	①	.0035–.0135	4	2.2490–2.2500	.0005–.0025	.008–.018
8—250	2.6378	.0004–.0027	.0010–.0070	3	1.9291	.0005–.0028	.008–.020

① Nos. 1,2,3: .0005–.0025
  No. 4: .0020–.0034

Piston and Ring Specifications

Engine	Ring Gap			Ring Side Clearance			Piston Clearance
	#1 Compr.	#2 Compr.	Oil Control	#1 Compr.	#2 Compr.	Oil Control	
6—181	.010–.020	.010–.020	.015–.055	.0030–.0050	.0030–.0050	.0035 max.	.0028–.0020
6—231	.010–.020	.010–.020	.015–.055	.0030–.0050	.0030–.0050	.0035 max.	.0008–.0020
6—263	.019–.027	.013–.021	.010–.022	.0050–.0070	.0030–.0050	.0010–.0050	.0035–.0045
8—250	.023–.026	.023–.026	.010–.050	.0016–.0037	.0016–.0037	self-sealing	.0010–.0018

Torque Specifications

Engine	Cyl. Head	Conn. Rod	Main Bearing	Crankshaft Damper	Flywheel	Manifold	
						Intake	Exhaust
6—181	80	40	100	200	60	47	25
6—231	80	40	100	200	60	47	25
6—263	①	42	89	300	76	41	31
8—250	②	22	85	18	63	③	18

① Bolts Nos. 5,6,11,12,13,14,59: 59 ft. lb.
　 All others: 142 ft. lb.
　 See the illustration in the text for bolt identification
② Tighten the bolts in two steps, first to 45 ft. lb.; then to 90 ft. lb.
③ Bolts nos. 1,2,3,4,: 15
　　　　　　5–16: 22
Then, tighten all bolts to 22 ft. lb. in the sequence shown in the text.
Then, go back and recheck all bolt torques.

INSPECTION TECHNIQUES

Procedures and specifications are given in this chapter for inspecting, cleaning and assessing the wear limits of most major components. Other procedures such as Magnaflux and Zyglo can be used to locate material flaws and stress cracks. Magnaflux is a magnetic process applicable only to ferrous materials. The Zyglo process coats the material with a flourescent dye penetrant and can be used on any material Check for suspected surface cracks can be more readily made using spot check dye. The dye is sprayed onto the suspected area, wiped off and the area sprayed with a developer. Cracks will show up brightly.

OVERHAUL NOTES

Aluminum has become extremely popular for use in engines, due to its low weight. Observe the following precautions when handling aluminum parts:

• Never hot tank aluminum parts (the caustic hot-tank solution will eat the aluminum.

• Remove all aluminum parts (identification tag, etc.) from engine parts prior to the tanking.

• Always coat threads lightly with engine oil or anti-seize compounds before installation, to prevent seizure.

• Never over-torque bolts or spark plugs especially in aluminum threads.

Stripped threads in any component can be repaired using any of several commercial repair kits (Heli-Coil, Microdot, Keenserts, etc.).

When assembling the engine, any parts that will be frictional contact must be prelubed to provide lubrication at initial start-up. Any product specifically formulated for this purpose can be used, but engine oil is not recommended as a prelube.

When semi-permanent (locked, but removable) installation of bolts or nuts is desired, threads should be cleaned and coated with Loctite® or other similar, commercial non-hardening sealant.

Standard Torque Specifications and Fastener Markings

In the absence of specific torques, the following chart can be used as a guide to the maximum safe torque of a particular size/grade of fastener.

- There is no torque difference for fine or coarse threads.
- Torque values are based on clean, dry threads. Reduce the value by 10% if threads are oiled prior to assembly.
- The torque required for aluminum components or fasteners is considerably less.

U.S. Bolts

SAE Grade Number	1 or 2			5			6 or 7		
Number of lines always 2 less than the grade number.									
Bolt Size (Inches)—(Thread)	Maximum Torque			Maximum Torque			Maximum Torque		
	Ft./Lbs.	Kgm	Nm	Ft./Lbs.	Kgm	Nm	Ft./Lbs.	Kgm	Nm
¼ — 20	5	0.7	6.8	8	1.1	10.8	10	1.4	13.5
— 28	6	0.8	8.1	10	1.4	13.6			
5/16 — 18	11	1.5	14.9	17	2.3	23.0	19	2.6	25.8
— 24	13	1.8	17.6	19	2.6	25.7			
⅜ — 16	18	2.5	24.4	31	4.3	42.0	34	4.7	46.0
— 24	20	2.75	27.1	35	4.8	47.5			
7/16 — 14	28	3.8	37.0	49	6.8	66.4	55	7.6	74.5
— 20	30	4.2	40.7	55	7.6	74.5			
½ — 13	39	5.4	52.8	75	10.4	101.7	85	11.75	115.2
— 20	41	5.7	55.6	85	11.7	115.2			
9/16 — 12	51	7.0	69.2	110	15.2	149.1	120	16.6	162.7
— 18	55	7.6	74.5	120	16.6	162.7			
⅝ — 11	83	11.5	112.5	150	20.7	203.3	167	23.0	226.5
— 18	95	13.1	128.8	170	23.5	230.5			
¾ — 10	105	14.5	142.3	270	37.3	366.0	280	38.7	379.6
— 16	115	15.9	155.9	295	40.8	400.0			
⅞ — 9	160	22.1	216.9	395	54.6	535.5	440	60.9	596.5
— 14	175	24.2	237.2	435	60.1	589.7			
1 — 8	236	32.5	318.6	590	81.6	799.9	660	91.3	894.8
— 14	250	34.6	338.9	660	91.3	849.8			

Metric Bolts

Relative Strength Marking	4.6, 4.8			8.8		
Bolt Markings						
Bolt Size Thread Size x Pitch (mm)	Maximum Torque			Maximum Torque		
	Ft./Lbs.	Kgm	Nm	Ft./Lbs.	Kgm	Nm
6 x 1.0	2–3	.2–.4	3–4	3–6	.4–.8	5–8
8 x 1.25	6–8	.8–1	8–12	9–14	1.2–1.9	13–19
10 x 1.25	12–17	1.5–2.3	16–23	20–29	2.7–4.0	27–39
12 x 1.25	21–32	2.9–4.4	29–43	35–53	4.8–7.3	47–72
14 x 1.5	35–52	4.8–7.1	48–70	57–85	7.8–11.7	77–110
16 x 1.5	51–77	7.0–10.6	67–100	90–120	12.4–16.5	130–160
18 x 1.5	74–110	10.2–15.1	100–150	130–170	17.9–23.4	180–230
20 x 1.5	110–140	15.1–19.3	150–190	190–240	26.2–46.9	160–320
22 x 1.5	150–190	22.0–26.2	200–260	250–320	34.5–44.1	340–430
24 x 1.5	190–240	26.2–46.9	260–320	310–410	42.7–56.5	420–550

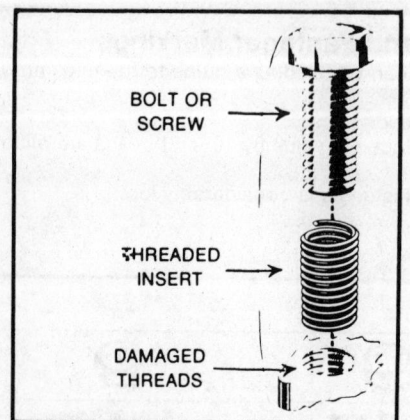

Damaged bolt holes can be repaired with thread repair inserts

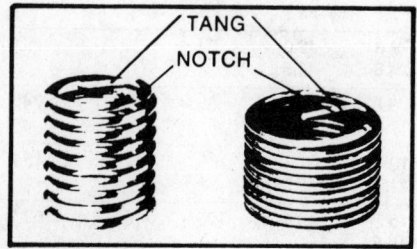

Standard thread repair insert (left) and spark plug thread insert (right)

REPAIRING DAMAGED THREADS

Several methods of repairing damaged threads are available. Heli-Coil® (shown here), Keen-serts® and Microdot® are among the most widely used. All involve basically the same principle—drilling out stripped threads, tapping the hole and installing a prewound insert—making welding, plugging and oversize fasteners unnecessary.

Two types of thread repair inserts are usually supplied—a standard type for most Inch Coarse, Inch Fine, Metric Course and Metric Fine thread sizes and a spark plug type to fit most spark plug port sizes. Consult the individual manufacturer's catalog to determine exact applications. Typical thread repair kits will contain a selection of prewound threaded inserts, a tap (corresponding to the outside diameter threads of the insert) and an installation tool. Spark plug inserts usually differ because they require a tap equipped with pilot threads and a combined reamer/tap section. Most manufacturers also supply blister-packed thread repair inserts separately in addition to a master kit containing a variety of taps and inserts plus installation tools.

Before effecting a repair to a threaded hole, remove any snapped, broken or damaged bolts or studs. Penetrating oil can be used to free frozen threads; the offending item can be removed with locking pliers or with a screw or stud extractor. After the hole is clear, the thread can be repaired, as follows:

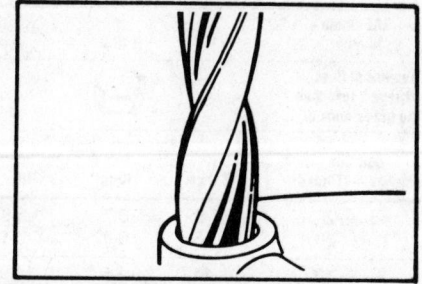

Drill out the damaged threads with specified drill. Drill completely through the hole or to the bottom of a blind hole

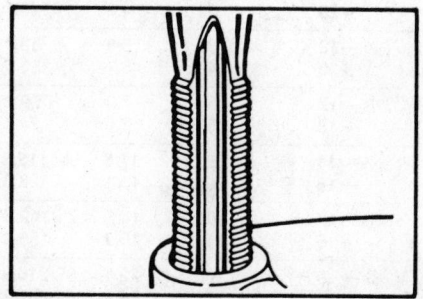

With the tap supplied, tap the hole to receive the thread insert. Keep the tap well oiled and back it out frequently to avoid clogging the threads

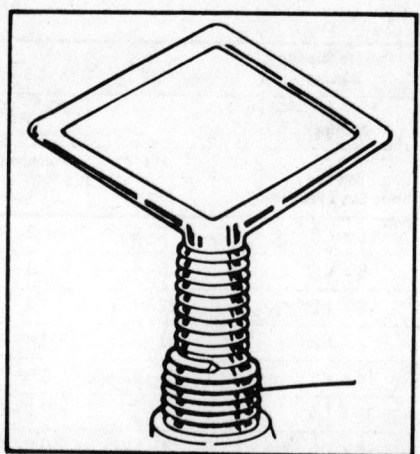

Screw the threaded insert onto the installation tool until the tang engages the slot. Screw the insert into the tapped hole until it is ¼–½ turn below the top surface, After installation break off the tang with a hammer and punch

CHECKING ENGINE COMPRESSION

A noticeable lack of engine power, excessive oil consumption and/or poor fuel mileage measured over an extended period are all indicators of internal engine war. Worn piston rings, scored or worn cylinder bores, blown head gaskets, sticking or burnt valves and worn valve seats are all possible culprits here. A check of each cylinder's compression will help you locate the problems.

As mentioned in the "Tools and Equipment" section of Chapter 1, a screw-in type compression gauge is more accurate that the type you simply hold against the spark plug hole, although it takes slightly longer to use. It's worth it to obtain a more accurate reading. Follow the procedures below for gasoline and diesel-engined cars.

Gasoline Engines

1. Warm up the engine to normal operating temperature.
2. Remove all spark plugs.
3. Disconnect the high-tension lead from the ignition coil.
4. On carbureted cars, fully open the throttle either by operating the carburetor throttle linkage by hand or by having an assistant "floor" the accelerator pedal. On fuel-injected cars, disconnect the cold start valve and all injector connections.
5. Screw the compression gauge into the No. 1 spark plug hole until the fitting is snug.
   NOTE: *Be careful not to crossthread the plug hole. On aluminum cylinder heads use extra care, as the threads in these heads are easily ruined.*
6. Ask an assistant to depress the accelerator pedal fully on both carbureted and fuel injected cars. Then, while you read the compression gauge, ask the assistant to crank the engine two or three times in short bursts using the ignition switch.
7. Read the compression gauge at the end of each series of cranks, and record the highest of these readings. Repeat this procedure for each of the engine's cylinders. Maximum compression should be 175-185psi. A cylinder's compression pressure is usually acceptable if it is not less than 80% of maximum. The difference between each cylinder should be no more than 12-14 pounds.

8. If a cylinder is unusually low, pour a tablespoon of clean engine oil into the cylinder through the spark plug hole and repeat the compression test. If the compression comes up after adding the oil, it appears that the cylinder's piston rings or bore are damaged or worn. If the pressure remains low, the valves may not be seating properly (a valve job is needed), or the head gasket may be blown near that cylinder. If compression in any two adjacent cylinders is low, and if the addition of oil doesn't help the compression, there is leakage past the head gasket. Oil and coolant water in the combustion chamber can result from this problem. There may be evidence of water droplets on the engine dipstick when a had gasket has blown.

Diesel Engines

Checking cylinder compression on diesel engines is basically the same procedure as on gasoline engines except for the following:

1. A special compression gauge adaptor suitable for diesel engines (because these engines have much greater compression pressures) must be used.
2. Remove the injector tubes and remove the injectors from each cylinder.
   NOTE: *Don't forget to remove the washer underneath each injector; it may get lost when the engine is cranked.*
3. When fitting the compression gauge adaptor to the cylinder head, make sure the bleeder of the gauge (if equipped) is closed.
4. When reinstalling the injector assemblies, install new washers underneath each injector.

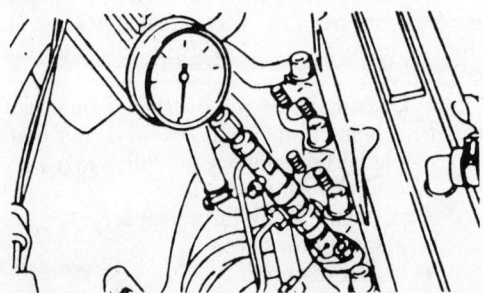

Diesel engines require a special compression gauge adaptor

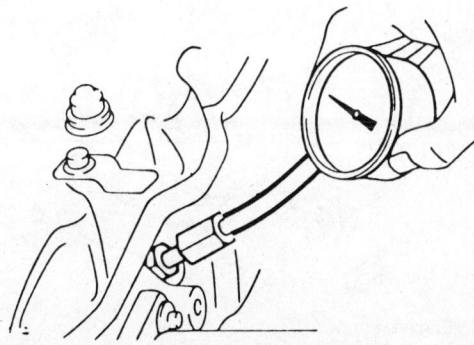

The screw-in type compression gauge is more accurate

Engine Removal and Installation

6–181 and 6–231

1. Disconnect the negative battery cable.
2. Tag and disconnect the air flow sensor wiring.
3. Disconnect the air intake duct. Drain the engine coolant.
4. Raise the front of the vehicle and support it on jackstands.
5. Unscrew the retaining bolts and separate the exhaust pipe from the manifold.
6. Loosen and remove the engine mount bolts.
7. Remove the bolts and then disconnect the driveline vibration absorber.
8. Tag and disconnect the starter wiring and then remove the starter.
9. Disconnect the A/C compressor and position it out of the way. DO NOT disconnect the refrigerant lines.
10. Disconnect the hydraulic lines at the power steering pump and wire them out of the way.
11. Loosen and remove the lower transaxle-to-engine bolts.

NOTE: *One bolt is situated between the transaxle case and the engine block. It is in-*

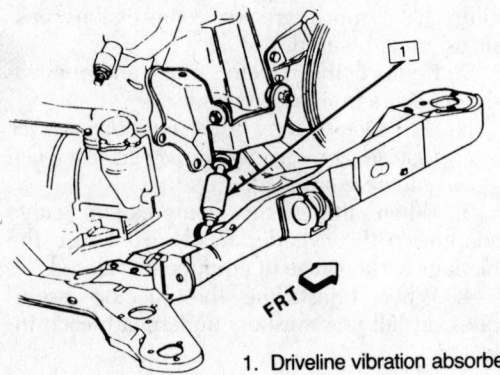

1. Driveline vibration absorber

Driveline vibration damper

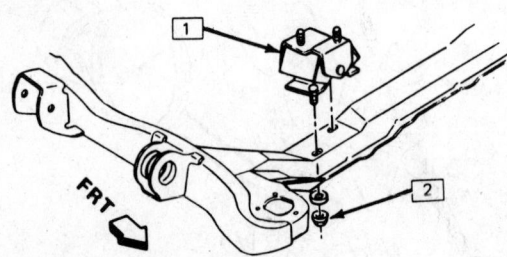

1. Engine mount
2. Nut 41 N·m (30 lbs. ft.)

3.0 & 3.8L right side engine mount

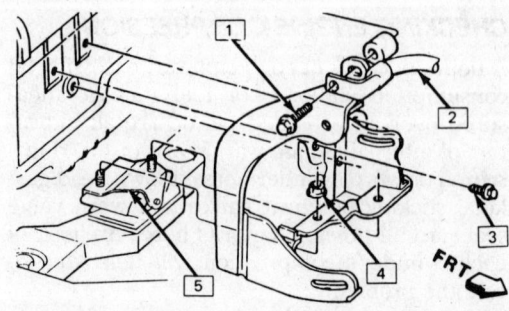

1. Bolt 50 N·m (37 lbs. ft.)
2. Negative battery cable
3. Bolt 95 N·m (70 lbs. ft.)
4. Nut 35 N·m (25 lbs. ft.)
5. Engine mount

3.0 & 3.8L left side engine mount

stalled in the opposite direction of the other bolts.

12. Remove the flexplate cover. Matchmark the flexplate-to-torque converter relationship to insure proper alignment upon installation. Remove the flexplate-to-torque converter bolts.
13. Disconnect the engine support bracket at the transaxle and then lower the vehicle.
14. Disconnect the radiator and heater hoses at the engine and position them out of the way.
15. Remove the alternator and rotate to the cowl.
16. Disconnect the engine wiring harness.
17. Remove the remaining upper transaxle-to-engine bolts.
18. Install a lifting fixture to the engine and remove the engine from the vehicle.
19. Installation is the reverse of the removal procedure.

8–250

1. Disconnect the negative battery cable. Drain the radiator coolant.
2. Remove the air cleaner. Matchmark the hood to the support brackets and remove the hood.
3. Disconnect the A/C hose strap from the strut tower. Disconnect The A/C accumulator from its bracket and position it out of the way.
4. Tag and disconnect the canister hoses and ground wire from the accumulator bracket and then remove the bracket itself from the inner strut tower.
5. Disconnect or remove the cooling fans, the drive belt and the radiator and heater hoses.
6. Tag and disconnect the following:
   a. Oil pressure switch
   b. Coolant temperature sensor
   c. Distributor wires
   d. EGR solenoid
   e. Engine temperature switch
   f. Accelerator cable

g. Cruise control linkage

h. Transmission TV cable

7. Remove the cruise control diaphragm and its bracket.

8. Remove the vacuum supply hose and the exhaust cross over pipe.

9. Disconnect the oil cooler lines at the oil filter adapter, unscrew their mounting bracket at the transaxle and position them out of the way.

10. Remove the air cleaner mounting bracket.

11. **CAREFULLY** bleed the fuel pressure at the Schraeder valve and then disconnect the fuel lines at the throttle body.

CAUTION: *When bleeding the fuel system, be sure to have a container or rags on hand to catch excess fuel.*

12. Unscrew the fuel line bracket at the transaxle and wire the fuel lines out of the way.

13. Tag and disconnect the small vacuum line at the brake booster.

14. Tag and disconnect the AIR solenoid electrical and hose connections. Remove the AIR valves and bracket.

15. Tag and disconnect the wires at the following:

a. ISC

b. TPS

c. Fuel injectors

d. MAT sensor

e. Oxygen sensor

f. Throttle bodybase warmer

g. Alternator

16. Remove the idler pulley. Remove the power steering pump hose strap from the stud-headed bolt in front of the right cylinder head. Remove the stud-headed bolt.

17. Remove the AIR pipe clip near the No. 2 spark plug.

18. Remove the power steering pump and belt tensioner (with bracket). Wire them out of the way.

19. Raise the vehicle and support it on jack stands.

20. Tag and disconnect the starter wires and the ground wire at the cylinder block.

21. Remove the two flex plate covers. Remove the starter. Remove the three flexplate-to-converter bolts.

22. Remove the A/C compressor lower dust shield.

23. Remove the right front wheel. Remove the outer wheelhouse plastic shield.

24. Remove the A/C compressor mounting bolt and lower the compressor out of the way.

25. Remove the lower radiator hose.

26. Remove the driveline vibration dampener and its brackets from the lower right front of the engine and cradle.

27. Remove the three right front engine-to-transaxle bracket bolts.

28. Disconnect the exhaust pipe at the manifold. Remove the AIR pipe-to-converter bracket from the exhaust manifold stud.

NOTE: *Be careful not to lose the springs when detaching the exhaust pipe.*

29. Remove the lower right hand bell housing-to-engine bolt. Support the engine with a jack.

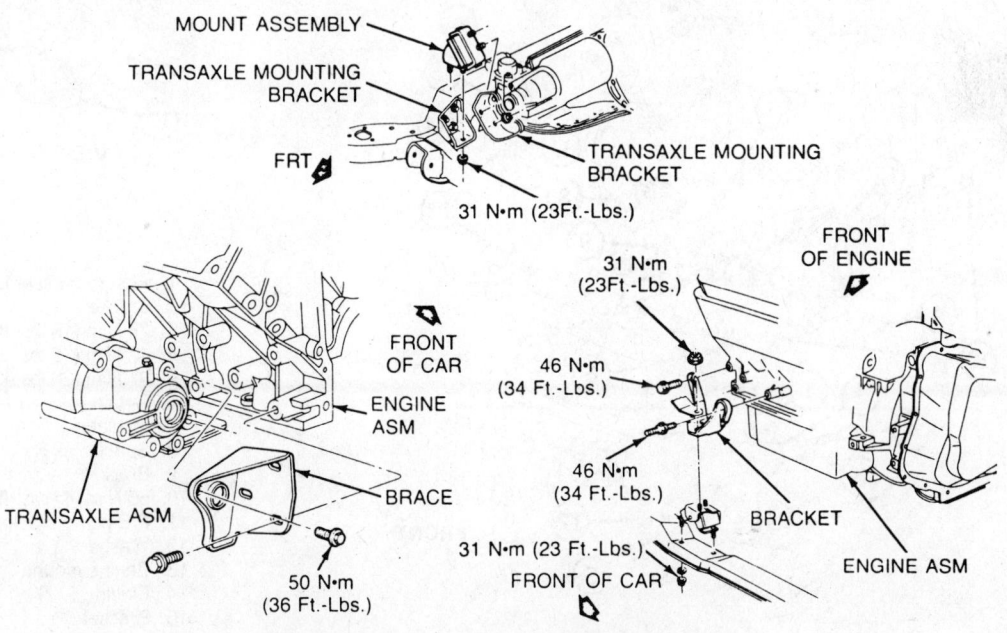

4.1L V8 engine/transaxle right side mounting points

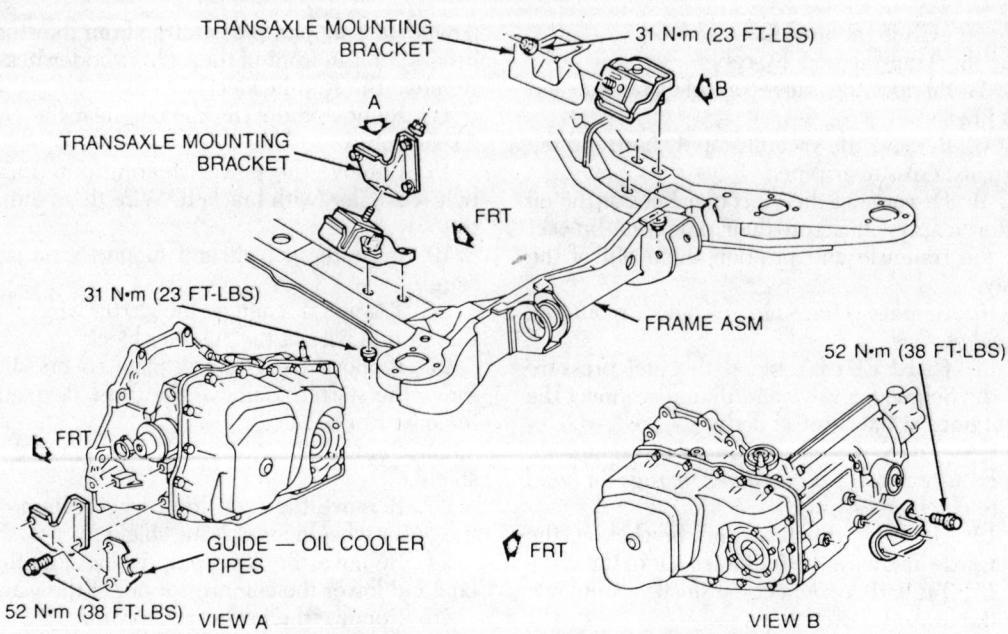

4.1L V8 engine/transaxle left side mounting points

30. Remove the five upper bell housing-to-engine bolts. Remove the three left front engine mount bracket-to-engine bolts.

31. Attach a suitable lifting fixture and remove the engine.

32. Installation is in the reverse order of removal.

6–263 Diesel

1. Disconnect the negative battery cable. Matchmark the hood to the support brackets and then remove the hood. Drain the cooling system.

2. Remove the serpentine drive belt. Remove the vacuum drive belt.

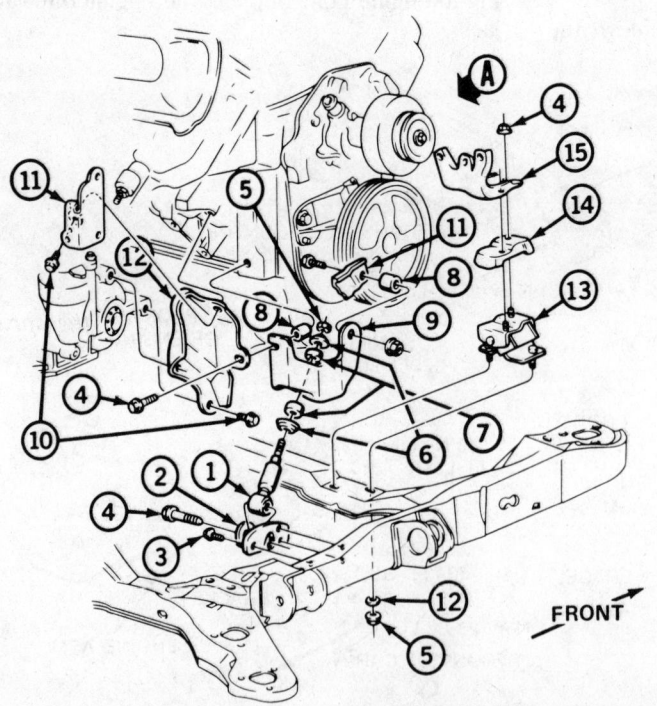

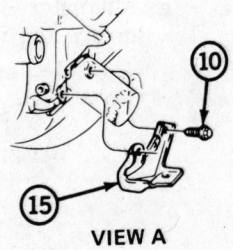

VIEW A

1. Absorber assembly
2. Bracket
3. 25 N·m (18 lbs. ft.)
4. 47 N·m (35 lbs. ft.)
5. 31 N·m (23 lbs. ft.)
6. Retainer
7. Insulator
8. Spacer
9. Bracket
10. 55 N·m (41 lbs. ft.)
11. P/S pump bracket
12. Washer
13. Engine mount
14. Shield
15. Bracket

V6 diesel engine mounting points

3. Remove the air cleaner. Install an air crossover screen cover (No. J-26996-1) or equivalent.

4. Tag and disconnect the ground wires at the inner fender panel and the engine ground strap.

5. Raise the vehicle and support it on jack stands.

6. Remove the engine-to-transaxle brace.

7. Remove the flywheel cover and then remove the flywheel-to-torque converter bolts.

8. Disconnect the exhaust pipe from the rear exhaust manifold.

9. Remove the engine mount-to-cradle retaining nuts and washers.

10. Remove the engine absorbers assembly from the frame bracket.

11. Tag and disconnect the following:
   a. Starter motor wires
   b. Glow plug wire at No. 2 cylinder
   c. Battery ground cable

12. Disconnect the lower oil cooler hose and cap the opening.

13. Remove the accessible power steering pump bracket fasteners. Lower the vehicle.

14. Remove the remaining power steering pump bracket/brace fasteners and lower the pump (with hoses connected) out of the way.

15. Disconnect the heater water return pipe.

16. Tag and disconnect the remaining glow plug leads and all other electrical leads connected to the engine.

17. Disconnect the engine harness at the cowl connector and body-mounted relays.

18. Remove the A/C compressor with the lines and brackets attached. Wire the compressor out of the way.

19. Disconnect all fuel and vacuum lines.
NOTE: *Cap all open fuel lines.*

20. Disconnect the throttle and TV cables at the injection pump and cable brackets.

21. Disconnect the upper oil cooler line. Cap the openings.

22. Remove the crossover pipe heat shield and the transaxle filler tube.

23. Remove the exhaust crossover pipe.

24. Install a suitable engine lifting device to the lift hooks on the bloc.

25. Use a floor jack to support the transaxle under the rear extension housing.

26. Remove the engine-to-transaxle bolts and remove the engine.

27. Installation is in the reverse order of removal.

Diesel Engine Valve Lifter Bleed-Down

If the intake manifold and valve rocker arms have been removed, it will be necessary to re-move, disassemble, drain and reassemble the lifters on that side. If the rocker arms have been loosened or removed, but the intake manifold was not removed, skip down to the Bleed-Down procedure.

REMOVAL

NOTE: *Keep lifters and pushrods in order! This is absolutely necessary for installation, since these parts have differences which could result in engine damage if not installed in their original positions!*

1. Remove intake manifold. Refer to "Intake Manifold".

2. Remove valve covers, rocker arm assemblies and pushrods.

3. Remove the valve lifter guide retainer bolts.

4. Remove the retainer guides and valve lifters.

DISASSEMBLY

1. Remove the retainer ring with a small screwdriver.

2. Remove pushrod seat and oil metering valve.

3. Remove plunger and plunger spring.

4. Remove check valve retainer from plunger, then remove valve and spring.

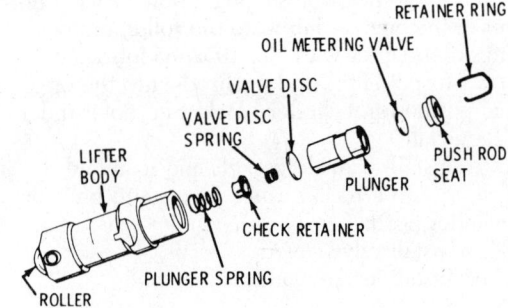

Diesel valve lifter disassembly

CLEANING AND INSPECTION

After lifters are disassembled, all parts should be cleaned in clean solvent. A small particle of foreign material under the check valve will cause malfunctioning of the lifter. Close inspection should be made for nicks, burrs or scoring of parts. If either the roller body or plunger is defective, replace with a new lifter assembly. Whenever lifters are removed, check as follows:

1. Roller should rotate freely, but without excessive play.

2. Check for missing or broken needle bearings.

3. Roller should be free of pits or rough-

ness. If present, check camshaft for similar condition. If pits or roughness are evident replace lifter and camshaft.

ASSEMBLY

1. Coat all lifter parts with a coating of clean kerosene or diesel fuel.
2. Assemble the ball check, spring and retainer into the plunger.
3. Install plunger spring over check retainer.
4. Hold plunger with spring up and insert into lifter body. Hold plunger vertically to prevent cocking spring.
5. Submerge the lifter in clean kerosene or diesel fuel.
6. Install oil metering valve and push rod seat into lifter and install retaining ring.

INSTALLATION

NOTE: *Prime new lifters by working lifter plunger while submerged in clean kerosene or diesel fuel. Lifter could be damaged when starting engine if dry.*

1. When the rocker arm is loosened or removed, valve lifter bleed down is required. Lifters must be bled down as possible valve to piston interference due to the close tolerances could exist. Before installing a new or used lifter in the engine, lubricate the roller and bearings of the lifter with No. 1052365 lubricant.
2. Install lifters and pushrods into the original position in cylinder block. See note under "Removal."
3. Install manifold gaskets and manifold.
4. Position rocker arms, pivots and bolts on cylinder head.
5. Install valve covers.
6. Install intake manifold assembly.

BLEED-DOWN

1. Before installing any removed rocker arms, rotate the engine crankshaft to a position of No. 1 cylinder being 32° BTDC. This is a 50 mm (2") counterclockwise from the 0° pointer. If only the right valve cover was removed, remove No. 1 cylinder's glow plug to determine if the position of the piston is the correct one. The compression pressure will tell you that you are in the right position.

If the left valve cover was removed, rotate the crankshaft until the number 5 cylinder intake valve pushrod ball is 7.0mm (.28 in.) above the number 5 cylinder exhaust valve pushrod ball.

NOTE: *Use only hand wrenches to torque*

the rocker arm pivot nuts to avoid engine damage.

2. If removed, install the No. 5 cylinder pivot and rocker arms. Torque the nuts alternately between the intake and exhaust valves until the intake valve begins to open, then stop.
3. Install remaining rocker arms except No. 3 exhaust valve (if this rocker arm was removed).
4. If removed, install but do not torque No. 3 valve pivots beyond the point that the valve would be fully open. This is indicated by strong resistance while still turning the pivot retaining bolts. Going beyond this would bend the pushrod. Torque the nuts SLOWLY allowing the lifter to bleed down.
5. Finish torquing No. 5 cylinder rocker arm pivot nut SLOWLY. Do not go beyond the point that the valve would be fully open. This is indicated by strong resistance while still turning the pivot retaining bolts. Going beyond this would bend the pushrod.
6. DO NOT turn the engine crankshaft for at least 45 minutes.
7. Finish reassembling the engine as the lifters are being bled.

NOTE: *Do not rotate the engine until the valve lifters have been bled down, or damage to the engine will occur.*

Rocker Arm, Shaft and Pushrod

REMOVAL AND INSTALLATION

6–181 and 6–231

1. Remove the rocker arm cover(s).
2. Remove the rocker arm shaft(s).
3. Place the shaft on a clean surface.
4. Remove the nylon rocker arm retainers. A pair of slip joint pliers is good for this.
5. Slide the rocker arms off the shaft and inspect them for wear or damage. Keep them in order!

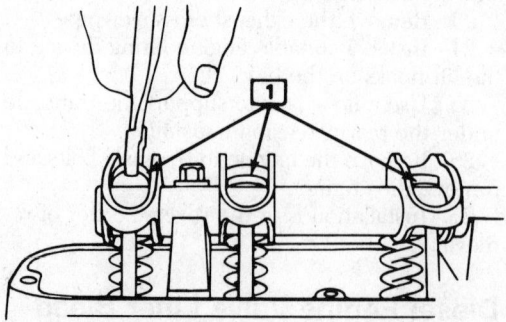

1. Rocker arm retainers
Removing nylon retainers from 3.0L or 3.8L V6

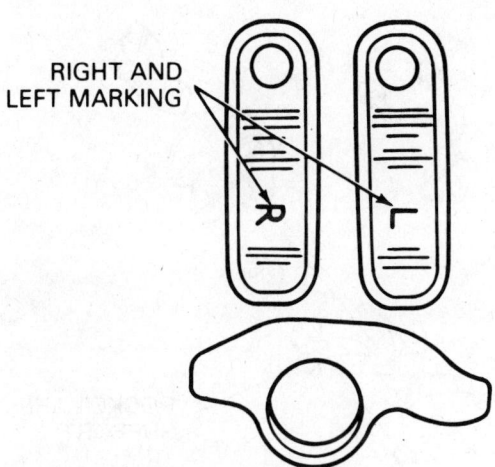

RIGHT AND
LEFT MARKING

3.0L & 3.8L rocker arm identification

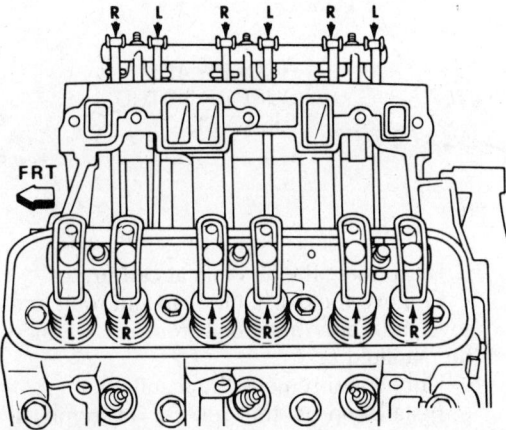

Positioning rocker arms on shaft

6. Installation is the reverse of removal. If new rocker arms are being installed, note that they are stamped R (right) or L (left). Each rocker arm must be centered over its oil hole. New nylon retainers must be used.

6-263

NOTE: *This procedure requires that the valve lifters be bled!*
1. Remove the valve cover(s). See the Valve Cover procedure.
2. Remove the rocker arm nuts, pivot and rocker arms.
3. If rocker arms are being replaced, they must be replaced in cylinder sets. Never replace just one rocker arm per cylinder! If a stud was replaced, coat the threads with locking compound and torque it to 11 ft. lb.
4. Installation is the reverse of removal. See the section on Valve lifter bleeddown. This is absolutely necessary! If lifters are not bled, en-

gine damage will be unavoidable! Torque the rocker arm nuts to 28 ft. lbs.; the cover to 5 ft. lb.

8-250

1. With the valve cover removed, remove the five nuts from the stud headed head bolts and remove the valve train support with the rocker arms and pivots attached as an assembly.
NOTE: *This method is preferred as the pivot assemblies may be damaged if pivot bolt torque is not removed evenly against valve spring pressure.*
2. Place the support in a vise and individually remove the rocker arms and pivot.
3. With the valve train support secured in a vise, position the rocker arms and pivots to the valve train support and loosely install the pivot bolts. Torque the pivot bolts to 20 ft. lbs.
NOTE: *When installing new parts, thoroughly lubricate all parts with an EP lubricant such as an axle lube.*
4. Position the valve train support with the rocker arms and pivots installed over the five stud headed head bolts.
5. Position the pushrod into the seat of each rocker arm and loosely install the five retaining nuts. Tighten the live nuts alternately and evenly while checking the positioning of the push rods from time to time. When the nuts are all the way down, tighten to 35 ft.lbs.
6. Install the rocker cover.

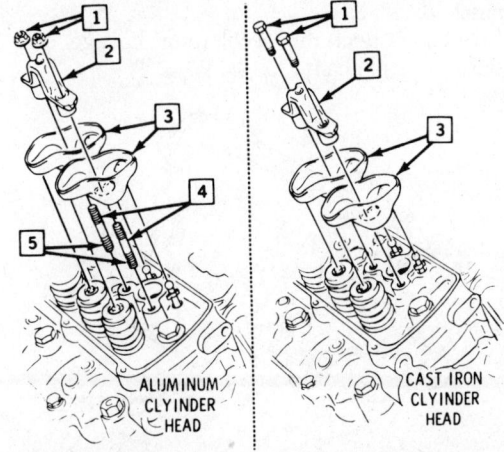

1. 37 N·m (28 lbs. ft.)
2. Pivot
3. Rocker arms
4. Studs—16 N·m (11 lbs. ft.)
5. Apply thread locking compound

Diesel engine rocker arms

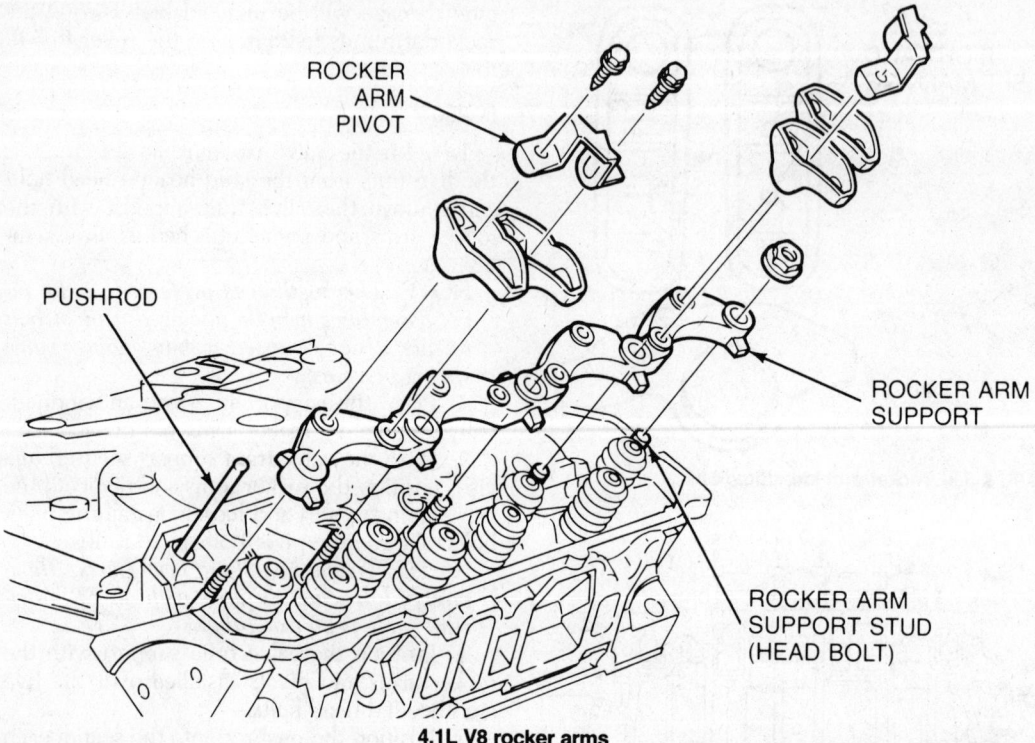

ROCKER
ARM
PIVOT

PUSHROD

ROCKER ARM
SUPPORT

ROCKER ARM
SUPPORT STUD
(HEAD BOLT)

4.1L V8 rocker arms

Intake Manifold

REMOVAL AND INSTALLATION

6–181 and 6–231

1. Disconnect the battery ground.
2. Drain the cooling system.
3. Remove the air cleaner.
4. Disconnect all hoses and wiring from the manifold.
5. Disconnect the accelerator linkage and cruise control chain.

6. Disconnect the fuel line at carburetor.
7. Remove the distributor cap and rotor and remove the Torx® head bolt from the left side of the manifold.
8. Unbolt and remove the manifold.
9. Installation is the reverse of removal. When installing the front and rear seals, make sure that the ends of the seals fit snugly against the block and head. Install nos. 1 & 2 bolts first and tighten them until snug, then install the other bolts in order.

6–263

NOTE: *This procedure requires the removal, disassembly draining and reassembly of the valve lifters. Read that procedure before continuing.*

1. Remove the air cleaner assembly.
2. Drain the radiator, then disconnect the upper radiator hose from the water outlet.
3. Disconnect the heater inlet hose from the outlet on the intake manifold and disconnect the heater outlet pipefrom the intake manifold attachments and move it aside.
4. Remove air crossover and the fuel injection pump.
5. Disconnect wiring as necessary at the generator, A/C compressor and switches, if so equipped.
6. Remove the cruise control servo if so equipped.

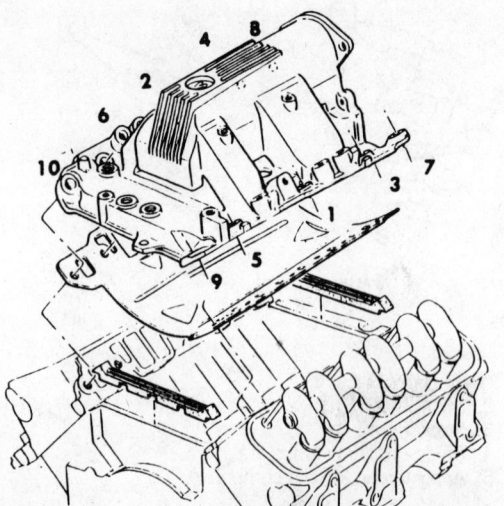

3.0L or 3.8L V6 intake manifold bolt torque sequence

7. Remove the A/C compressor bracket and brace bolts and position the compressor (if so equipped) with lines attached out of the way.

8. Remove the generator assembly.

9. Disconnect the engine mounting strut.

10. Remove the fuel lines, filter and brackets. Cap all openings.

11. Disconnect the electrical leads to the glow plug controller and sending units.

12. Disconnect the exhaust crossover pipe head shield.

13. Remove the left (forward) injection lines and cap all openings. Use a backup wrench on the nozzles.

14. Disconnect the throttle and T.V. cables from the bracket.

15. Remove the drain tube.

16. Remove the intermediate pump adapter.

17. Remove pump adapter and seal.

18. Remove the intake manifold.

19. Clean the machined surfaces of cylinder head and intake manifold with a putty knife. Use care not to gouge or scratch the machined surfaces. Clean all bolts and bolt holes.

20. Coat both sides of gasket sealing surface that seal the intake manifold to the head with 1050026 sealer or equivalent and position intake manifold gasket. Install end seals, making sure that ends are positioned under the cylinder heads. The seals and mating surfaces must be dry. Any liquid, including sealer will act as a lubricant and cause the seal to move during assembly. Use RTV sealer only on each end of the seal.

21. Position intake manifold on engine. Lubricate the entire intake manifold bolt (all) with lubricant 1052080 or equivalent.

22. Torque the bolts in sequence shown to 20 ft. lb. Then retorque to 41 ft. lbs.

23. Install the drain tube.

24. Install the pump adapter.

25. Apply chassis lube to seal area of intake manifold and pump adapter.

26. Apply chassis lube to inside and outside diameter of seal and seal area of tool J-28425.

27. Install seal on tool and install the seal.

28. Install intermediate pump adapter.

29. Reverse the order of removal and install all other removed parts except the air crossover.

30. Fill the cooling system.

31. Install manifold covers, J-29657.

32. Start engine and check for leaks.

33. Check and if necessary, reset the injection pump timing.

34. Remove screen covers from manifold.

35. Install air crossover.

36. Install the air cleaner.

37. Road test car and inspect for leaks.

8–250

1. Drain the coolant from the radiator and disconnect the upper radiator hose from the upper thermostat housing.

2. Disconnect the electrical connections from the following: Coolant sensor, MAT sensor, throttle position sensor, the 12 volt feed wire and the 4 way connector, ISC motor and the injectors.

3. Disconnect the heater hose from the rear of the intake manifold.

4. Disconnect the fuel lines from the throttle body.

5. Remove the distributor.

6. Remove the rocker arm covers then remove the rocker arm support with the rocker arms intact by removing the five nuts which attach the support to the studheaded head bolts. Keep the pushrods in sequence so they may be reassembled in their original position.

7. Partially disconnect the A/C compressor and move to one side without discharging the A/C system.

8. Remove the vacuum harness connections from the TVS at the rear of the intake manifold.

9. Remove the 16 intake manifold bolts and remove the two bolts securing the lower thermostat housing to the front cover.

10. Bend the front and rear engine lift brackets out of the way then remove the intake manifold and lower thermostat housing as as assembly by lifting straight up.

11. Installation is the reverse of removal. Make sure the pushrods are properly seated and tighten the retaining nuts alternately and evenly until the nuts are all the way down, then tighten to 35 ft. lbs. Use new gaskets and O-rings. With the new intake gaskets in position, apply RTV sealant to the four corners where the end seals meet the side gaskets. Refer to the intake manifold torque sequence illustration and tighten the bolts to the torque values given in the "Torque Specifications" chart. Check the ignition timing.

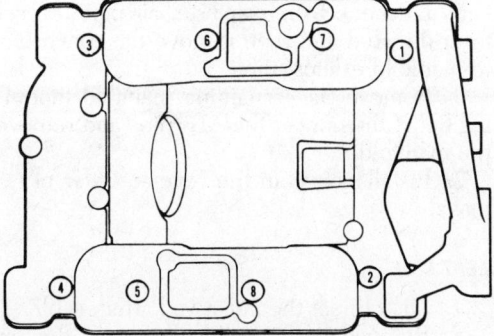

Diesel intake manifold torque sequence

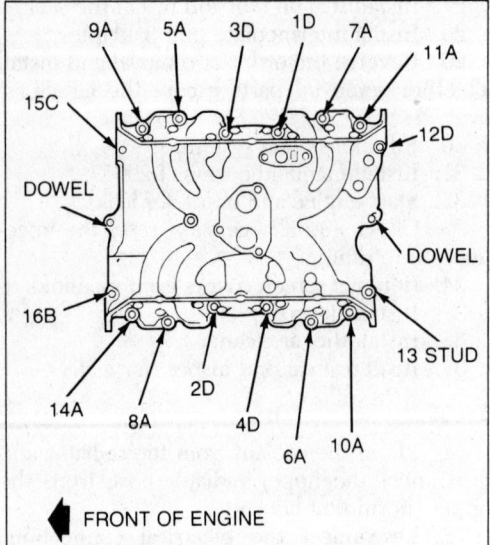

FRONT OF ENGINE

BOLT TIGHTENING SEQUENCE

1. TIGHTEN BOLTS 1, 2, 3, & 4 IN SEQUENCE TO 20.0 N•m (15 FT-LBS).

2. TIGHTEN BOLTS 5 THRU 16 IN SEQUENCE TO 30.0 N•m (22 FT-LBS).

3. RETIGHTEN ALL BOLTS IN SEQUENCE TO 30.0 N•m (22 FT-LBS).

4. REPEAT STEP 3.

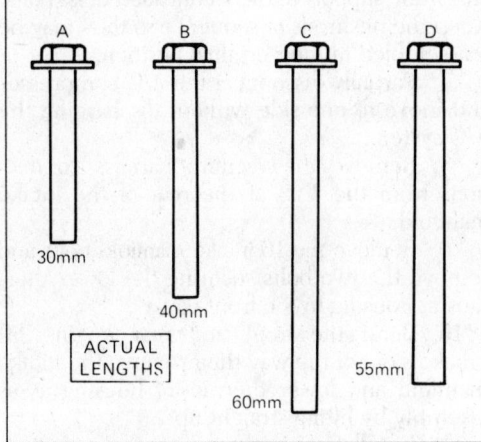

4.1L V8 intake manifold installation

Exhaust Manifold
REMOVAL AND INSTALLATION
6–181 and 6–231

LEFT SIDE

1. Disconnect the negative battery cable.

2. Remove the mass air flow sensor, air intake duct and crankcase ventilation pipe.

3. Remove the two bolts attaching the exhaust crossover pipe to the manifold.

4. Tag and disconnect the spark plug wires.

5. Remove the mounting bolts and remove the manifold.

NOTE: *The oil dipstick tube may need to be removed to provide access to the manifold bolts.*

6. Installation is in the reverse order of removal.

RIGHT SIDE

1. Disconnect the negative battery cable.

2. Repeat Step 2 of the "Left Side" procedure.

3. Disconnect the IAC connector at the throttle body (6–231 only).

4. Tag and disconnect the spark plug wires and the oxygen sensor lead.

5. Disconnect the heater inlet pipe from the manifold studs.

6. Remove the exhaust crossover pipe.

7. Remove the front alternator support bracket.

8. Remove the exhaust manifold mounting bolts. Raise and support the front of the vehicle.

9. Disconnect the exhaust pipe from the manifold.

10. Remove the front exhaust pipe. Remove the manifold.

11. Installation is in the reverse order of removal.

8–250
RIGHT SIDE

1. Disconnect the negative battery cable. Remove the air cleaner.

2. Remove the exhaust crossover pipe. Disconnect the oxygen and coolant temperature sensors.

3. Remove the catalytic converter-to-AIR pipe clip bolt.

4. Remove the two front manifold mounting bolts. Raise and support the front of the car.

5. Disconnect the converter air pipe bracket from the stud and then remove the converter-to-manifold exhaust pipe.

6. Remove the remaining manifold mounting bolts, disconnect the AIR pipe and remove the manifold.

7. Installation is in the reverse order of removal.

LEFT SIDE

1. Disconnect the negative battery cable.

2. Remove both cooling fans and the exhaust crossover pipe.

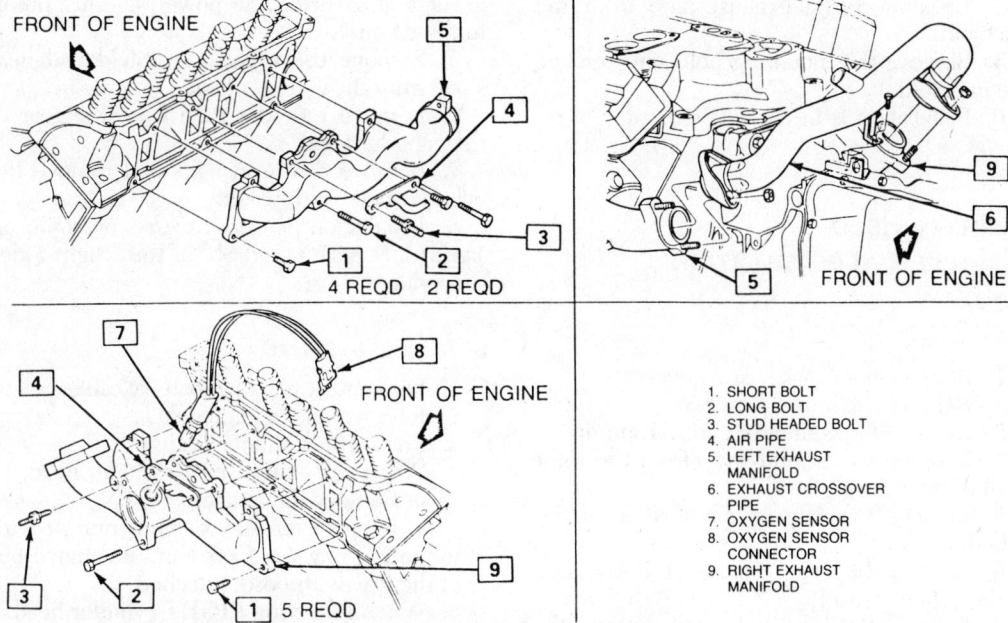

1. SHORT BOLT
2. LONG BOLT
3. STUD HEADED BOLT
4. AIR PIPE
5. LEFT EXHAUST MANIFOLD
6. EXHAUST CROSSOVER PIPE
7. OXYGEN SENSOR
8. OXYGEN SENSOR CONNECTOR
9. RIGHT EXHAUST MANIFOLD

4.1L V8 exhaust manifold installation

3. Remove the drive belt and the AIR pump pivot bolt.

4. Remove the belt tensioner and the power steering pump brace.

5. Remove the manifold mounting bolts. Disconnect the air pipe and remove the manifold.

6. Installation is in the reverse order of removal.

6–263 Diesel

LEFT (FRONT) SIDE

1. Disconnect the negative battery cable.
2. Remove the exhaust crossover pipe.

3. Raise and support the front of the car.

4. Remove the right engine splash shield.

5. Remove the vacuum pump-to-exhaust manifold brace.

6. Remove the mounting bolts and remove the manifold.

7. Installation is in the reverse order of removal.

RIGHT (REAR) SIDE

1. Disconnect the negative battery cable.
2. Remove the exhaust crossover pipe.
3. Raise and support the front of the car.

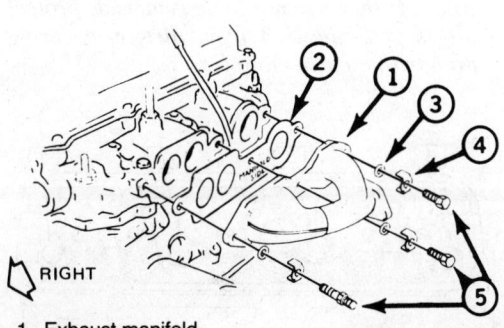

1. Exhaust manifold
2. Gasket
3. Washer (3)
4. Lock (3)
5. Bolt–38 N·m (28 lbs. ft.)

APPLY LUBRICANT TO ENTIRE BOLT

Diesel left (front) side exhaust manifold

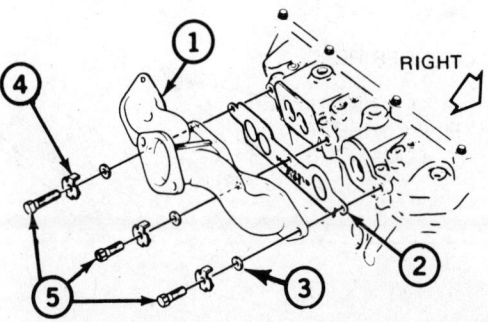

1. Exhaust manifold
2. Gasket
3. Washer (3)
4. Lock (3)
5. Bolt–39 N·m (28 lbs. ft.) (3)

Diesel right (rear) side exhaust manifold

4. Disconnect the exhaust pipe from the manifold.

5. Remove the mounting bolts and remove the manifold.

6. Installation is in the reverse order of removal.

Cylinder Head

REMOVAL AND INSTALLATION

8-250

RIGHT SIDE

1. Remove both rocker arm covers.
2. Remove the intake manifold.
3. Remove the generator and AIR pump.
4. Remove the ground strap from the right front of the cylinder head.
5. Remove the exhaust manifold from the cylinder head.
6. Remove the screw retaining the AIR pipe to the cylinder head.
7. Remove the head bolts and remove the cylinder head.
8. Installation is the reverse of removal with the exception of the following precautions:
    a. Apply graphite lubricant to the exhaust faces of the cylinder head.
    b. Install the cylinder head bolts finger tight with the studheaded bolts in the upper row and the conventional bolts in the lower row, then tighten all bolts in sequence to 45 ft. lbs. and again in sequence to 90 ft. lbs. as shown in illustration.

LEFT SIDE

1. Remove both rocker arm covers.
2. Remove the intake manifold.
3. Remove the vacuum pump and mounting bracket.
4. Remove the upper bolt and loosen the

lower bolt securing the power steering pump to the engine.

5. Remove the exhaust manifold and heat stove from the cylinder head.
6. Remove the AIR pipe from the rear of the cylinder head.
7. Remove the head bolts and remove the cylinder head and gasket.
8. Installation is the reverse of removal. Perform Steps 8a, and 8b, of the "Right Side" procedure.

6-181 and 6-231

1. Disconnect negative battery cable.
2. Remove intake manifold.
3. Loosen and remove belt(s).
4. When removing LEFT cylinder head;
    a. Remove oil dipstick.
    b. Remove air and vacuum pumps with mounting bracket if present, and move out of the way with hoses attached.
5. When removing RIGHT cylinder head:
    a. Remove alternator.
    b. Disconnect power steering gear pump and brackets attached to cylinder head.
6. Disconnect wires from spark plugs, and remove the spark plug wire clips from the rocker arm cover studs.
7. Remove exhaust manifold bolts from head being removed.
8. With air hose and cloths, clean dirt off cylinder head and adjacent area to avoid getting dirt into engine. It is extremely important to avoid getting dirt into the hydraulic valve lifters.
9. Remove rocker arm cover and rocker arm and shaft assembly from cylinder head. Lift out pushrods.

NOTE: *If lifters are to be serviced, remove them at this time and place them in a container with numbered holes or a similar device, to keep them identified as to engine position. If they are not to be removed, protect lifters and camshaft from dirt by covering area with a clean cloth.*

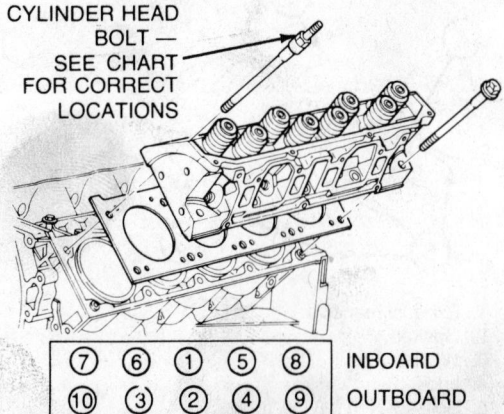

CYLINDER HEAD BOLT — SEE CHART FOR CORRECT LOCATIONS

⑦	⑥	①	⑤	⑧	INBOARD
⑩	③	②	④	⑨	OUTBOARD

4.1L V8 head bolt torque sequence

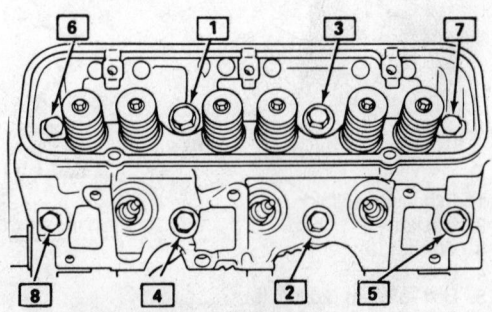

3.0L or 3.8L V6 cylinder head torque sequence

10. Loosen all cylinder head bolts, then remove bolts and lift off the cylinder head.

11. With cylinder head on bench, remove all spark plugs for cleaning and to avoid damaging them during work on the head.

12. Installation is the reverse of removal. Clean all gasket surfaces thoroughly. Always use a new head gasket. The head gasket is installed with the bead downward. Coat the head bolt threads with thread sealer. Torque the head bolts in three equal stages. Recheck head bolt torque after the engine has been warmed to operating temperature.

6–263

NOTE: *This procedure requires the complete disassembly of the valve lifters as explained under "Diesel Engine Valve Lifter Bleed-Down".*

1. Remove intake manifold.

2. Remove valve cover. Loosen or remove any accessory brackets or pipe clamps which interfere.

3. Disconnect glow plug wiring (and block heater lead if so equipped on rear bank).

4. Remove the ground strap from right (rear) cylinder head.

5. Remove rocker arm nuts, pivots, rocker arms and pushrods. Scribe pivots and keep rocker arms separated so they can be installed in their original locations.

6. Disconnect the exhaust crossover pipe from the exhaust manifold on the side being worked on and loosen it on the other.

7. Remove engine block drain plug, from side of the block where head is being removed.

8. Remove the pipe plugs covering the upper cylinder head bolts.

9. Remove all the cylinder head bolts and remove the cylinder head.

10. If necessary to remove the prechamber, remove the glow plug and injection nozzle, then tap out with a small blunt 1/8 in. drift. Do NOT use a tapered drift.

11. Installation is the reverse of removal. Do not use sealer on the head gasket. If a prechamber was replaced, measure the chamber height and grind the new one to within 0.001

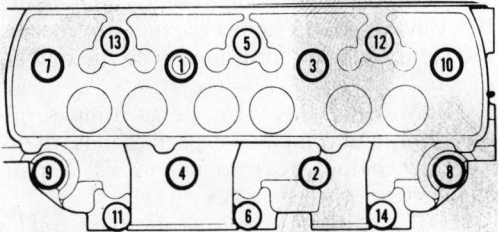

Diesel cylinder head torque sequence

in. of the old chamber's height, using #80 grit wet sandpaper to polish it. Coat the head bolts with sealer.

CLEANING AND INSPECTION

1. Remove all traces of carbon from the head, using a decarbon-type wire brush mounted in an electric drill. Do not use a motorized brush on any gasket mating surface.

2. Lay a straight edge across the cylinder head face and check between the straight edge and the head with feeler gauges. Make the check at six points minimum. Cylinder head flatness should be within .003-.006 inch. These surfaces may be reconditioned by parallel grinding. If more than 10% must be removed, the head should be replaced.

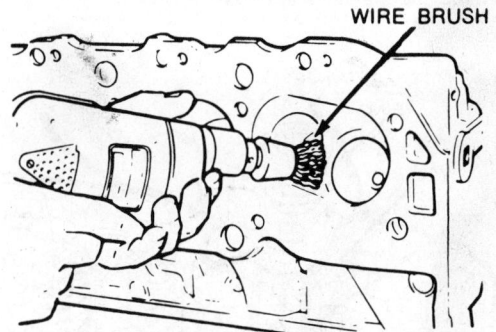

Use a wire brush and electric drill to remove carbon from the combustion chambers and exhaust ports

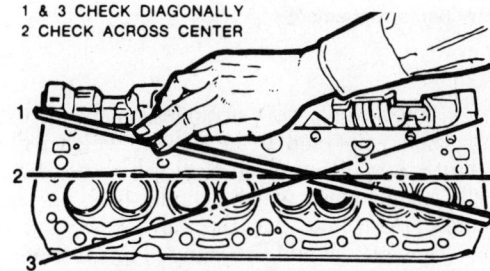

Check the cylinder head mating surface for warpage with a precision straight edge

Valves

REMOVAL AND INSPECTION

1. Remove the cylinder head(s) from the vehicle as previously outlined.

2. Using a suitable valve spring compressor, compress the valve spring and remove the valve keys using a magnetic retrieval tool.

3. Slowly release the compressor and remove the valve spring caps (or rotors) and the valve springs.

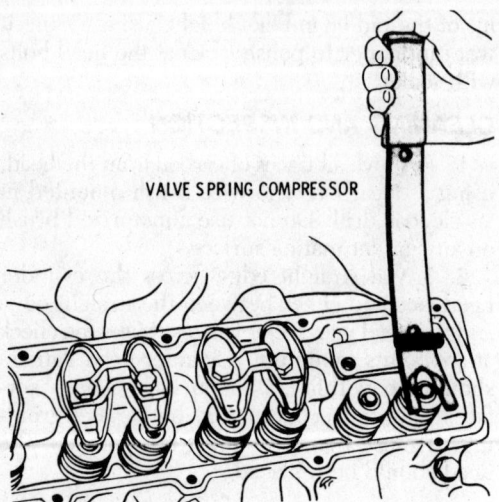

VALVE SPRING COMPRESSOR

Removing the valve springs

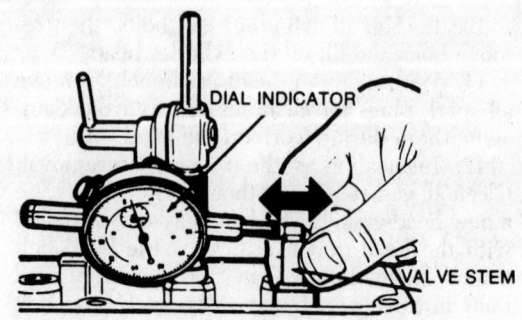

DIAL INDICATOR

VALVE STEM

Checking the valve stem-to-guide clearance

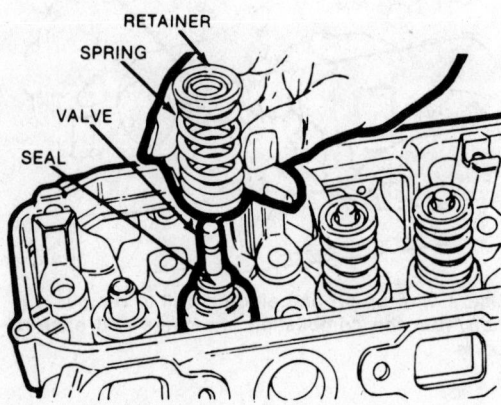

RETAINER

SPRING

VALVE

SEAL

Valve part arrangement

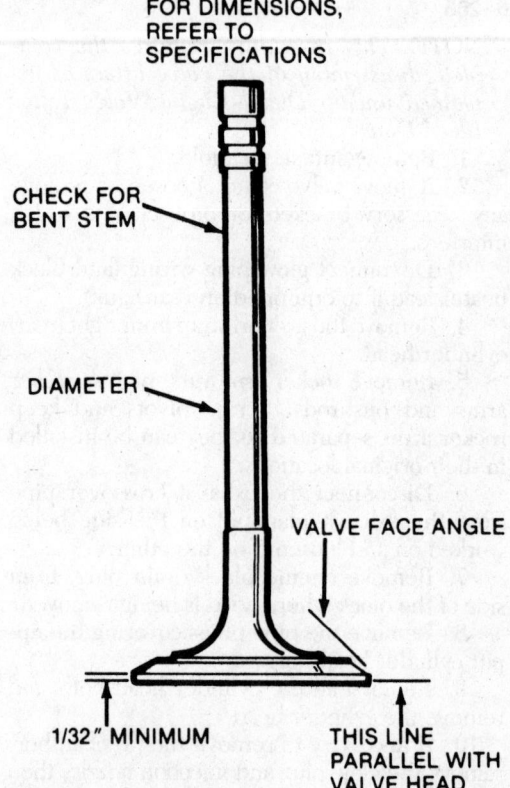

FOR DIMENSIONS, REFER TO SPECIFICATIONS

CHECK FOR BENT STEM

DIAMETER

VALVE FACE ANGLE

1/32" MINIMUM

THIS LINE PARALLEL WITH VALVE HEAD

Critical valve dimensions

4. Fabricate a valve arrangement board to use when you remove the valves, which will indicate the port in which each valve was originally installed (and which cylinder head on V6 models). Also note that the valve keys, rotators, caps, etc. should be arranged in a manner which will allow you to install them on the valve on which they were originally used.

5. Remove and discard the valve seals. On models using the umbrella type seals, note the location of the large and small seals for assembly purposes.

6. Thoroughly clean the valves on the wire wheel of a bench grinder, then clean the cylinder head mating surface with a soft wire wheel, a soft wire brush, or a wooden scraper. Avoid using a metallic scraper, since this can cause damage to the cylinder head mating surface, especially on models with aluminum heads.

7. Using a valve guide cleaner chucked into a drill, clean all of the valve guides.

8. Install each valve into its respective port (guide) of the cylinder head.

9. Mount a dial indicator so that the stem is at 90° to the valve stem, as close to the valve guide as possible.

10. Move the valve off its seat, and measure the valve guide-to-stem clearance by rocking the stem back and forth to actuate the dial indicator.

11. Measure the valve stems using a micrometer, and compare to specifications, to determine whether stem or guide wear is responsible for excessive clearance.

NOTE: *Consult the Specifications tables earlier in this chapter.*

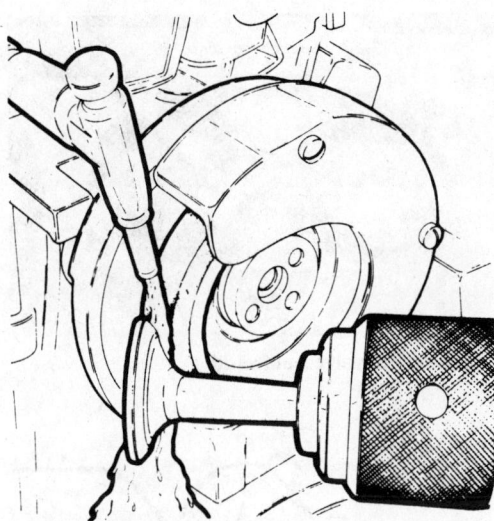

Valve grinding by machine

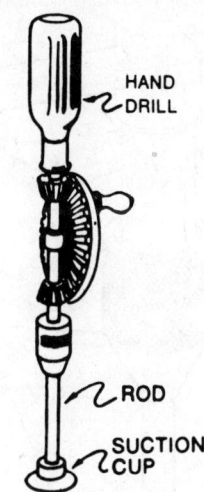

Home-made valve lapping tool

REFACING

Using a valve grinder, resurface the valves according to specifications in this chapter.

NOTE: *All machine work should be performed by a competent, professional machine shop.*

CAUTION: *Valve face angle is not always identical to valve seat angle.*

A minimum margin of ⅟₃₂ in. should remain after grinding the valve. The valve stem top should also be squared and resurfaced, by placing the stem in the V-block of the grinder, and turning it while pressing lightly against the grinding wheel. Be sure to chamfer the edge of the tip so that the squared edges don't dig into the rocker arm.

LAPPING

This procedure should be performed after the valves and seats have been machined, to insure that each valve mates to each seat precisely.

Lapping the valves by hand

1. Invert the cylinder head, lightly lubricate the valve stems, and install the valves in the head as numbered.

2. Coat valve seats with fine grinding compound, and attach the lapping tool suction cup to a valve head.

NOTE: *Moisten the suction cup.*

3. Rotate the tool between your palms, changing position and lifting the tool often to prevent grooving.

4. Lap the valve until a smooth, polished seat is evident.

5. Remove the valve and tool, and rinse away all traces of grinding compound.

Valve Guide Service

The valve guides used in these engines are integral with the cylinder head, that is, they cannot be replaced.

NOTE: *Refer to the previous "Valves—Removal and Installation" to check the valve guides for wear.*

Valve guides are most accurately repaired using the bronze wall rebuilding method. In this operation, "threads" are cut into the bore of the valve guide and bronze wire is turned into the threads. The bronze "wall" is then reamed to the proper diameter.

This method is well received for a number of reasons: it is relatively inexpensive, it offers better valve lubrication (the wire forms channels whichretain oil), it offers less valve friction, and it preserves the original valve guide-to-seat relationship.

Another popular method of repairing valve guides is to have the guides "knurled." Knurling entails cutting into the bore of the valve guide with a special tool The cutting action

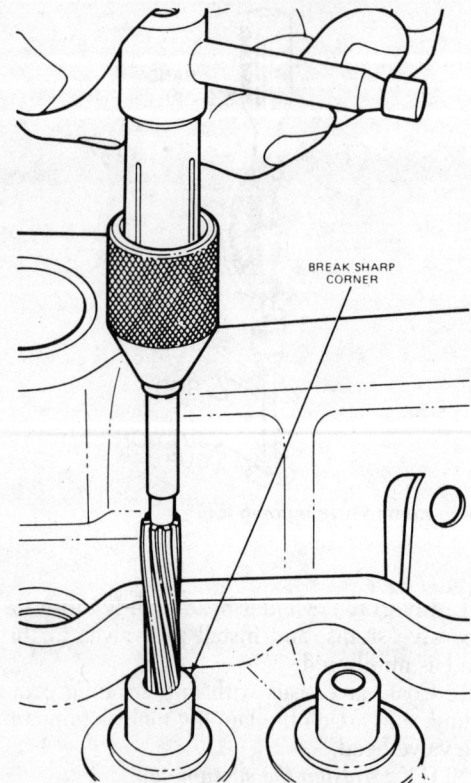

Reaming the valve guides—typical

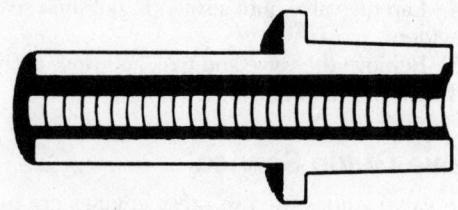

Cutaway of a knurled valve guide

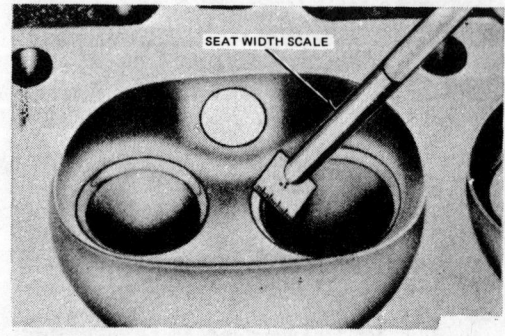

Checking the valve seat width

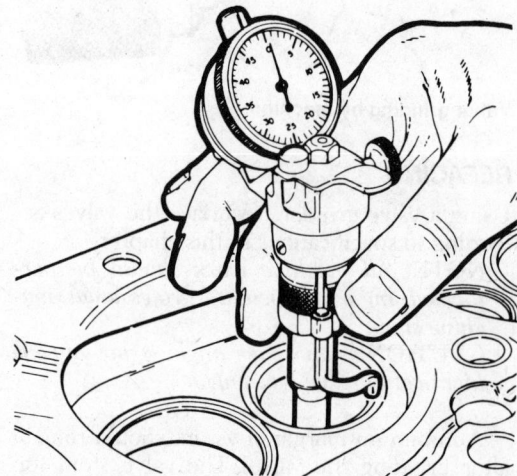

Have the valve seat concentricity checked at a machine shop

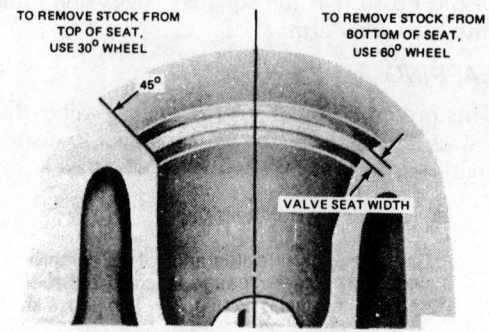

Facing the valve seat

"raises" metal off of the guide bore which actually narrows the inner diameter of the bore, thereby reducing the clearance between the valve guide bore and the valve stem. This method offers the same advantages as the bronze wall method, but will generally wear faster.

Either of the above services must be performed by a professional machine shop which has the specialized knowledge and tools necessary to perform the service.

Valve Seat Service

The valve seats are integral with the cylinder head on all engines. On all engines the seats are machined into the cylinder head casting itself.

VALVE SPRING TESTING

Place the spring on a flat surface next to a square. Measure the height of the spring, and rotate it against the edge of the square to measure distortion. If spring height varies (by comparison) by more than 1/16 in. or if distortion exceeds 1/16 in., replace the spring.

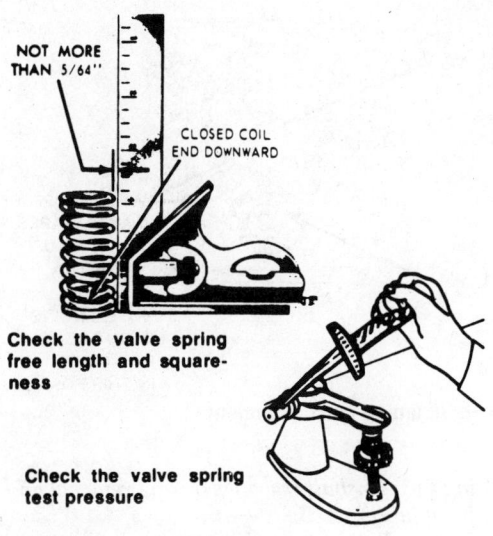

Check the valve spring free length and square-ness

Check the valve spring test pressure

Checking the valve springs

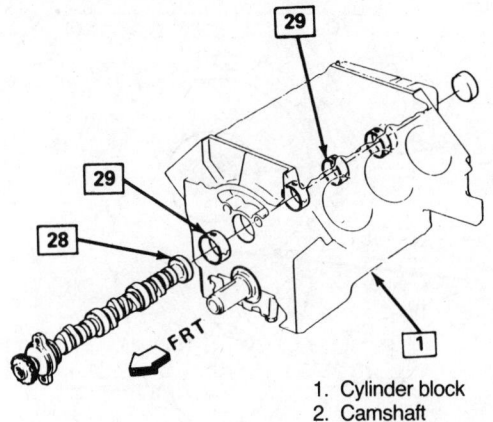

1. Cylinder block
2. Camshaft
3. Camshaft bearing

3.0L or 3.8L camshaft

In addition to evaluating the spring as above, test the spring pressure at the installed and compressed (installed height minus valve lift) height using a valve spring tester. Spring pressure should be ± 1 lb. of all other springs in either position.

VALVE AND SPRING INSTALLATION

NOTE: *Be sure that all traces of lapping compound have been cleaned off before the valves are installed.*

1. Lubricate all of the valve stems with a light coating of engine oil, then install the valves into the proper ports/guides.

2. If umbrella-type valve seals are used, install them at this time. Be sure to use a seal protector to prevent damage to the seals as they are pushed over the valve keeper grooves. If O-ring seals are used, don't install them yet.

3. Install the valve springs and the spring retainers (or rotators), and using the valve compressing tool, compress the springs.

4. If umbrella-type seals are used, just install the valve keepers (white grease may be used to hold them in place) and release the pressure on the compressing tool. If O-ring type seals are used, carefully work the seals into the second groove of the valve (closest to the head), install the valve keepers and release the pressure on the tool.

NOTE: *If the O-ring seals are installed BE-FORE the springs and retainers are compressed, the seal will be destroyed.*

5. After all of the valves are installed and retained, tap each valve spring retainer with a rubber mallet to seat the keepers in the retainer.

Camshaft

REMOVAL AND INSTALLATION

6–181 and 6–231

1. Remove the engine as described earlier.
2. Remove the intake manifold.
3. Remove the rocker arm covers.
4. Remove the rocker arm assemblies, pushrods and lifters.
5. Remove the timing chain cover.

NOTE: *Align the timing marks of the camshaft and crankshaft sprockets to avoid burring the camshaft journals by the crankshaft.*

6. Remove the timing chain and camshaft sprocket as described earlier.
7. Installation is the reverse of removal.

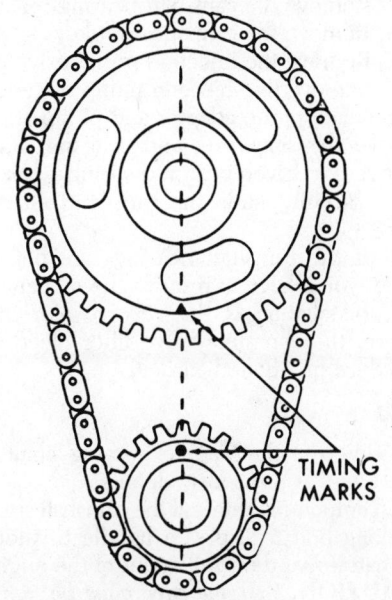

3.0L or 3.8L timing mark alignment

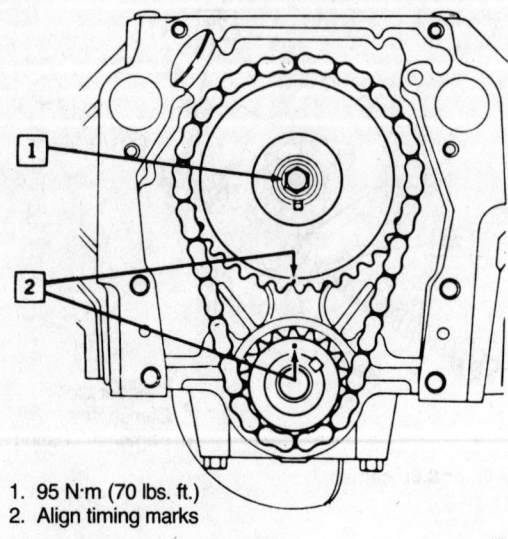

1. 95 N·m (70 lbs. ft.)
2. Align timing marks

Diesel timing mark alignment

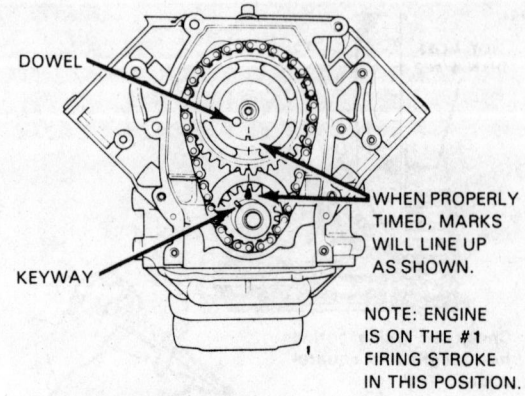

DOWEL

KEYWAY

WHEN PROPERLY
TIMED, MARKS
WILL LINE UP
AS SHOWN.

NOTE: ENGINE
IS ON THE #1
FIRING STROKE
IN THIS POSITION.

4.1L V8 timing mark alignment

6–263

NOTE: *This procedure requires the removal, disassembly, cleaning, reassembly and bleed-down of all the valve lifters. Read that procedure, described earlier, before proceeding.*

1. Remove the engine as described earlier.
2. Remove the intake manifold.
3. Remove the oil pump drive assembly.
4. Remove the timing chain cover.
5. Align the timing marks.
6. Remove the rocker arms, pushrods and lifters, keeping them in order for reassembly.
7. Remove the timing chain and camshaft sprocket as described earlier.
8. Remove the camshaft bearing retainer.
9. Remove the cam sprocket key.
10. Remove the injection pump drive gear.
11. Remove the injection pump driven gear, intermediate pump adapter and pump adapter. Remove the snap ring and selective washer. Remove the driven gear and spring.
12. Carefully slide the camshaft out of the block.
13. If the camshaft bearings are being replaced, you'll have to remove the oil pan.
14. Installation is the reverse of removal. Perform the complete valve lifter bleed-down procedure mentioned earlier.

8–250

1. Remove the engine, timing chain and valve lifters as previously detailed.
2. Temporarily reinstall the camshaft sprocket or a long bolt to use as a handle to slide the camshaft forward until it is out of the engine.
CAUTION: *Extreme care must be exercised to prevent the camshaft lobes from scratch-*ing *the camshaft bearings during removal and installation.*

3. Installation is in the reverse order of removal. Apply a thin coat of rear axle lubricant or the equivalent to the camshaft lobes, distributor gear teeth and bearing journals.

Timing Gear Cover
REMOVAL AND INSTALLATION
6–181 and 6–231

1. Drain the cooling system.
2. Disconnect the lower radiator hose and the heater hose at the water pump.
3. Remove the two nuts from the front engine mount at the cradle and raise the engine using a suitable lifting device.
4. Remove the water pump pulley and all drive belts.
5. Remove the alternator and brackets.
6. Remove the distributor
NOTE: *If the timing chain and sprockets are not going to be disturbed, note the position of the distributor rotor for reinstallation in the same position.*

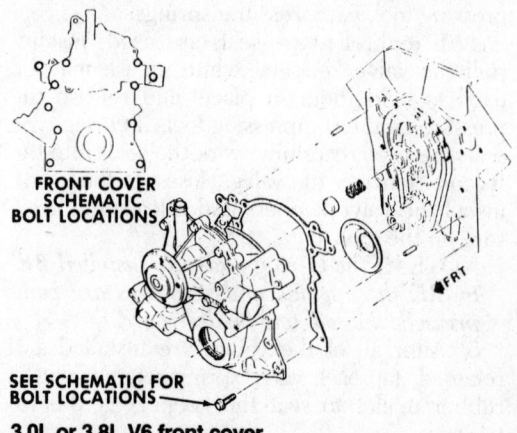

FRONT COVER
SCHEMATIC
BOLT LOCATIONS

◄FRT

SEE SCHEMATIC FOR
BOLT LOCATIONS

3.0L or 3.8L V6 front cover

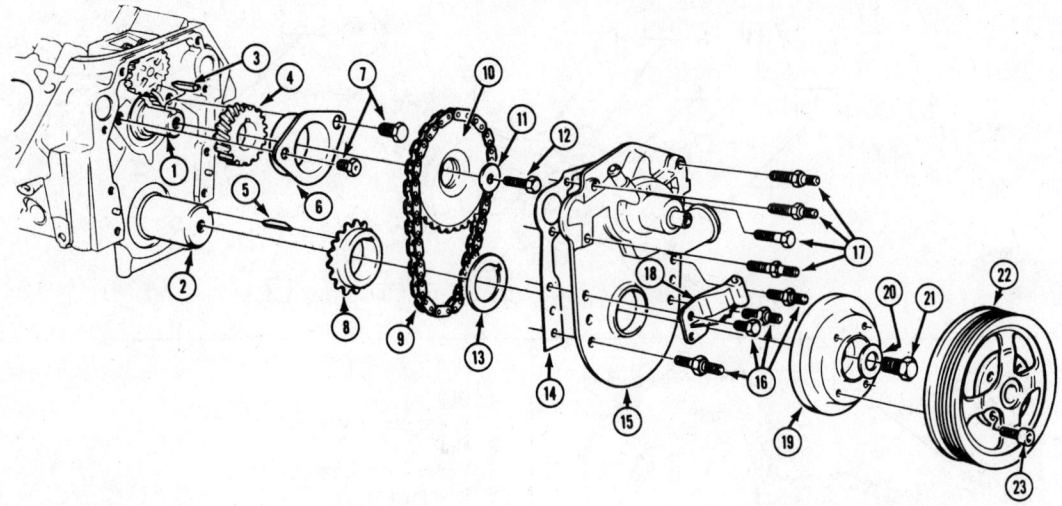

1. Camshaft
2. Crankshaft
3. Camshaft sprocket key
4. Injection pump drive gear
5. Crankshaft sprocket key
6. Front camshaft bearing retainer
7. 65 N·m (48 ft. lbs.)
8. Crankshaft sprocket
9. Timing chain
10. Camshaft sprocket
11. Washer
12. 95 N·m (70 ft. lbs.)
13. Slinger
14. Gasket
15. Front cover
16. 57 N·m (42 ft. lbs.)—apply adhesive
17. 28 N·m (21 ft. lbs.)—apply adhesive
18. Probe holder (rpm counter)
19. Crankshaft balancer
20. Washer
21. 275–475 N·m (203–350 ft. lbs.)
22. Pulley assembly
23. 40 N·m (30 ft. lbs.)

Front end of the diesel engine

7. Remove the balancer bolt and washer, and using a puller, remove the balancer.

8. Remove the cover-to-block bolts. Remove the two oil pan-to-cover bolts.

9. Remove the cover and gasket.

10. Installation is the reverse of removal. Always use a new gasket coated with sealer. Remove the oil pump cover and pack the area around the gears with petroleum jelly so that no air space is left within the pump. Apply sealer to the cover bolt threads.

6–263

1. Drain the cooling system.

2. Disconnect the lower radiator hose and the heater hose at the water pump. Disconnect the heater outlet pipe at the manifold.

3. Disconnect the power steering pump, vacuum pump, belt tensioner, air conditioning compressor and alternator brackets.

CAUTION: *Do not disconnect any refrigerant lines.*

4. Remove the crankshaft balancer using a puller.

5. Unbolt and remove the front cover and gasket.

6. Installation is the reverse of removal. Grind a chamfer on the end of each dowel pin to aid in cover installation. Trim ⅛ inch from the ends of the new front pan seal. Apply RTV

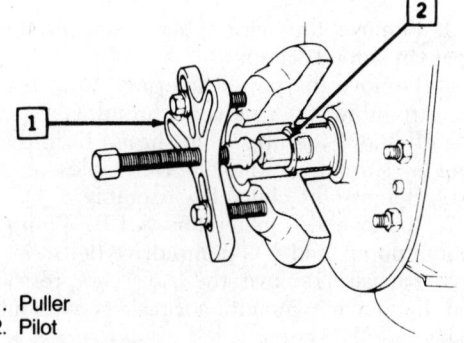

1. Puller
2. Pilot

Typical balancer removal

sealer to the oil pan seal retainer. After the cover gasket is in place, apply sealer to the junction of the pan, gasket and block. When installing the cover, rotate it right and left while guiding the pan seal into place with a small screwdriver.

8–250

1. Disconnect the negative battery cable and drain the radiator.

2. On the Fleetwood and Deville models, remove the two screws on each side of the radiator securing the support rod. Move the support rods out of the way.

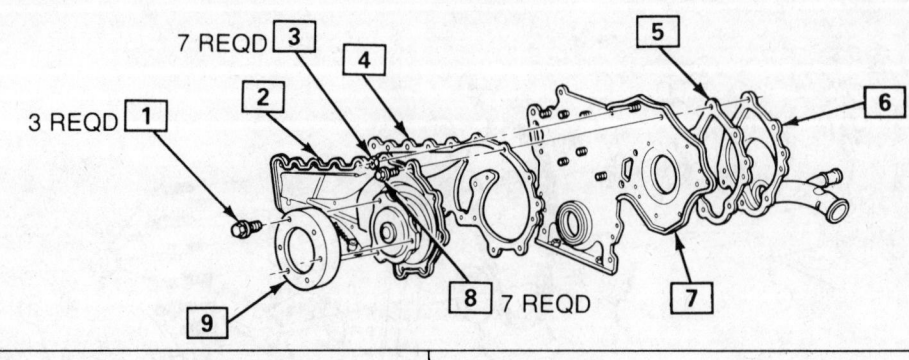

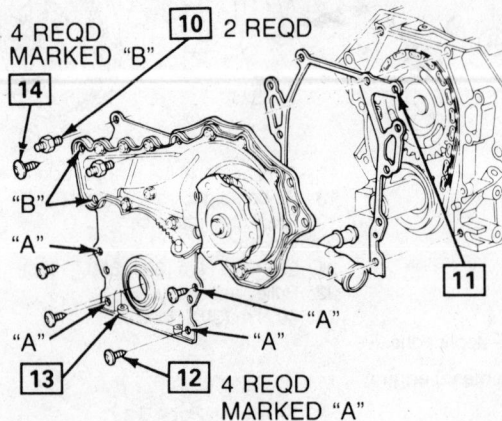

1. Bolt
2. Water pump assembly
3. Nut
4. Water pump gasket
5. Inlet gasket
6. Inlet
7. Front cover
8. Bolt
9. Water pump pulley
10. Stud headed bolt
11. Front cover gasket
12. Torx® screw
13. Front cover/water pump/inlet assembly
14. Torx® screw

4.1L V8 front cover

3. Remove the wiring harness from the upper fan shroud clamps.

4. Remove the power steering pump reservoir from the upper radiator shroud.

5. Remove the upper fan shroud from the lower fan shroud by removing the staples.

6. Remove the clutch fan assembly.

7. Remove the generator, A.I.R. Pump, vacuum pump, and A/C pump drive belts.

8. Partially remove the A/C compressor from the engine mounting brackets without discharging the system.

9. Remove the alternator and support bracket from the engine.

10. Loosen the clamp and disconnect the coolant reservoir to water pump hose at the pump.

11. Disconnect the inlet and outlet hoses at the water pump.

12. Drain the crankcase by either removing the crankcase plugs (one on each side) or by elevating the rear wheels. This will prevent coolant from draining into the oil pan as the front cover is removed.

13. Remove the water pump and crankcase pulleys.

14. Remove the A/C bracket at the water pump.

15. Remove the timing mark tab from the front cover.

16. Remove the crankcase pulley to hub bolts and separate the pulley from the hub.

17. Remove the plug from the end of the crankshaft. Install a puller and remove the hub or balancer.

18. Remove the remaining front cover attaching screws and remove the cover with the water pump and lower thermostat housing as an assembly.

19. Install the engine front cover by reversing the above removal procedure.

Timing Gear and/or Chain
REMOVAL AND INSTALLATION
6–181 and 6–231

1. Remove the timing chain cover as outlined earlier.

2. Turn the crankshaft so that the timing marks are aligned.

3. Remove the crankshaft oil slinger.

4. Remove the camshaft sprocket bolts.

5. Use two prybars to alternately pry the camshaft and crankshaft sprocket free along with the chain.

6. Installation is the reverse of removal. If the engine was turned, make sure that the No. 1 cylinder is at TDC.

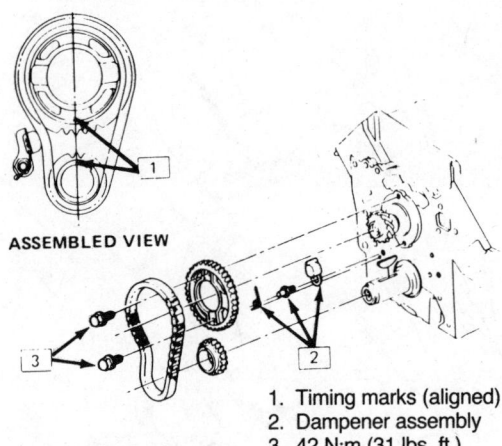

ASSEMBLED VIEW

1. Timing marks (aligned)
2. Dampener assembly
3. 42 N·m (31 lbs. ft.)

3.0L or 3.8L V6 timing chain

6-263

NOTE: *The following procedure requires the bleed-down of the valve lifters. Readthat procedure before proceeding.*

1. Remove the front cover.
2. Loosen all the rocker arms. See Rocker Arm Removal and Installation.
3. Remove the crankshaft oil slinger.
4. Remove the camshaft sprocket bolt.
5. Using two prybars, work the camshaft and crankshaft sprockets alternately off their shafts along with the chain. It may be necessary to remove the crankshaft sprocket with a puller.
6. Installation is the reverse of removal. If the engine was turned, make sure that the No. 1 piston is at TDC. Bleed the lifters following the procedure under "Diesel Engine Valve Lifter Bleed-Down."

8-250

1. Remove the front cover.
2. Remove the oil slinger from the crankshaft.
3. Rotate the engine and line up the timing marks as shown in the illustration.
4. Remove the screw securing the camshaft sprocket to the camshaft, then remove the camshaft and crankshaft sprocket with the chain attached.
5. Installation is the reverse of removal. After installing the timing chain over the camshaft sprocket rotate the crankshaft until the timing mark on the crank sprocket is positioned straight up.
6. Install the cam sprocket and timing chain over the crankshaft so that the timing marks are aligned as the illustration.
7. Hold the camshaft sprocket in position against the end of the camshaft and press the sprocket on the camshaft by hand. Make sure

the index pin in the camshaft is lined up with the index hole in the sprocket.

8. If necessary, keep the engine from rotating while torquing the camshaft sprocket screw to 37 ft. lbs.

NOTE: *Engine timing has been set so that the No. 1 cylinder is in the T.D.C. firing position. If for some reason the distributor was removed make sure the rotor is set so that cylinder No. 1 is in the firing position.*

9. Install the oil slinger on the crankshaft with the smaller end of the slinger against the crankshaft sprocket.

Timing Cover Oil Seal
REMOVAL AND INSTALLATION
Except 8-250

1. After removing the timing cover, pry oil seal out of front of cover.
2. Install new lip seal with lip (open side of seal) inside and drive or press seal carefully into place.

NOTE: *The timing cover oil seal can be replaced without removing the cover. Remove the fan belts, crankshaft pulley and harmonic balancer. Pry the oil seal out the cover working carefully to prevent damage to the seal mating surface. Lubricate the new seal and drive it into place with the open side toward the engine. Use a seal installer to avoid damaging or cocking the seal.*

8-250

NOTE: *The engines are equipped with a molded-type front cover crankshaft oil seal. The seal may be replaced without removing the engine front cover.*

1. Disconnect the battery and remove the air cleaner.
2. Remove the power steering pump drive belt.
3. Remove the alternator drive belt.
4. On air conditioned cars, and cars equipped with the A.I.R. system, remove the pump drive belts.
5. Reuse and support the front of the car on jack stands. Remove the fan.
6. Remove pulley and harmonic balancer, as outlined in Timing Chain and Sprocket Removal.
7. With a suitable tool, pry out front cover oil seal.
8. Lubricate new oil seal with wheel bearing grease. Position the seal on the end of the crankshaft with the garter spring side toward the engine.
9. Using a seal installer, drive the front seal into the front cover until it bottoms.

10. Assemble and install the remaining parts in reverse order of disassembly.

Pistons and Connecting Rod Assemblies

REMOVAL

1. Remove the engine assembly from the car, see "Engine Removal and Installation".
2. Remove the intake manifold, cylinder head or heads.
3. Remove the oil pan.
4. Remove the oil pump assembly.
5. Stamp the cylinder number on the machine surfaces of the bolt bosses of the connecting rod and cap for identification when reinstalling. If the pistons are to be removed from the connecting rod, mark the cylinder number on the piston with a silver pencil or quick-drying paint for proper cylinder indentification and cap-to-rod location. Engines are numbered oddly on the right bank; evenly on the left.
6. Examine the cylinder bore above the ring travel. If a ridge exists, remove the ridge with

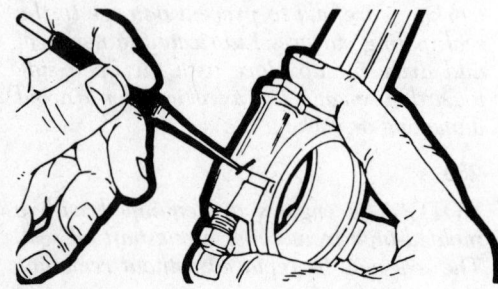

Match the connecting rods to their caps with a scribe mark

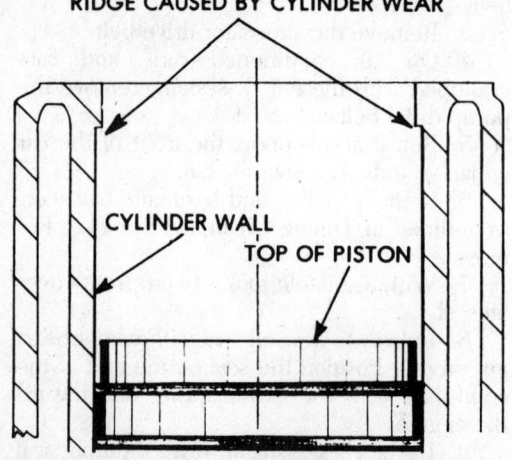

RIDGE CAUSED BY CYLINDER WEAR

CYLINDER WALL

TOP OF PISTON

Ridge formed by piston rings at the top of their travel

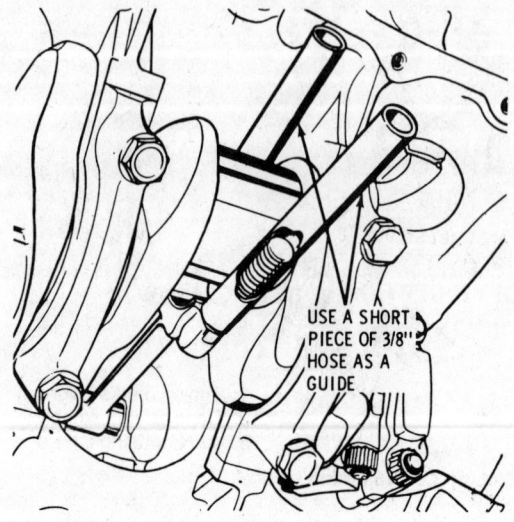

USE A SHORT PIECE OF 3/8" HOSE AS A GUIDE

Connecting rod bolt guide

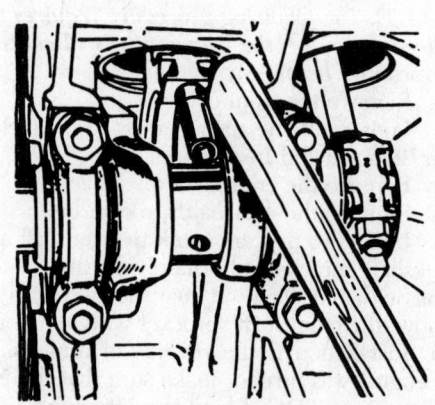

Push the piston and rod out with a hammer handle

a ridge reamer before attempting to remove the piston and rod assembly.

7. Remove the rod bearing cap and bearing.
8. Install a guide hose over threads of rod bolts. This is to prevent damage to bearing journal and rod bolt threads.
9. Remove the rod and piston assembly through the top of the cylinder bore.
10. Remove the other rod and piston assemblies in the same manner.

PISTON PIN REMOVAL AND INSTALLATION

Use care at all times when handling and servicing connecting rods and pistons. To prevent possible damage to these units, do not clamp the rod or piston in a vise since they may become distorted. Do not allow the pistons to strike against one another, against hard objects or bench surfaces, since distortion of the piston contour or nicks in the soft aluminum material may result.

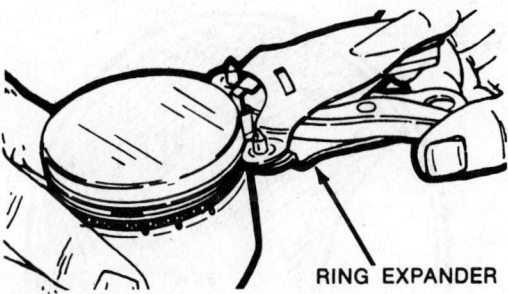

Remove the piston rings

RING EXPANDER

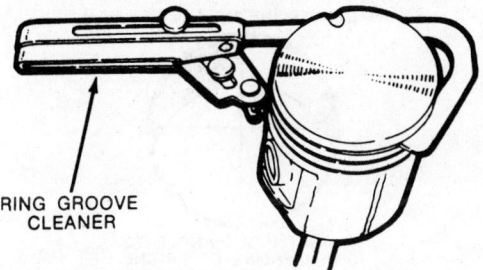

RING GROOVE
CLEANER

Cleaning the piston ring grooves

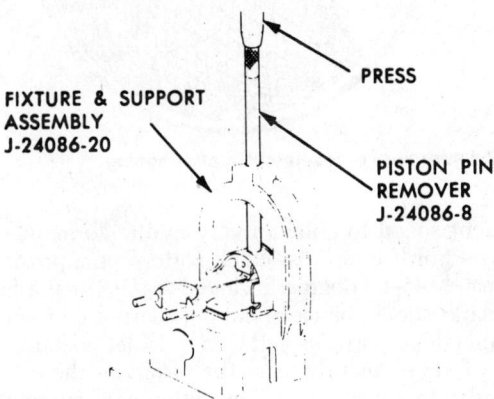

FIXTURE & SUPPORT
ASSEMBLY
J-24086-20

PRESS

PISTON PIN
REMOVER
J-24086-8

Removing piston pin

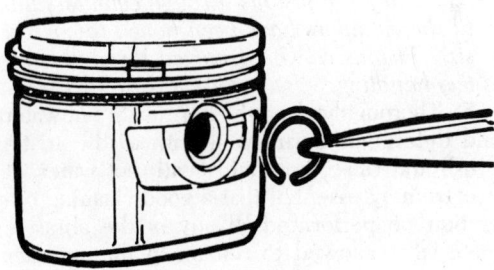

Install the piston lock-rings, if used

1. Remove the piston rings using a suitable piston ring remover.
2. Remove the piston pin lockring, if used. Install the guide bushing of the piston pin removing and installing tool.
3. Install the piston and connecting rod assembly on a support, and place the assembly in an arbor press. Press the pin out of the connecting rod, using the appropriate piston pin tool.
4. Assembly is the reverse of disassembly. Use new lockrings where needed.

CLEANING AND INSPECTION

Connecting Rods

Wash connecting rods in cleaning solvent and dry with compressed air. Check for twisted or bent rods and inspect for nicks or cracks. Replace connecting rods that are damaged.

Pistons

Clean varnish from piston skirts and pins with a cleaning solvent. DO NOT WIRE BRUSH ANY PART OF THE PISTON. Clean the ring grooves with a groove cleaner and make sure oil ring holes and slots are clean. Inspect the piston for cracked ring lands, skirts or pin bosses, wavy or worn ring lands, scuffed or damaged skirts, eroded areas at the top of the piston. Replace pistons that are damaged or show signs of excessive wear. Inspect the grooves for nicks or burrs that might cause the rings to hang up.

Measure piston skirt (across center line of piston pin) and check piston clearance.

MEASURING THE OLD PISTONS

Check used piston-to-cylinder bore clearance as follows:
1. Measure the cylinder bore diameter with a telescope gauge.
2. Measure the piston diameter. When measuring the pistons for size or taper, measurements must be made with the piston pin removed.
3. Subtract the piston diameter from the cylinder bore diameter to determine piston-to-bore clearance.

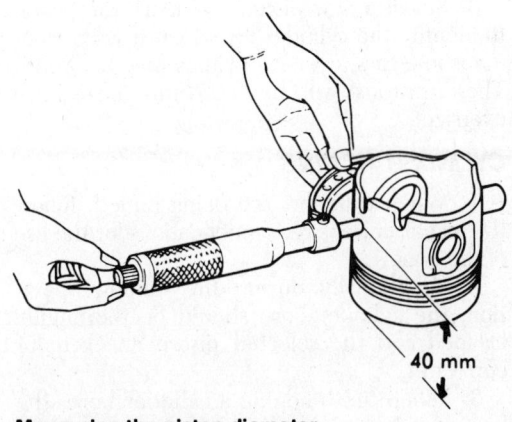

40 mm

Measuring the piston diameter

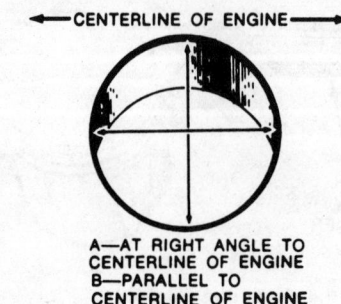

A—AT RIGHT ANGLE TO CENTERLINE OF ENGINE
B—PARALLEL TO CENTERLINE OF ENGINE

Cylinder bore measuring points

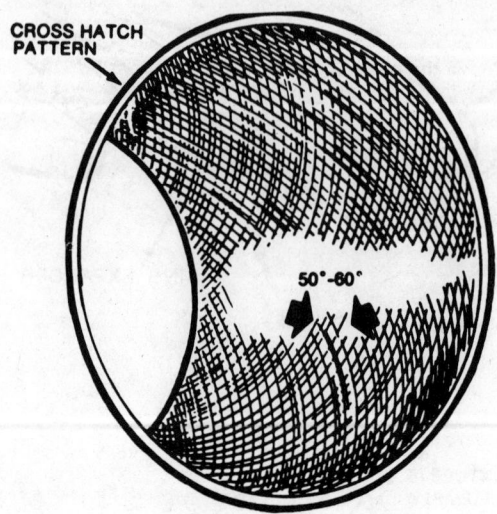

Cylinder bore cross-hatching after honing

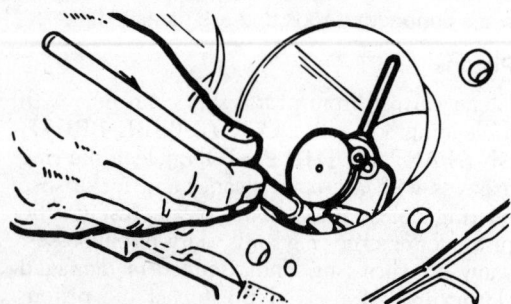

Measuring cylinder bore with a dial gauge

4. Compare the piston-to-bore clearances obtained with those clearances recommended. Determine if the piston-to-bore clearance is in the acceptable range.

5. When measuring taper, the largest reading must be at the bottom of the skirt.

SELECTING NEW PISTONS

1. If the used piston is not acceptable, check the service piston size and determine if a new piston can be selected. (Service pistons are available in standard, high limit and standard 0.254mm (0.010 in.) oversize.).

2. If the cylinder bore must be reconditioned, measure the new piston diameter, then hone the cylinder bore to obtain the prefered clearance.

3. Select a new piston and mark the piston to identify the cylinder for which it was fitted. (On some cars, oversize pistons may be found. These pistons will be 0.254mm (0.010 in.) oversize).

CYLINDER HONING

1. When cylinders are being honed, follow the manufacturer's recommendations for the use of the hone.

2. Occasionally during the honing operation, the cylinder bore should be thoroughly cleaned and the selected piston checked for correct fit.

3. When finish-honing a cylinder bore, the hone should be moved up and down at a suffi-

cient speed to obtain a very fine uniform surface finish in a cross-hatch pattern of approximately 45–65 degrees included angle. The finish marks should be clean but not sharp, free from imbedded particles and torn or folded metal.

4. Permanently mark the piston for the cylinder to which it has been fitted and proceed to hone the remaining cylinders.

NOTE: *Handle pistons with care. Do not attempt to force pistons through cylinders until the cylinders have been honed to correct size. Pistons can be distorted through careless handling.*

5. Thoroughly clean the bores with hot water and detergent. Scrub well with a stiff bristle brush and rinse thoroughly with hot water. It is extremely essential that a good cleaning operation be performed. If any of the abrasive material is allowed to remain in the cylinder bores, it will rapidly wear the new rings and cylinder bores. The bores should be swabbed several times with light engine oil and a clean cloth and then wiped with a clean dry cloth. CYLINDERS SHOULD NOT BE CLEANED WITH KEROSENE OR GASOLINE. Clean the remainder of the cylinder block to remove the excess material spread during the honing operation.

CHECKING CYLINDER BORE

Cylinder bore size can be measured with inside micrometers or a cylinder gauge. The most wear will occur at the top of the ring travel.

Reconditioned cylinder bores should be held to not more than 0.025mm (0.001 in.) taper.

If the cylinder bores are smooth, the cylinder walls should not be deglazed. If the cylinder walls are scored, the walls may have to be honed before installing new rings. It is important that reconditioned cylinder bores be thoroughly washed with a soap and water solution to remove all traces of abrasive material to eliminate premature wear.

Piston Rings

The pistons have three rings (two compression rings and one oil ring). The oil ring consists of two rails and an expander. Pistons do not have oil drain holes behind the rings.

RING TOLERANCES

When installing new rings, ring gap and side clearance should be checked as follows:

Piston Ring and Rail Gap

Each ring and rail gap must be measured with the ring or rail positioned squarely and at the bottom of the ring-travel area of the bore.

Side Clearance

Each ring must be checked for side clearance in its respective piston groove by inserting a feeler gauge between the ring and its upper land. The piston grooves must be cleaned before checking the ring for side clearance specifications. To check oil ring side clearance, the oil rings must be installed on the piston.

RING INSTALLATION

For service ring specifications and detailed installation productions, refer to the instructions furnished with the parts package.

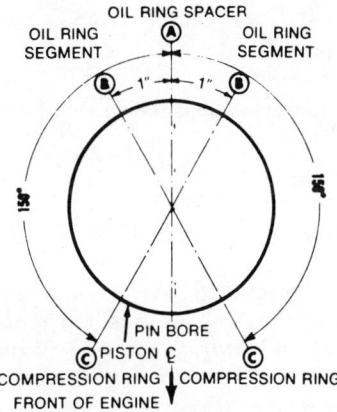

Piston ring spacing

Connecting Rod Bearings

If you have already removed the connecting rod and piston assemblies from the engine, follow only Steps 3–7 of the following procedure.

REMOVAL, INSPECTION, INSTALLATION

The connecting rod bearings are designed to have a slight projection above the rod and cap faces to insure a positive contact. The bearings can be replaced without removing the rod and piston assemblies from the engine.

1. Remove the oil pan. See the Oil Pan procedures, below. It may be necessary to remove the oil pump to provide access to rear connecting rod bearings.

2. With the the connecting rod journal at the bottom, stamp the cylinder number on the machined surfaces of the connecting rod and cap for identification when installing, then remove the caps.

3. Inspect journals for roughness and wear. Slight roughness may be removed with a fine grit polishing cloth saturated with engine oil. Burrs may be removed with a fine oil stone by moving the stone on the journal circumference. Do not move the stone back and forth across the journal. If the journals are scored or ridged, the crankshaft must be replaced.

4. The connecting rod journals should be

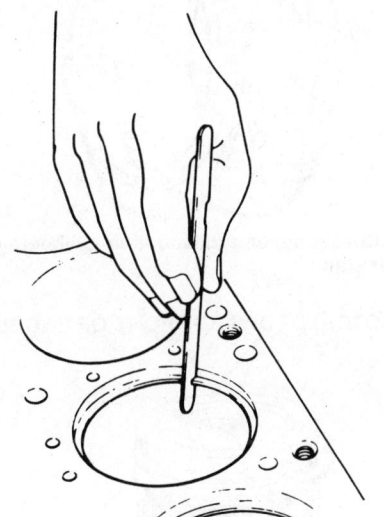

Measuring piston ring gap

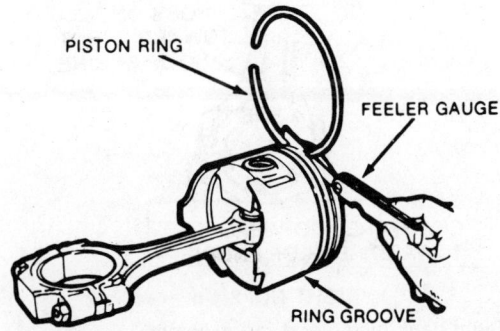

Checking ring side clearance

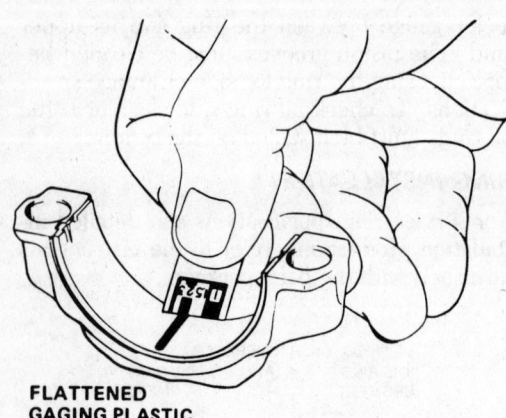

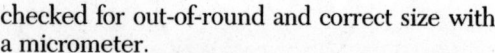

**FLATTENED
GAGING PLASTIC**

Checking rod bearing clearance with Plastigage®
or equivalent

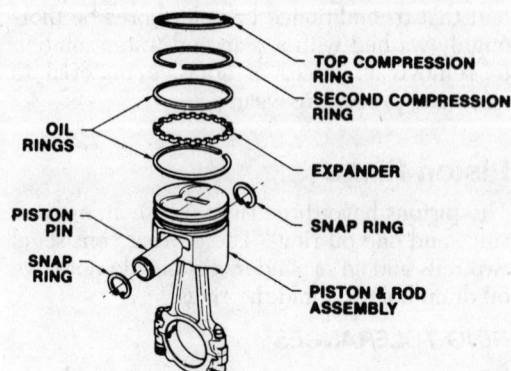

Piston rings and wrist pin

checked for out-of-round and correct size with
a micrometer.

> NOTE: *Crankshaft rod journals will nor-
> mally be standard size. If any undersized
> bearings are used, all will be 0.254mm un-
> dersize and 0.254mm will be stamped on the
> number 4 counterweight.*

If plastic gauging material is to be used:

5. Clean oil from the journal bearing cap,
connecting rod and outer and inner surfaces of
the bearing inserts. Position the insert so that
the tang is properly aligned with the notch in
the rod and cap.

6. Place a piece of plastic gauging material
in the center of lower bearing shell.

7. Remove the bearing cap and determine
the bearing clearances by comparing the width
of the flattened plastic gauging material at its
widest point with the graduation on the container. The number within the graduation on
the envelope indicates the clearance in thousandths of an inch or millimeters. If this clearance is excessive, replace the bearing and recheck the clearance with the plastic gauging
material. Lubricate the bearing with engine oil
before installation. Repeat Steps 2–7 on the remaining connecting rod bearings. All rods must
be connected to their journals when rotating
the crankshaft, to prevent engine damage.

Piston and Connecting Rod
INSTALLATION

1. Install some lengths of rubber tubing over
the connecting rod bolts to prevent damage to
the journals.

2. Apply engine oil to the rings and piston,
then install a piston ring compressing tool on
the piston.

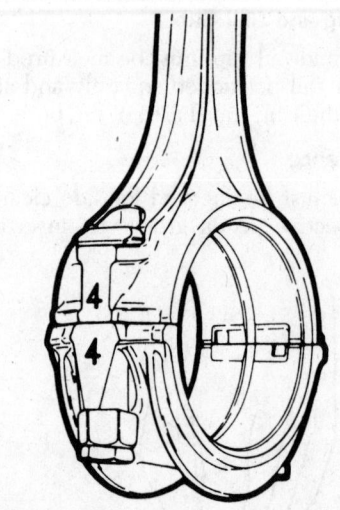

Match the connecting rods to their cylinders with a
number stamp

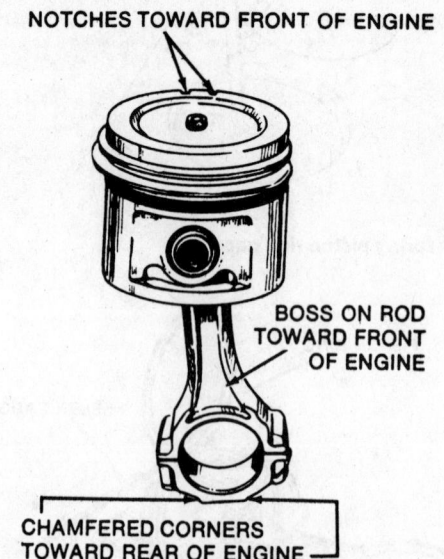

RIGHT NO. 2-4-6

Right bank piston and rod assembly,
V6s

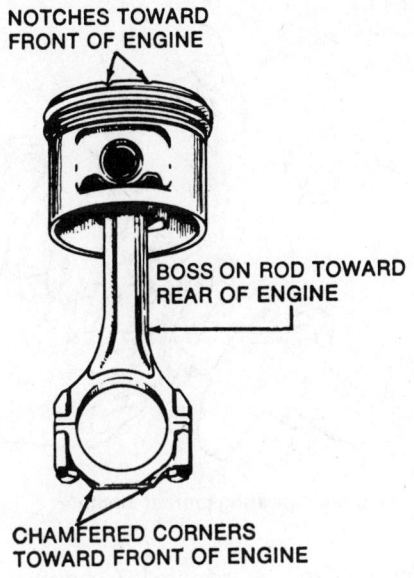

NOTCHES TOWARD
FRONT OF ENGINE

BOSS ON ROD TOWARD
REAR OF ENGINE

CHAMFERED CORNERS
TOWARD FRONT OF ENGINE

LEFT NO. 1-3-5

Left bank piston and rod assembly, V6s

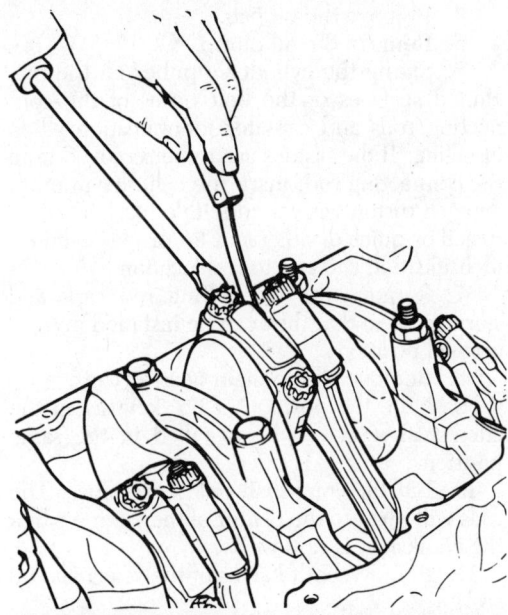

Checking connecting rod side clearance with a feeler gauge. Use a small pry bar to carefully spread the connecting rods

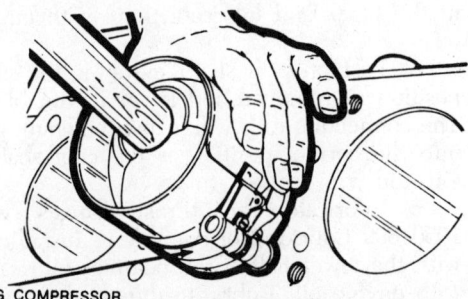

RING COMPRESSOR

Using a wooden hammer handle, tap the piston down through the ring compressor and into the cylinder

3. Install the assembly in its respective cylinder bore.

4. Lubricate the crankshaft journal with engine oil and install the connecting rod bearing and cap, with the bearing index tang in rod and cap on same side.

NOTE: *When more than one rod and piston assembly is being installed, the connecting rod cap attaching nuts should be tightened only enough to keep each rod in position until all have been installed. This will aid installation of the remaining piston assemblies.*

5. Torque the rod bolt nuts to specification. Using a feeler gauge and small prybar, check connecting rod side clearance.

6. Install all other parts in reverse order of removal.

7. Install the engine in the car. See Engine Removal and Installation.

Crankshaft

REMOVAL

1. Remove the engine assembly as previously outlined.

2. Remove the engine front cover.

3. Remove the timing chain and sprockets.

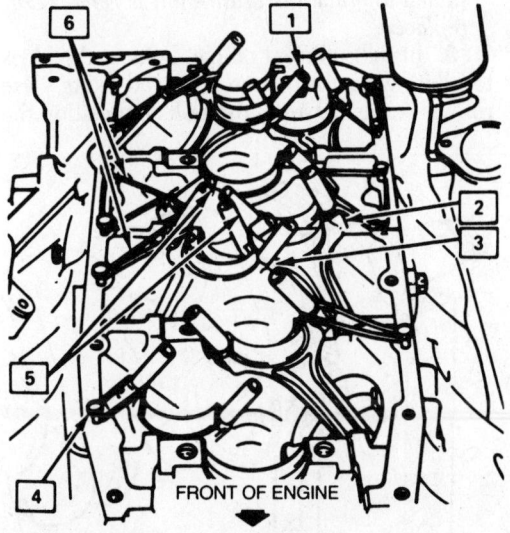

FRONT OF ENGINE

1. Rubber hose
2. # 4 Rod
3. # 3 Rod
4. Oil pan bolt
5. Note overlap of adjacent rods
6. Rubber bands

Crankshaft removal showing hose lengths on rod bolts

4. Remove the oil pan.

5. Remove the oil pump.

6. Stamp the cylinder number on the machined surfaces of the bolt boses of the connecting rods and caps for identification when installing. If the pistons are to be removed from the connecting rod, mark the cylinder number on each piston with an indelible marker, silver pencil or quick drying paint for proper cylinder identification and cap to rod location.

7. Remove the connecting rod caps and store them so that they can be installed in their original positions.

8. Remove all the main bearing caps.

9. Note the position of the keyway in the crankshaft so it can be installed in the same position.

10. Lift the crankshaft out of the block. The rods will pivot to the center of the engine when the crankshaft is removed.

11. Remove both halves of the rear main oil seal.

INSPECTION AND INSTALLATION

1. Using a dial indicator, check the crankshaft journal runout. Measure the crankshaft journals with a micrometer to determine the correct size rod and main bearings to be used. Whenever a new or reconditioned crankshaft is installed, new connecting rod bearings and main bearings should be installed. See Main Bearings and Rod Bearings.

2. Clean all oil passages in the block (and crankshaft if it is being reused).

NOTE: *A new rear main seal should be installed anytime the crankshaft is removed or replaced.*

3. Install sufficient oil pan bolts in the block to align with the connecting rod bolts. Use rubber bands between the bolts to position the

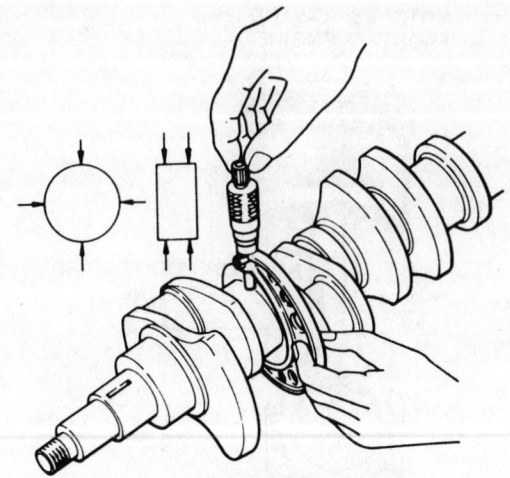

Checking main bearing journal diameter

connecting rods as required. Connecting rod position can be adjusted by increasing the tension on the rubber bands with additional turns around the pan bolts or thread protectors.

4. Position the upper half of main bearings in the block and lubricate them with engine oil.

5. Position crankshaft keyway in the same position as removed and lower it into block. The connecting rods will follow the crank pins into the correct position as the crankshaft is lowered.

6. Lubricate the thrust flanges with 10501609 Lubricant or equivalent. Install caps with the lower half of the bearings lubricated with engine oil. Lubricate the cap bolts with engine oil and install, but do not tighten.

7. With a block of wood, bump the shaft in each direction to align the thrust flanges of the main bearing. After bumping the shaft in each direction, wedge the shaft to the front and hold it while torquing the thrust bearing cap bolts.

NOTE: *In order to prevent the possibility of cylinder block and/or main bearing cap damage, the main bearing caps are to be tapped into their cylinder block cavity using a wood or rubber mallet before the bolts are installed. Do not use attaching bolts to pull the main bearing caps into their seats. Failure to observe this information may damage the cylinder block or a bearing cap.*

8. Torque all main bearing caps to specification. Check crankshaft endplay, using a flat feeler gauge.

9. Remove the connecting rod bolt thread protectors and lubricate the connecting rod bearings with engine oil.

10. Install the connecting rod bearing caps in their original position. Torque the nuts to specification.

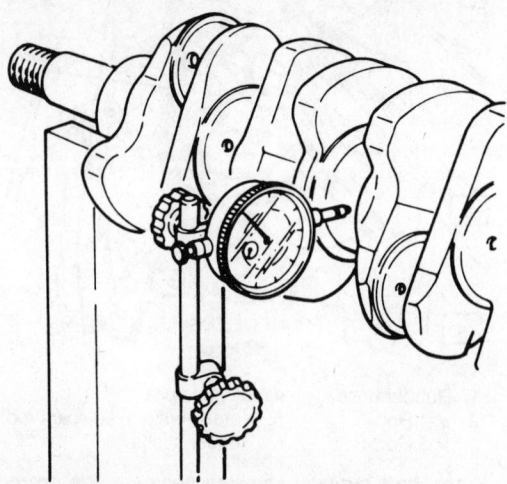

Checking crankshaft journal runout

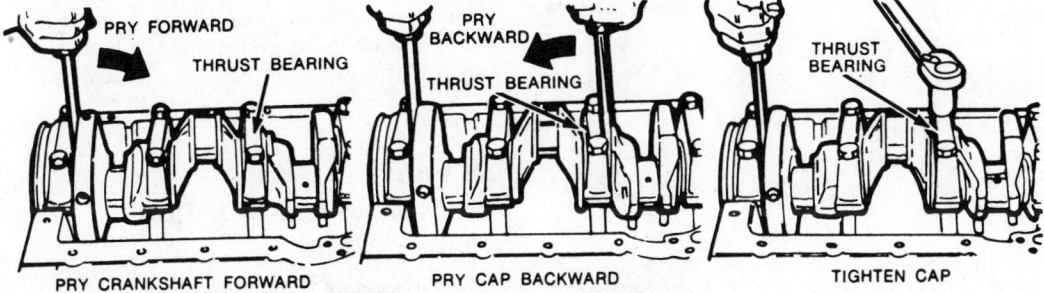

PRY FORWARD   THRUST BEARING

PRY BACKWARD   THRUST BEARING

THRUST BEARING

PRY CRANKSHAFT FORWARD   PRY CAP BACKWARD   TIGHTEN CAP

Aligning the crankshaft thrust bearing

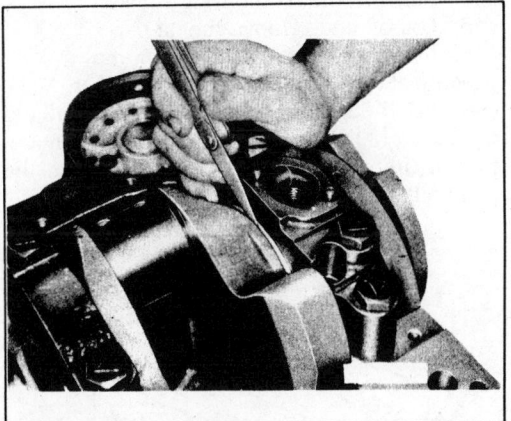

Measuring crankshaft end-play

11. Complete the installation by reversing the removal steps.

Main Bearings

CHECKING BEARING CLEARANCE

1. Remove the bearing cap and wipe the oil from the crankshaft journal and the outer and inner surfaces of the bearing shell.

2. Place a piece of plastic gauging material in the center of the bearing.

3. Use a floor jack or other means to hold the crankshaft against the upper bearing shell. This is necessary to obtain accurate clearance readings when using plastic gauging material.

4. Install the bearing cap and bearing. Place engine oil on the cap bolts and install. Torque the bolts to specification.

5. Remove the bearing cap and determine the bearing clearance by comparing the width of the flattened plastic gauging material at its widest point with the graduations on the gauging material container. The number within the graduation on the envelope indicates the clearance in millimeters or thousandths of an inch. If the clearance is greater than allowed, RE-PLACE BOTH BEARING SHELLS AS A SET. Recheck the clearance after replacing the shells. (Refer to Main Bearing Replacement).

REPLACEMENT

Main bearing clearances must be corrected by the use of selective upper and lower shells. UNDER NO CIRCUMSTANCES should the use of shims behind the shells to compensate for wear be attempted. To install the main bearing shells, proceed as follows:

1. Remove the oil pan as outlined below. On some models, the oil pump may also have to be removed.

2. Loosen all main bearing caps.

3. Remove the bearing cap and remove the lower shell.

4. Insert a flattened cotter pin or roll pin in the oil passage hole in the crankshaft, then rotate the crankshaft in the direction opposite

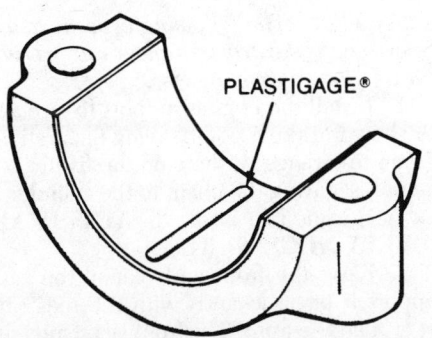

PLASTIGAGE®

Plastigauge installed on the lower bearing shell— typical

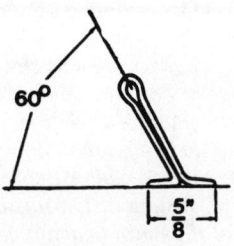

60°

$\frac{5"}{8}$

Home-made bearing roll-out pin

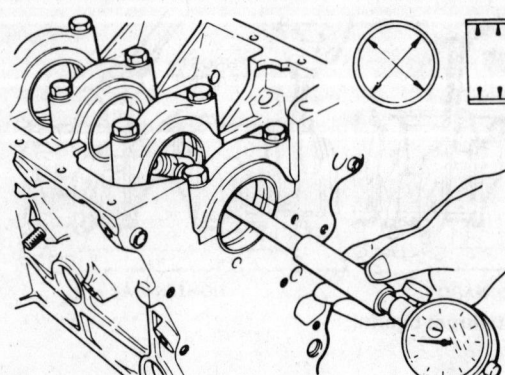

Measuring main bearing diameter

to cranking rotation. The pin will contact the upper shell and roll it out.

5. The main bearing journals should be checked for roughness and wear. Slight roughness may be removed with a fine grit polishing cloth saturated with engine oil. Burrs may be removed with a fine oil stone. If the journals are scored or ridged, the crankshaft must be replaced.

The journals can be measured for out-of-round with the crankshaft installed by using a crankshaft caliper and inside micrometer or a main bearing micrometer. The upper bearing shell must be removed when measuring the crankshaft journals. Maximum out-of-round of the crankshaft journals must not exceed 0.037mm (0.0015 in.).

6. Clean the crankshaft journals and bearing caps thoroughly for installing new main bearings.

7. Apply special lubricant, No. 1050169 or equivalent, to the thrust flanges of bearing shells.

8. Place a new upper shell on the crankshaft journal with locating tang in the correct position and rotate the shaft to turn it into place using a cotter pin or roll pin as during removal.

9. Place a new bearing shell in the bearing cap.

10. Install a new oil seal in the rear main bearing cap and block.

11. Lubricate the main bearings with engine oil. Lubricate the thrust surface with lubricant 1050169 or equivalent.

12. Lubricate the main bearing cap bolts with engine oil.

NOTE: *In order to prevent the possibility of cylinder block and/or main bearing cap damage, the main bearing caps are to be tapped into their cylinder block cavity using a wood or rubber mallet before the attaching bolts are installed. Do not use attaching bolts to pull the main bearing caps into their seats. Failure to observe this information may*

damage the cylinder block or a bearing cap.

13. Torque the main bearing cap bolts to 145 Nm (107 ft. lbs.).

Oil Pan

REMOVAL AND INSTALLATION

6–181 and 6–231

1. Disconnect the battery ground.
2. Raise and support the car on jackstands.
3. Drain the oil.
4. Remove the bellhousing cover.
5. Unbolt and remove the oil pan.
6. Installation is the reverse of removal. RTV gasket material is used in place of a gasket. Make sure that the sealing surfaces are free of all old RTV material. Use a ⅛ in. bead of RTV material on the oil pan sealing flange. Torque the pan bolts to 10–14 ft. lb.

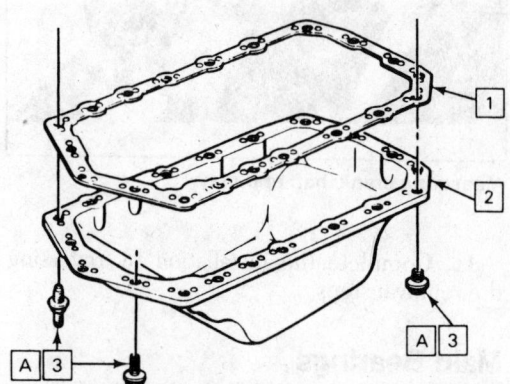

1. Formed rubber gasket
2. Oil pan
3. 10 N·m (88 lbs. in.)
4. Do not over tighten

3.0L or 3.8L V6 oil pan

6–263

CAUTION: *The following procedure will be personally hazardous unless the procedures are followed exactly.*

1. Install the engine support fixture assembly shown in the accompanying illustration. Be certain to arrange washers on the fixture so that the bolt securing the chain to the cylinder head can be torqued to 20 ft. lb. **THIS IS ABSOLUTELY NECESSARY!**

2. Raise the front and rear of the car and support it on jackstands with the rear slightly lower than the front. The front jackstands should be located at the front lift points shown in your owner's manual.

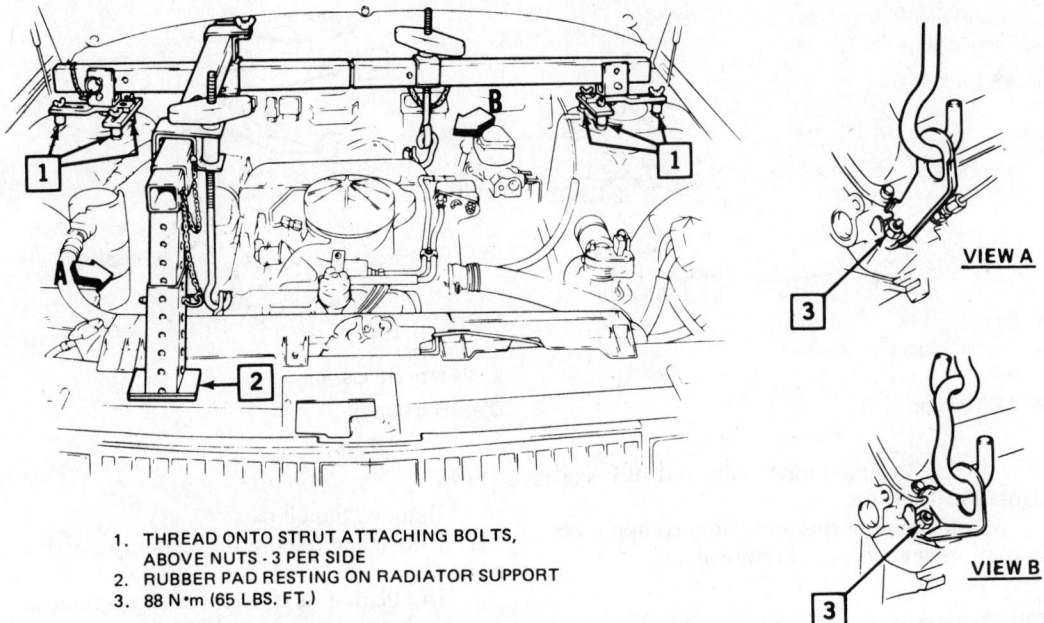

1. THREAD ONTO STRUT ATTACHING BOLTS, ABOVE NUTS - 3 PER SIDE
2. RUBBER PAD RESTING ON RADIATOR SUPPORT
3. 88 N•m (65 LBS. FT.)

VIEW A

VIEW B

Diesel engine holding fixture which MUST be used when removing the oil pan

3. Drain the oil.

4. Remove the leftside steering gear cradle bolt and loosen the right side cradle bolts.

5. Remove the front stabilizer bar.

6. Using a ½ in. drill bit, drill through the spot weld located between the rear holes at the left front stabilizer bar mounting.

7. Remove the nuts securing the engine and transaxle to its cradle.

8. Disconnect the left lower ball joint from the knuckle.

9. Place a wood block on a floor jack and raise the transaxle under the pan until the mount studs clear the cradle.

10. Remove the bolts securing the front crossmember to the right side of the cradle.

11. Remove the bolts from the left side front body mounts.

12. Remove the left side and front crossmember assemblies. It will be necessary to lower the rear crossmember below the left side of the body through the careful use of a large prybar.

13. Remove the bellhousing cover.

14. Remove the starter.

15. Remove the engine front mount bracket.

16. Unbolt and remove the oil pan.

17. Installation is the reverse of removal. Apply sealer to both sides of the oil pan gasket and make sure that the tabs on the gaskets are installed in the seal notches. Apply RTV sealer to the front cover oil pan seal retainer, and to each seal where it contacts the block. Wipe the seal area of the pan with clean engine oil before installing the pan. Torque the pan bolts to 10 ft. lb.

8–250

1. Disconnect the negative battery cable.

2. Remove the two flywheel covers.

3. Drain the oil.

4. Remove the mounting bolts and nuts and then remove the oil pan.

NOTE: *If the pan is difficult to remove, try tapping the edges lightly with a rubber mallet.*

5. Seal the oil pan to the block with RTV sealant.

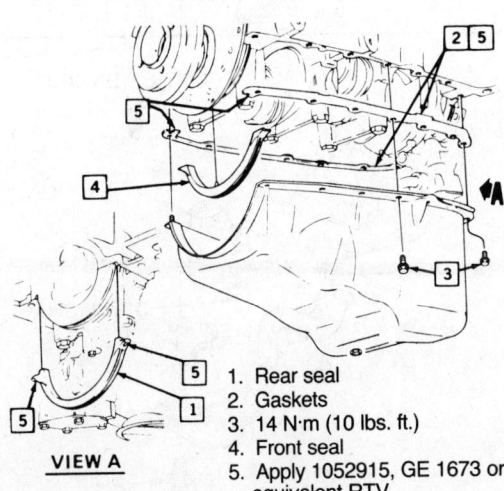

1. Rear seal
2. Gaskets
3. 14 N·m (10 lbs. ft.)
4. Front seal
5. Apply 1052915, GE 1673 or equivalent RTV.

VIEW A

Diesel engine oil pan

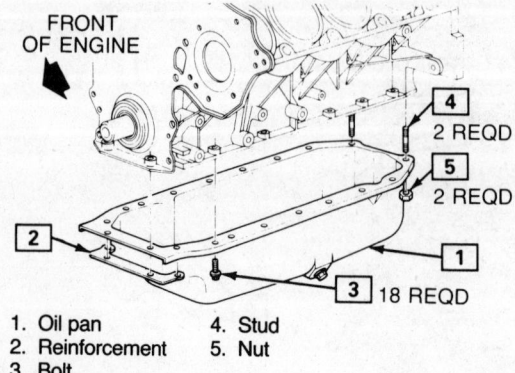

1. Oil pan   4. Stud
2. Reinforcement   5. Nut
3. Bolt

4.1L V8 oil pan

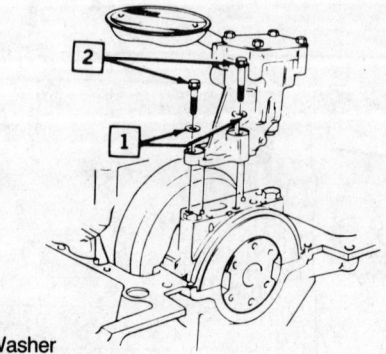

1. Washer
2. 24 N·m (18 lbs. ft.)

Diesel oil pump

6. Install the mounting bolts and nuts and tighten to 12 ft. lbs.

7. Installation of the remaining components is in the reverse order of removal.

Oil Pump

REMOVAL AND INSTALLATION

6–181 and 6–231

1. Remove the oil filter.

2. Unbolt the oil pump cover from the timing chain cover.

3. Slide out the oil pump gears. Clean all parts thoroughly in solvent and check for wear. Remove the oil pressure relief valve cap, spring and valve.

4. Installation is the reverse of removal. Torque the pressure relief valve cap to 35 ft. lb. Install the pump gears and check their clearances:
- End clearance: 0.002–0.006 in.
- Side clearance: 0.002–0.005 in.5.

Place a straightedge across the face of the pump cover and check that it is flat to within 0.001 in. Pack the oil pump cavity with petroleum jelly so that there is no air space. Install the cover and torque the bolts to 10 ft. lb.

6–263

1. Remove the oil pan.

2. Unbolt and remove the oil pump and drive extension.

3. Installation is the reverse of removal. Torque the pump bolts to 18 ft. lb.

8–250

1. Jack up the car and support it with jack stands.

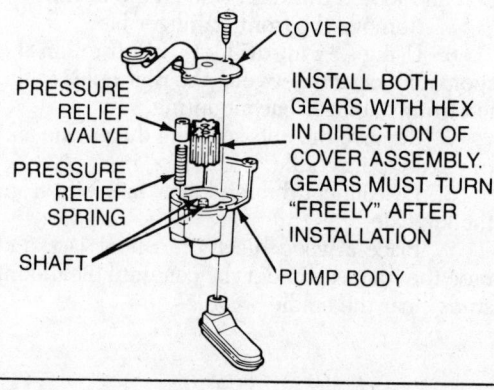

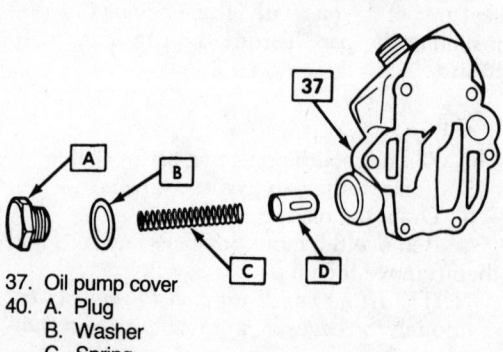

37. Oil pump cover
40. A. Plug
   B. Washer
   C. Spring
   D. Relief valve

3.0L or 3.8L V6 oil pump

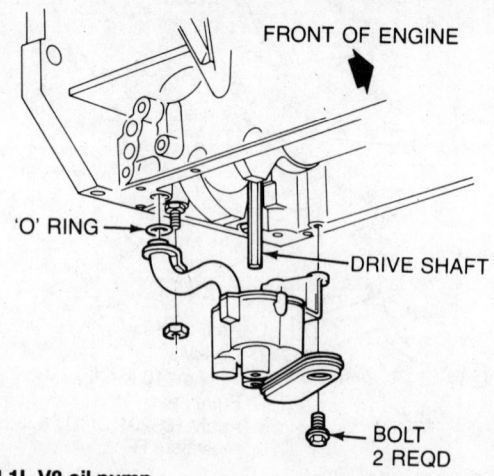

4.1L V8 oil pump

2. Remove the oil pan.

3. Remove the two screws and one nut securing the oil pump to the engine.

4. To disassemble, remove the four screws holding the oil pump cover to the housing, then slide the drive shaft, drive gear and driven gear out of the pump housing.

5. Remove the oil pressure regulator valve and spring from the bore in the housing assembly.

6. Inspect the oil pressure regulator valve for nicks and burrs.

7. Measure the free length of the regulator valve spring. It should be 2.57–2.69 in.

8. Inspect the drive gear and driven gear for nicks and burrs.

9. Assemble the pump drive gear over the drive shaft so that the retaining ring is inside the gear. Position the drive gear over the pump housing shaft closest to the pressure regulator bore.

10. Slide the driven gear over the remaining shaft in the pump housing, meshing the driven gear with the drive gear.

11. Install the oil pressure regulator spring and valve in the bore of the pump housing assembly.

12. Install the pump cover and four retaining screws.

13. Install the oil pump assembly to the block, engaging the drive shaft to the distributor gear. Tighten the nut to 22 ft. lbs. and the two screws to 15 ft. lbs.

14. Install the oil pan and lower the car.

Rear Main Seal

REMOVAL AND INSTALLATION

6–181, 6–231 and 6–263

Braided fabric seals are pressed into grooves formed in crankcase and rear bearing cap to rear of the oil collecting groove, to seal against leakage of oil around the crankshaft.

A new braided fabric seal can be installed in crankcase only when crankshaft is removed, but it can be repaired while crankshaft is installed, as outlined under "Rear Main Bearing Upper Oil Seal Repair". The seal can be replaced in cap whenever the cap is removed. Remove old seal and place new seal in groove with both ends projecting above parting surface of cap. Force seal into groove rubbing down with hammer handle or smooth stick until seal projects above the groove not more than ¹⁄₁₆ in. Cut ends off flush with surface of cap, using sharp knife or razor blade.

The engine must be operated at slow speed when first started after a new braided seal is installed.

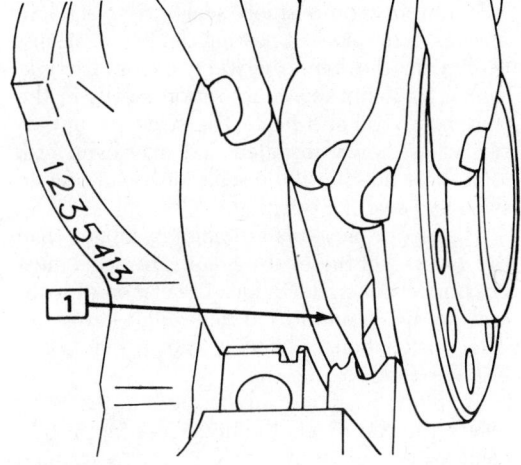

1. Packing tool J-21526-2

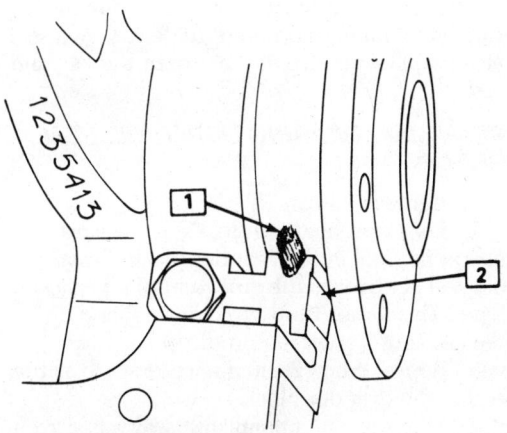

1. Short piece of rope seal
2. Guide tool J-21526-1 installed

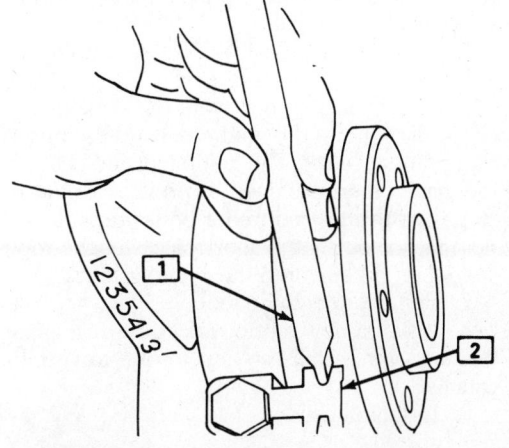

1. Packing tool
2. Guide tool

Rear main seal installation on all but the V8

Neoprene composition seals are placed in grooves in the sides of bearing cap to seal against leakage in the joints between cap and crankcase. The neoprene composition swells in the presence of oil and heat. The seals are undersize when newly installed and may even leak for a short time until the seals have had time to swell and seal the opening.

The neoprene seals are slightly longer than the grooves in the bearing cap. The seals must not be cut to length. Before installation of seals, soak for 1 to 2 minutes in light oil or kerosene. After installation of bearing cap in crankcase, install seal in bearing cap.

To help eliminate oil leakage at the joint where the cap meets the crankcase, apply silicone sealer, or equivalent, to the rear main bearing cap split line. When applying sealer, use only a thin coat as an over abundance will not allow the cap to seat properly.

After seal is installed, force seals up into the cap with a blunt instrument to be sure of a seal at the upper parting line between the cap and case.

REAR MAIN BEARING UPPER OIL SEAL REPAIR

1. Remove oil pan.
2. Insert packing tool (J-21526-2) against one end of the seal in the cylinder block. Drive the old seal gently into the groove until it is packed tight. This varies from 1/4–3/4 in. depending on the amount of pack required.
3. Repeat Step 2 on the other end of the seal in the cylinder block.
4. Measure the amount the seal was driven up on one side and add 1/16 in. Using a single edge razor blade, cut that length from the old seal removed from the rear main bearing cap. Repeat the procedure for the other side. Use the rear main bearing cap as a holding fixture when cutting the seal.
5. Install Guide Tool (J-21526-1) onto cylinder block.
6. Using packing tool, work the short pieces cut in Step 4 into the guide tool and then pack into cylinder block. The guide tool and packing tool have been machined to provide a built-in stop. Use this procedure for both sides. It may help to use oil on the short pieces of the rope seal when packing into the cylinder block.
7. Remove the guide tool.
8. Install a new fabric seal in the rear main bearing cap. Install cap and torque to specifications.
9. Install oil pan.

8–250

1. Remove the oil pan (See oil pan removal).

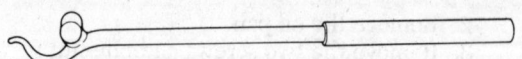

V8 rear main seal remover tool

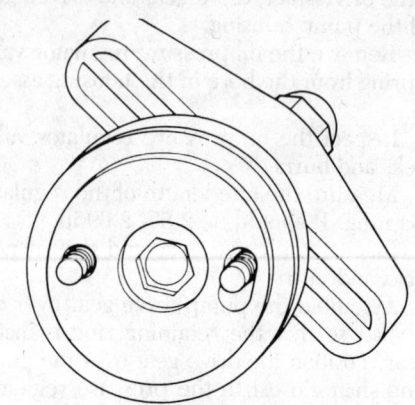

V8 rear main seal installation tool

2. Remove the rear main bearing cap and loosen the bolts holding the other four bearings about three turns each. Remove the old rear main bearing seals.
3. Clean the groove in the cap and in the block. Lubricate seals with engine oil.
4. Make an installation tool.
5. Start the upper half into the groove in the block with the lip facing forward and rotate it into position, using the tool as a guide. Press firmly on both ends to be sure it is protruding uniformly on each side.
6. Install the lower half of the seal into the bearing cap with the lip facing forward and one end of the seal over the ridge and flush with the split line. Hold one finger over this end to prevent it from slipping, and push the seal into seated position by applying pressure to the otherend. Be sure the seal is firmly seated and protrudes evenly on each side. Do not apply pressure to the lip. This may damage the effectiveness of the seal.

NOTE: *Vehicles equipped with neoprene type seals, make sure that the seal is flush at the split line to avid leaks.*

7. Apply rubber cement to the mating surfaces of the block and cap being careful not to get any cement on the bearing, the crankshaft or the seal. The cement coating should be about 0.010 in. thick.
8. Tighten the bearing bolts to 89 ft. lbs. for the 250 and 90–100 ft. lbs. for all others. Be sure to tighten the bolts of the other four bearings also. Rotate the crankshaft one full turn to check for binding.
9. Reinstall the oil pan.

Radiator

REMOVAL AND INSTALLATION

1. Disconnect the negative battery cable.
2. Drain the coolant.
3. On the 8–250 , detach the electrical connectors, remove the mounting bolts and then remove the left and right cooling fans.
4. Loosen the clamp-screws and remove the coolant reservoir and upper radiator hoses.
5. Disconnect the engine, transaxle and auxiliary oil cooler lines at the radiator. Wire the lines out of the way.
6. Disconnect the lower radiator hose at the radiator.
7. Remove the mounting bolts and lift out the radiator.
8. On all other engines, remove the upper fan mounting bolts (the 6–263 diesel engine has two cooling fans like the 8–250).
9. Disconnect the upper air cleaner duct and/or silencer on the Ninety Eight.
10. Unscrew the mounting bolts and remove the upper radiator valance panel.
11. Unscrew the clamp-screws and disconnect the coolant recovery tank hose and the upper radiator hose from the radiator.
12. Disconnect the transaxle and engine (diesel only) oil cooler lines from the radiator side tank. Wire them out of the way.
13. Unscrew the mounting bolts and then lift the radiator from the engine compartment.
14. Installation is in the reverse order of removal.
NOTE: *When installing the engine oil cooler lines on the diesel, always use new O-rings. Tighten to 26 ft. lbs.*

Water Pump

REMOVAL AND INSTALLATION

6–181

1. Disconnect the negative battery cable.
2. Remove accessory drive belts.
3. Remove water pump attaching bolts.
4. Remove the engine support strut.
5. Place a floor jack under the front crossmember of the cradle and raise the jack until the jack just starts to raise the car.
6. Remove the front two body mount bolts with the lower cushions and retainers.
7. Thread the body mount bolts with retainers a minimum of three turns into the cage so that the bolts restrain cradle movement.
8. Release the floor jack slowly until the crossmember contacts the body mount bolt retainers. As the jack is being lowered watch and correct any interference with hoses, lines, pipes and cables.

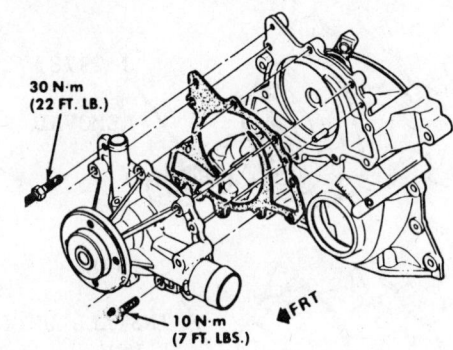

30 N·m
(22 FT. LB.)

10 N·m
(7 FT. LBS.)

◀FRT

3.0L or 3.8L water pump

NOTE: *Do not lower the cradle without its being restrained as possible damage can occur to to the body and underhood items.*
9. Remove water pump from engine.
10. Reverse removal procedure.
11. Install pump and torque to 25 ft. lbs.
12. Connect negative battery cable.
13. Fill with coolant and check for leaks.

6–231

1. Drain the cooling system. Remove the fan shroud, if necessary for clearance.
2. Loosen the belt or belts, then remove the fan blades and pulley or pulleys from the hub on the water pump shaft. Remove the belt or belts.
3. Disconnect the hose from the water pump inlet and the heater hose from the nipple. Remove the bolts, then remove the pump and gasket from the timing case cover or engine block.
4. Install the pump assembly with a new gasket. Bolts and lock washers must be torqued evenly.
5. Connect the radiator hose to the pump inlet and the heater hose to the nipple. Fill the cooling system and check all points of possible coolant leaks.
6. Install the fan pulley or pulleys and the fan blade. Install the belt or belts and adjust for correct tension.

6–263 Diesel

1. Drain the radiator.
2. Disconnect lower radiator hose and water pump.
3. Disconnect the heater return hose at the water pump, remove the bolt retaining the heater water return pipe to the intake manifold and position the pipe out of the way.
4. If equipped with A/C, remove the vacuum pump drive belt.
5. Remove the serpentine drive belt.

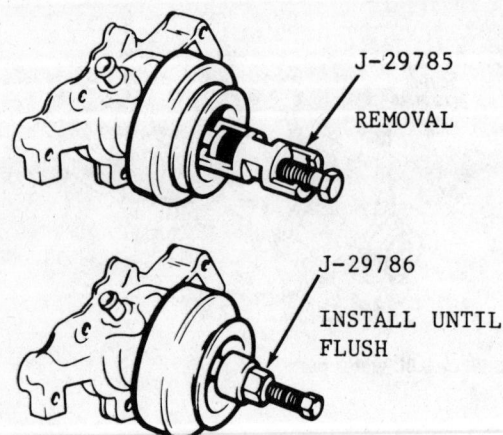

J-29785

REMOVAL

J-29786

INSTALL UNTIL FLUSH

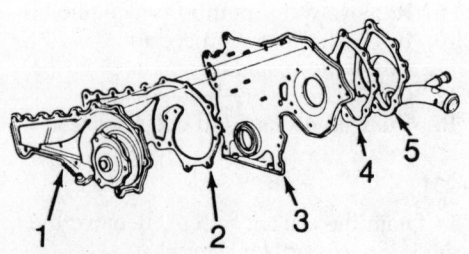

Removal and installation tools for the diesel water pump

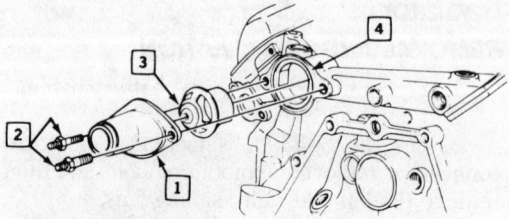

1. Water outlet
2. 24 N·m (18 lbs. ft.) dip threads of studs into 1050026 sealer or equivalent
3. Thermostat—install with pointed end out
4. Apply RTV sealer

Diesel thermostat housing

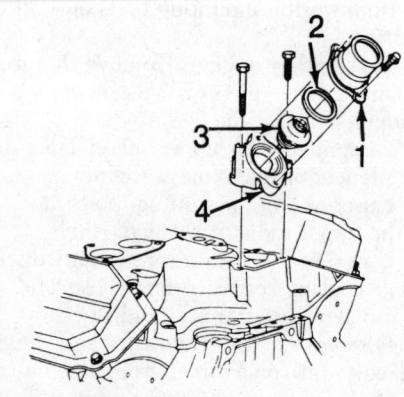

1. Upper housing
2. Gasket
3. Thermostat assembly
4. Lower housing

4.1L thermostat housing

1. Water pump assembly
2. Water pump gasket
3. Front cover
4. Water pump inlet gasket
5. Water pump inlet

4.1L V8 water pump

6. Remove the generator, A/C compressor vacuum pump brackets.

7. Remove the water pump attaching bolts and remove the water pump assembly.

8. Remove the water pump pulley using tool J-29785

9. Clean gasket material from engine block.

10. Installation is the reverse of removal. The pulley is installed using tool J-29786. Apply a thin coat of 1050026 sealer or equivalent to the water pump housing to retain the gasket, then position new gasket on the housing. Also apply sealer 1052624, or equivalent, to water pump mounting bolts. Torque bolts to 12–15 ft. lb.

8–250

1. Disconnect the negative battery terminal.

2. Drain the coolant.

3. Disconnect the A/C accumulator from the bracket and then position it out of the way. Disconnect the bracket from the wheel arch.

4. Remove the right side cross-car brace.

5. Remove the drive belt, the idler pulley and the bracket.

6. Unscrew the three mounting bolts and remove the water pump pulley.

7. Remove the water pump and gasket.

8. Installation is in the reverse order of removal. Always use a new water pump gasket.

Thermostat

REMOVAL AND INSTALLATION

To replace the thermostat, drain the cooling system below the level of the thermostat and remove the two bolts holding the water neck in place. Remove the water neck and the thermostat will lift out. Clean the mating surfaces of both the intake manifold and the water neck. Use a new gasket when installing a new thermostat. If only silicone sealer was used from the factory, use only sealer during assembly.

Be sure the thermostat is not reversed in its installed position. The spring should be installed toward the engine.

Emission Controls and Fuel Systems

4

EMISSION CONTROLS

Positive Crankcase Ventilation

Checking crankcase vacuum is the most effective way to test any PCV sytem. If there is a vacuum in the crankcase, then the major part of the system has to be working.

Inspect the system to find out where the fresh air enters the engine. This is usually through a hose attached to the air cleaner, but is may be through the oil filler cap on some models. If the fresh air entry is separate from the oil filler cap, simply remove the cap.

On all models, use a piece of paper or a PCV tester to measure the crankcase vacuum at the oil filler cap, with the cap removed and the engine idling in Park or Neutral. It may take a few seconds for the vacuum to build up enough to suck the piece of paper against the oil filler hole. If the vacuum does not build up, check to be sure you have plugged the fresh air entry. An alternate method on some cars is to use the piece of paper or PCV tester on the end of the fresh air entry hose. When you do it that way, the oil filler cap must be the solid type and you must leave it in place.

If there is no crankcase vacuum, pull the PCV valve from the crankcase and hold your finger over the end of it. You should feel full manifold vacuum with the engine idling. If not, the valve is plugged or there is an obstruction in a hose or passageway. On some designs the valve may be screwed into its mounting, with a hose leading to the rocker cover or crankcase. If the valve has good suction, but there is no crankcase vacuum, check the hose to be sure it is open. PCV valves that are restricted or plugged must be replaced, unless they are the type that will come apart for cleaning. Lack of crankcase vacuum can also be caused by vacuum leaks at rocker cover, oil pan, or other engine gaskets. Usually, tightening the bolts will stop the leak.

In some extreme cases, usually on high mileage engines, the PCV system is in good shape, but the blow-by past the rings is so much that the system can't handle it, and the engine will blow smoke out the oil filler hole. Switching to a PCV valve with a higher flow may temporarily correct the problem. But the only good solution is to overhaul the engine. If the motor oil is contaminated with gasoline, the PCV system will pick up the unburned vapors, add them to the intake mixture and cause the engine to run excessively rich.

After checking crankcase vacuum, always check the condition of the fresh air filter and hose, to be sure they are clean and not clogged.

NOTE: *The PCV system operation is not computer controlled, but if inoperative it will directly affect the operation of any computerized emission system. If poor performance is a problem, check the PCV system first.*

Evaporative Emission Control
CHARCOAL VAPOR SYSTEM

Most fuel evaporation losses come from the fuel tank. On an uncontrolled car the vapors go out through the tank vent, which may be in several places at the top of the tank or in the cap. There are also some losses through the bowl vent on the carburetor, but these are minor compared to the tank.

Evaporation controls are made up of hoses which allow the tank and carburetor vapors to go to a canister filled with charcoal. When the engine is running, a hose to the intake manifold or carburetor base allows engine vacuum to pull fresh air through the canister, drawing the vapors into the engine where they are burned. Fresh air enters the canister through a filter, which keeps the charcoal clean.

When the engine is running, air must enter the tank to replace the fuel that is used up and prevent a vacuum. On all makes of canister storage models, air enters the tank through the filter in the canister, but air can also enter the tank through the pressure/vacuum tank cap.

All evaporation control systems use some sort of vapor separator at the fuel tank to prevent liquid fuel from traveling along the vent line to the canister. The early models had very elaborate separators mounted separately from the tank, but now they are simpler and usually attached to the top of the tank. The only periodic servicing required on evaporation controls is replacement of the canister filter on those models on which it is replaceable.

NOTE: *If the vent lines become blocked, it is possible for some evaporation control systems to pull liquid fuel from the tank into the charcoal canister. If any charcoal canister is found to be fuel-soaked it should be replaced and all hoses checked for obstructions.*

Early Fuel Evaporation System

The electrically operated EFE system used on GM engines performs the same function as the vacuum operated heat riser on other engines, which is to preheat the engine induction system during cold driveway. Rapid heating is desirable because it provides quick fuel evaporation and more uniform fuel distribution to aid cold driveability.

The electrically heated EFE system has a ceramic heater grid located underneath the primary bore(s) of the carburetor which is part of the carburetor insulator. When the ignition is turned on and engine coolant temperature is low, voltage is applied to the EFE relay, which in turn transfers the voltage to the EFE heater in the ceramic grid. When temperature in-

creases, a thermal valve switch de-energizes the relay and the heater is turned off.

TESTING EFE HEATER

To check the resistance of the heater, turn the ignition off, disconnect the heater electrical connector, using a ohmmeter, measure the resistance across the two terminals of the heater connector. If resistance is under 2 ohms, the heater is good. If not, replace the heater.

Catalytic Converter

The catalytic converters are mounted in the engine exhaust stream and works as a gas reactor in which its major function is to speed up the heat producing chemical reaction between the exhaust gas components, in order to reduce the air pollutants in the engine exhaust.

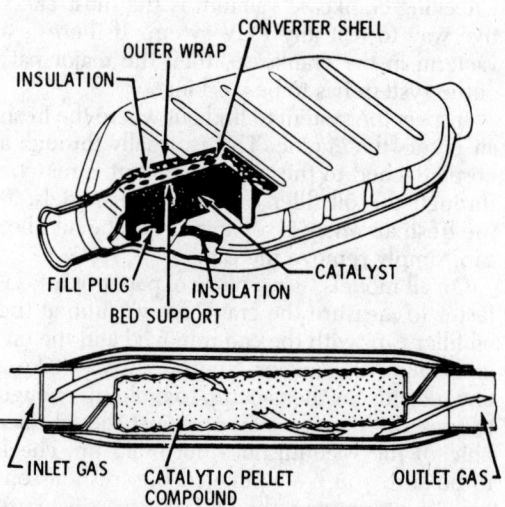

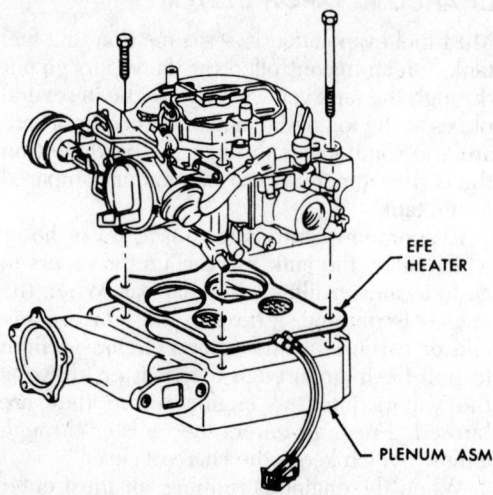

Electrically-heated type EFE

Bead type catalytic converter

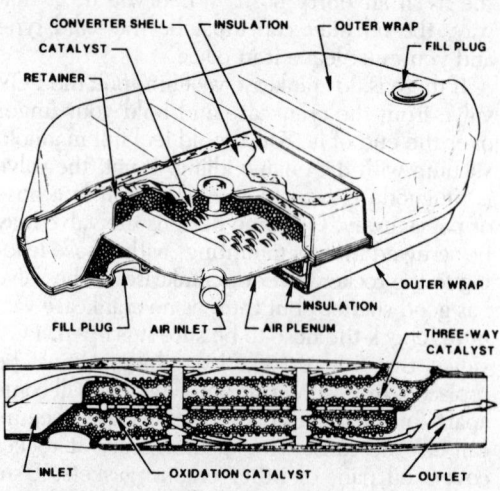

Dual bed type catalytic converter

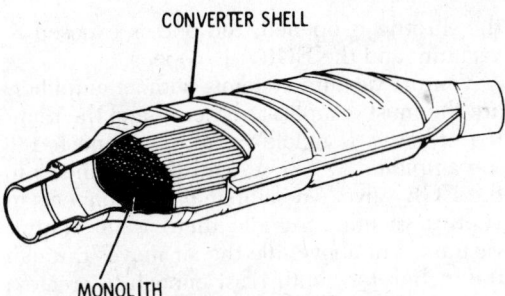

CONVERTER SHELL

MONOLITH

Single bed monolith catalytic converter

The converters produce CO_2 and water when operating.

The catalyst material is either a ceramic substrate or pellets that are coted with a base of alumina and then impregnated with catalytically active, precious (noble) metals. It is the surface of the catalyst material that controls the heat producing chemical reaction.

Two main types of converters are used on today's vehicles, one, an oxidation type converter containing two precious (noble) metals, platinum and palladium to effectively catalyze the oxidation of the hydrocarbons (HC) and carbon monoxide (CO). The second type converter used is considered a three-way catalyst, containing platinum and rhoduim in the front part of the converter to reduce the oxides of nitrogen (NOx), while platinum and palladium are used in the rear section to oxidize the hydrocarbons (HC) and carbon monoxide (CO), as was done in the two-way converters.

THREE-WAY CATALYTIC CONVERTERS

The three-way catalytic converters use a combination of catalysts which produce two different chemical reactions, oxidation and reduction. By adding fresh air to the unburned hydrocarbons (HC) and carbon monoxide (CO) within the converter, the oxidizing or combustion process takes place.

Just the reverse process is required to lower the oxides of nitrogen (NOx) emissions. The oxides of nitrogen (NOx) already contains excessive oxygen and the process of separating the excess oxygen from the nitrogen is called a reducing reaction.

This reducing or reduction process is done in the front section of the converter while the oxidizing process is accomplished in the rear section A fresh air connector is located on the center of the converter shell, to add fresh air from the air system as required.

To enable the three-way converter to operate properly, the engine's air/fuel ratio must be held within a tight range, called a "Stoichiometric" range. This is accomplished with the use of the closed loop, feedback fuel management systems incorporating the latest electronic controls.

Different controls are used by the vehicle manufacturers to prevent converter damage and/or burnout. Unleaded fuels must be used in the vehicles when equipped with the catalytic converters to prevent contamination failure.

The catalyst inside the converter is made in two forms. General Motors converters use the pellet form, in which loose pellets are packed into the converter and can be emptied out and changed, if necessary. GM models also use small honeycomb catalysts in both one and two-converter applications.

Testing

There is no way to test a converter in the field to see if it is actually working. Tailpipe readings may be used to set carburetor idle mixtures, when the car maker requires it, but taking a tailpipe reading to determine if the converter is working is not possible.

The one field check that is recommended in all cases is to inspect for mechanical damage. If a converter gets overheated, the catalyst can melt and block the exhaust. Pellets or pieces of the catalyst may even come flying out the tailpipe while the engine is running. If this happens, the pellets or the entire converter must be changed.

Checking for a melted converter that restricts the exhaust can be done with a vacuum gauge connected to the engine. Run the engine at about 2500 rpm in Park or Neutral. If the vacuum reading slowly drops, it indicates a buildup of pressure in the exhaust.

The use of leaded fuel will slowly destroy the efficiency of the catalyst. If used long enough, leaded fuel can even cause catalyst plugging to the point where the engine will not run. If you know that a car has been run on several tanks or leaded fuel, then you can be sure that the catalyst is ruined. The only thing you can do is change the catalyst or install a new converter.

NOTE: *Do not change the catalyst if the car has been run on only one tank or less of leaded fuel. Switching back to unleaded will allow the catalyst to recover and be almost as efficient as it was.*

Exhaust Gas Recirculation
GASOLINE ENGINES

NOx (oxides of nitrogen) is a tailpipe emission caused by the oxidation of nitrogen in the combustion chamber. When the peak combustion temperatures go over 2500°F., NOx is formed

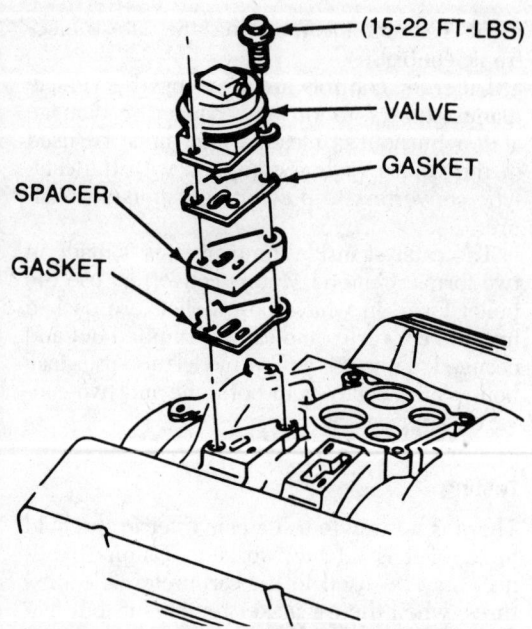

— (15-22 FT-LBS)

— VALVE

— GASKET

SPACER —

GASKET —

Typical EGR valve location, gasoline engines

in excessive amounts. To keep the combustion temperatures down, exhaust gas is recirculated on most cars. Recirculation is accomplished by allowing intake manifold vacuum to draw exhaust gas into the intake manifold. The lower combustion temperatures also help control spark knock (ping).

An EGR valve is used to control the flow of exhaust gas into the intake manifold. All EGR valve look alike, and are operated by vacuum. When the vacuum is off, the valve is closed. Several different types of controls are used to turn the vacuum to the EGR valve on and off. Most of them have to do with engine temperature, as described later. On computerized control systems, EGR operation is regulated by the electronic control unit.

NOTE: *All EGR systems are designed to cut off exhaust recirculation when the engine is cold, at idle, or under hard acceleration. If the EGR valve is stuck open, the engine won't idle.*

Ported vacuum EGR systems are the simplest. When the EGR valve hose is connected to the base of the carburator, without a separate amplifier, the system is operated by ported vacuum. The hose may not run directly from the EGR valve to the carburetor, but may go through a temperature control valve of some sort. In a ported vacuum system, the vacuum to operate the EGR valve is taken from a port that is above the throttle plate at idle, and thus not subject to vacuum. Because there is no vacuum, the spring in the EGR valve closes it, and the exhaust gas does not recirculate. As

the throttle is opened, the port is exposed to vacuum, and the EGR valve opens.

Venturi vacuum systems, with an amplifier, are the most complicated, because of the number of hoses. Manifold vacuum is connected to the amplifier by a hose, and then connects to the EGR valve. The amplifier also connects to venturi vacuum. At idle there is no venturi vacuum, but above idle the air moves through the carburetor venturi fast enough to create a vacuum. This slight amount of vacuum opens the amplifier, which then allows manifold vacuum to open the EGR valve.

Temperature controls for EGR systems come in many different designs. They are all made so that the EGR valve stays closed when the engine is cold. After the engine warms up, the temperature control allows the EGR valve to operate normally.

Testing EGR System

Testing of EGR systems should verify that when the engine is at normal operating temperature, the EGR valve is closed at idle, open above idle, and that the exhaust gas is actually recirculating. If the EGR valve sticks open at idle, the engine will run very rough, or may not even start. If this happens the valve should be removed and cleaned, or replaced. To check for valve operating above idle, check with a mirror or your fingers to see if the diaphragm or stem moves when the engine is at fast idle in Park or Neutral. If the diaphragm does not move when the throttle is opened, there is either a problem with vacuum, or the valve is stuck closed. With a vacuum gauge hooked up to the EGR port, you should see vacuum on the gauge when the throttle is opened. EGR valves should not leak when tested with a hand vacuum pump. If they do they must be replaced.

CAUTION: *The EGR valve gets hot during normal operation. Take normal precautions to avoid accidental burns.*

NOTE: *The EGR valve should open when about 3–5 in. Hg. is applied with the hand vacuum pump. Back pressure operated EGR valves cannot be vacuum tested.*

To find out if the exhaust gas is actually recirculating, use a hand vacuum pump or mouth suction through a hose to open the EGR valve with the engine idling. If the engine runs rough or dies, you know the exhaust gas is recirculating. If the engine does not run rough, make a second test of 2500 rpm. Opening the EGR valve at that rpm should cause a change in engine speed. If it does, you know the exhaust gas is recirculating. To make the 2500 rpm test, remove and plug the hose from the EGR port. Attach your suction hose to the EGR valve before running the engine at 2500 rpm. Simply pulling off the EGR hose at 2500 rpm is not a valid test, because the extra air entering the engine through the hose could cause a speed change all by itself. On most engines you won't have to go this far, because opening the EGR valve at idle will prove that the exhaust is recirculating.

If the exhaust is not recirculating, it means that a passageway or the valve itself is clogged up. The only way to fix it is to clean out the clogging as best you can, replace the clogged part, or replace the EGR valve.

Many EGR valves have a back pressure sensor built into the valve. This sensor is a pressure operated bleed that disables the EGR valve and keeps it closed when there is no exhaust pressure. This type of valve cannot be tested with a hand vacuum pump with the engine off because the bleed is open. The only practical way to test these new valves is by substitution of a known good valve. If a valve is not available, the suspect valve can be removed, and the mounting holes temporarily taped shut. If this corrects the problem, then a new valve should be installed.

DIESEL EGR SYSTEMS

General Information

GM has equipped its V6 diesel engines with EGR systems. The diesel EGR systems work in the same basic manner as gasoline engine EGR systems: exhaust gases are introduced into the combustion chambers to reduce combustion temperatures. On all V6 diesel EGR systems, the vacuum from the vacuum pump is modulated by the Vacuum Regulator Valve (VRV) mounted on the injection pump. The amount of EGR valve opening is further modulated by a Vacuum Modulator Valve (VMV). The VMV allows for an increase in vacuum to the EGR valve as the throttle is closed, up to the switching point of the VMV. The system also employs an VRV valve.

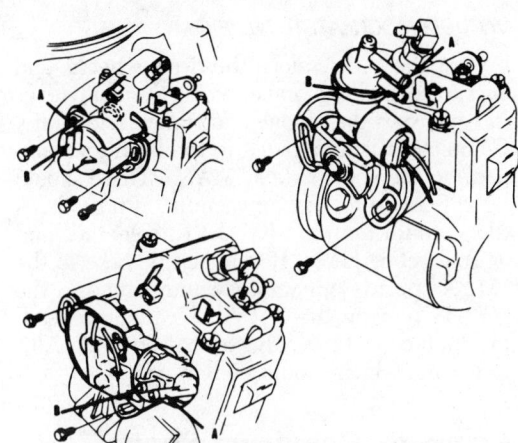

Diesel vacuum regulator valve, mounted to injection pump

Testing Diesel EGR System

VACUUM REGULATOR VALVE (VRV)

The VRV is attached to the side of the injection pump and regulates vacuum in proportion to throttle angle. Vacuum from the vacuum pump is supplied to port A and vacuum at port B (see illustration) is reduced as the throttle is opened. At closed throttle the vacuum is 15 in. Hg.; at half throttle, 6 in. Hg.; at wide open throttle there should be zero vacuum.

EXHAUST GAS RECIRCULATION (EGR) VALVE

Apply vacuum to the vacuum port. The valve should be fully open at 12 in. Hg. and closed below 6 in. Hg.

RESPONSE VACUUM REDUCER (RVR)

Connect a vacuum gauge to the port marked "To EGR valve or TCC solenoid". Connect a hand operated vacuum pump to the VRV port. Draw 15 in. of vacuum on the pump and the reading on the vacuum gauge should be .75 in. Hg. lower than the vacuum pump reading.

EXHAUST PRESSURE REGULATOR VALVE

Apply vacuum to the vacuum port of the valve. The valve should be fully closed at 12 in. Hg. and open below 6 in. Hg.

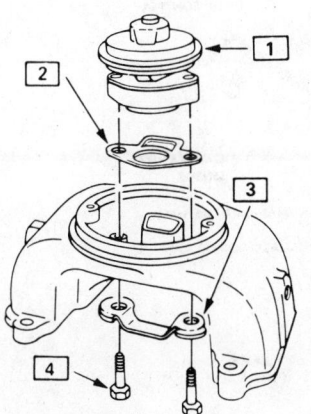

1. EGR valve
2. Gasket
3. Bolt lock
4. 24 N.M. (18 ft. lbs.)

AFTER BOLTS ARE TORQUED TO SPECS BEND LOCK TABS AROUND BOLT HEADS

Diesel EGR valve location

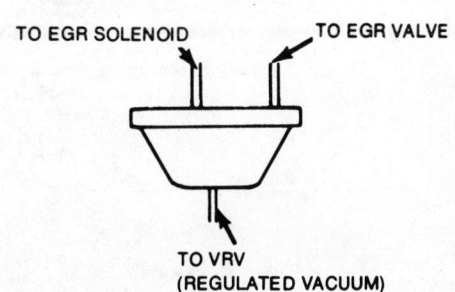

TO EGR SOLENOID   TO EGR VALVE

TO VRV
(REGULATED VACUUM)

Diesel EGR vacuum reduceer, except California

VACUUM MODULATOR VALVE (VMV)

To test the VMV, block the drive wheels, and apply the parking brake. With the shift lever in Park, start the engine and run at a slow idle. Connect a vacuum gauge to the hose that connects to the port marked "MAN". There should be at least 14 in. Hg. of vacuum. If not, check the vacuum pump, VRV, RVR, solenoid, and all connecting hoses. Reconnect the hose to the "MAN" port. Connect a vacuum gauge to the "DIST" port on the VMV. The vacuum reading should be 12 in. Hg. except on High Altitude cars, which should be 9 in. Hg.

Computer Command Control

The CCC System monitors up to fifteen engine engine/vehicle operating conditions which it uses to control up to nine engine and emission control systems. In addition to maintaining the ideal fuel/ratio for the catalytic converter and adjusting ignition timing, the CCC System also controls the AIR Management System so that the catalytic converter can operate at the highest efficiency possible. The system also controls the lockup on the transmission torque converter clutch (certain automatic transmission models only), adjusts idle speed over a wide range of conditions, purges the evaporative emissions charcoal canister, controls the EGR valve operation and operates the early fuel evaporative (EFE) system. Not all engines use all of the above sub-systems.

There are two operation modes: closed loop and open loop fuel fuel control. Closed loop fuel control means the oxygen sensor is controlling the carburetor's air/fuel mixture ratio. Under open loop fuel control operating conditions (wide open throttle), engine and/or oxygen sensor cold), the oxygen sensor has no effect on the air/fuel mixture.

NOTE: *On some engines, the oxygen sensor will cool off while the engine is idling, putting the system into open loop operation. To restore closed loop operation, run the engine at part throttle and accelerate from idle to part throttle a few times.*

CCC SYSTEM OPERATION

The CCC System ECM, in addition to monitoring sensors and sending a control signal to the carburetor, also controls the following components or sub-systems; charcoal canister purge, AIR Management System, idle speed control, automatic transmission converter lockup, distributor ignition timing, EGR valve control, EFE control and the air conditioner compressor clutch operation. The CCC ECM is equipped with a PROM assembly.

SERVICE PRECAUTIONS

To prevent ECM damage on CCC, DFI and DEFI equipped vehicles, the power supply feeding the ECM must not be interrupted with the ignition switch in the run, start or ACC position. The ignition switch must be placed in the off position when performing the following service operations:

- Disconnecting or connecting either battery cable.
- Remove or replacing the fuse provided continuous battery power to the ECM.

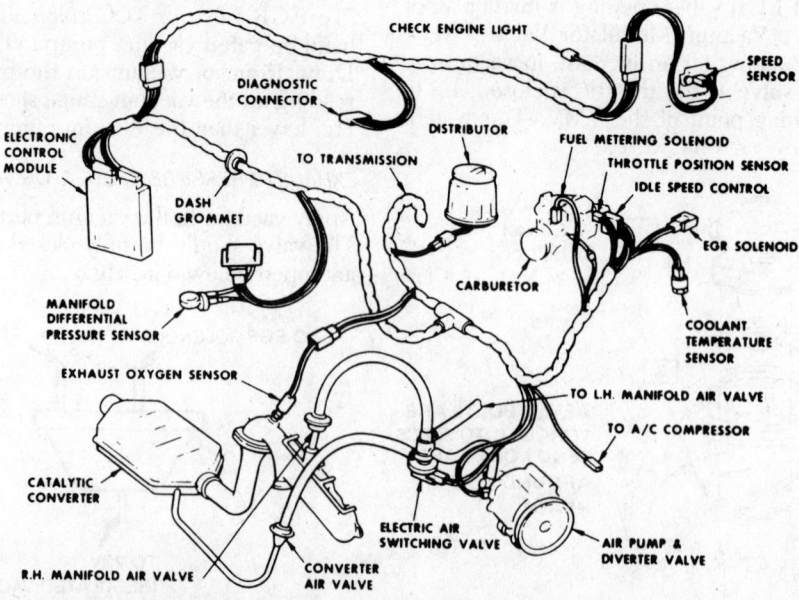

Computer Command Control (CCC) system schematic

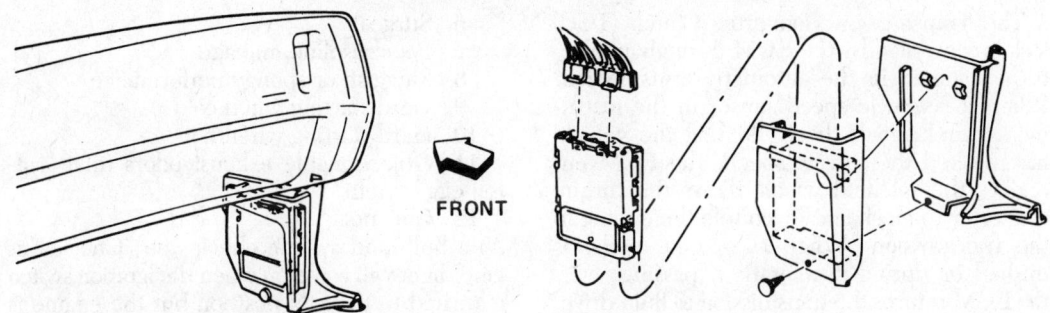

Electronic Control Module (ECM) location, all models similar

- Disconnecting or connecting any ECM connectors.
- Disconnecting (unless the engine is running) or connecting jumper cables (the jumper polarity must be correct, as even a momentary reversal of cable polarity may cause ECM damage).
- Disconnecting or connecting a battery charger (it is recommended that the battery be charged with both battery cables disconnected using the battery side terminal adapter, AC Delco ST-1201 or equivalent).

The AIR Management System is an emission control which provides additional oxygen either to the catalyst or the cylinder head ports (in some cases exhaust manifold). An AIR Management System, composed of an air switching valve and/or an air control valve, controls the air pump flow and is itself controlled by the ECM. A complete description of the AIR system is given towards the front of this unit repair section. The major difference between the CCC AIR System and the systems used on other cars is that the flow of air from the air pump is controlled electrically by the ECM, rather than by vacuum signal.

The charcoal canister purge control is an electrically operated solenoid valve controlled by the ECM. When energized, the purge control solenoid blocks vacuum from reaching the canister purge control solenoid, vacuum is allowed to reach the canister and operate the purge valve. This releases the fuel vapors collected in the canister into the induction system.

The EGR valve control solenoid is activated by the ECM in similar fashion to the canister purge solenoid. When the engine is cold, the ECM energizes the solenoid, which blocks the vacuum signal to the EGR valve. When the engine is warm, the ECM de-energizes the solenoid and the vacuum signal is allowed to reach and activate the EGR valve.

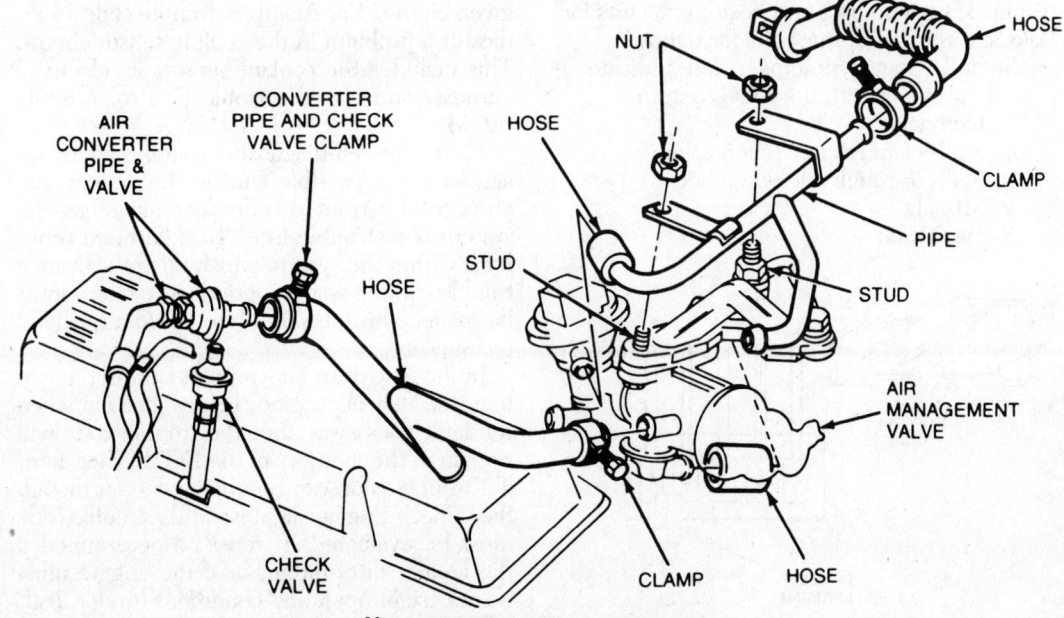

Air management system components

The Transmission Converter Clutch (TCC) lock is controlled by the ECM through an electrical solenoid in the automatic transmission. When the vehicle speed sensor in the instrument panel signals the ECM that the vehicle has reached the correct speed, the ECM energizes the solenoid which allows the torque converter to mechanically couple the engine to the transmission. When the brake pedal is pushed or during deceleration, passing, etc., the ECM returns the transmission to fluid drive.

The idle speed control adjusts the idle speed to load conditions, and will lower the idle speed under no-load or low-load conditions to conserver gasoline.

The Early Fuel Evaporative (EFE) system is used on some engines to provide rapid heat to the engine induction system to promote smooth start-up operation. There are two types of system: vacuum servo and electrically heated. They use different means to achieve the same end, which is to preheat the incoming air/fuel mixture. They are controlled by the ECM.

DIAGNOSIS AND TROUBLESHOOTING

NOTE: *The following explains how to activate the Trouble Code signal light in the instrument cluster and gives an explanation of what each code means. This is not a full CCC System troubleshooting and isolation procedure.*

Before suspecting the CCC System or any of its components as faulty, check the ignition system including distributor, timing, spark plugs and wires. Check the engine compression, air cleaner, and emission control components not controlled by the ECM. Also check the intake manifold, vacuum hose and hose connectors for leaks and the carburetor bolts for tightness.

The following symptoms could indicate a possible problem with the CCC System.

1. Detonation
2. Stalls or rough idle when cold
3. Stalls or rough idle when hot
4. Missing
5. Hesitation
6. Surges
7. Poor gasoline mileage
8. Sluggish or spongy performance
9. Hard starting when cold
10. Hard starting when hot
11. Objectionable exhaust odors (that "rotten egg" smell)
12. Cuts out

As a bulb and system check, the "Check Engine" light will come on when the ignition switch is turned to the ON position but the engine is not started.

The "Check Engine" light will also produce the trouble code or codes by a series of flashes which translate as follows. When the diagnostic test terminal under the dash is grounded, with the ignition in the ON position and the engine not running, the "Check Engine" light will flash once, pause, then flash twice is rapid succession. This is a code 12, which indicates that the diagnostic system is working. After a longer pause, the code 12 will repeat itself two or more times. The cycle will then repeat itself until the engine is started or the ignition is turned off.

When the engine is started, the "Check Engine" light will remain on for a few seconds, then turn off. If the "Check Engine" light remains on, the self-diagnostic system has detected a problem. If the test terminal is then grounded, the trouble code will flash three times. If more than one problem is found, each trouble code will flash three times. Trouble codes will flash in numerical order (lowest code number to highest). The trouble codes series will repeat as long as the test lead or terminal is grounded.

A trouble code indicates a problem with a given circuit. For example, trouble code 14 indicates a problem in the cooling sensor circuit. This includes the coolant sensor, its electrical harness, and the Electronic Control Module (ECM).

Since the self-diagnostic system cannot diagnose every possible fault in the system, the absence of a trouble code does not mean the system is in trouble-free. To determine problems within the system which do not activate a trouble code, a system performance check must be made. This job should be left to a qualified technician.

In the case of an intermittant fault in the system, the "Check Engine" light will go out when the fault goes away, but the trouble code will remain in the memory of the ECM. Therefore, if a trouble code can be obtained even though the "Check Engine" light is on, the trouble code must be evaluated. It must be determined if the fault is intermittant or if the engine must be at certain operating conditions (under load, etc.) before the "Check Engine" light will come on. Some trouble codes will not be recorded in

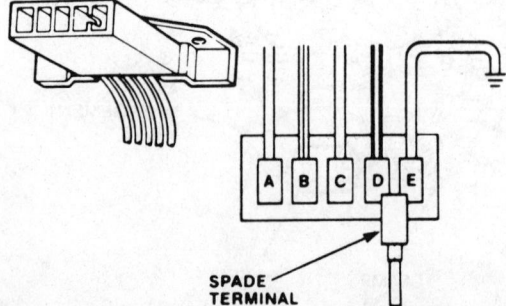

SPADE TERMINAL

CCC system diagnostic test terminal located underneath the left side of the instrument panel

CHILTON'S
FUEL ECONOMY & TUNE-UP TIPS

Tune-up • Spark Plug Diagnosis • Emission Controls

Fuel System • Cooling System • Tires and Wheels

General Maintenance

55 WAYS TO IMPROVE FUEL ECONOMY

CHILTON'S FUEL ECONOMY & TUNE-UP TIPS

Fuel economy is important to everyone, no matter what kind of vehicle you drive. The maintenance-minded motorist can save both money and fuel using these tips and the periodic maintenance and tune-up procedures in this Repair and Tune-Up Guide.

There are more than 130,000,000 cars and trucks registered for private use in the United States. Each travels an average of 10-12,000 miles per year, and, and in total they consume close to 70 billion gallons of fuel each year. This represents nearly ⅔ of the oil imported by the United States each year. The Federal government's goal is to reduce consumption 10% by 1985. A variety of methods are either already in use or under serious consideration, and they all affect you driving and the cars you will drive. In addition to "down-sizing", the auto industry is using or investigating the use of electronic fuel delivery, electronic engine controls and alternative engines for use in smaller and lighter vehicles, among other alternatives to meet the federally mandated Corporate Average Fuel Economy (CAFE) of 27.5 mpg by 1985. The government, for its part, is considering rationing, mandatory driving curtailments and tax increases on motor vehicle fuel in an effort to reduce consumption. The government's goal of a 10% reduction could be realized — and further government regulation avoided — if every private vehicle could use just 1 less gallon of fuel per week.

How Much Can You Save?

Tests have proven that almost anyone can make at least a 10% reduction in fuel consumption through regular maintenance and tune-ups. When a major manufacturer of spark plugs sur-

TUNE-UP

1. Check the cylinder compression to be sure the engine will really benefit from a tune-up and that it is capable of producing good fuel economy. A tune-up will be wasted on an engine in poor mechanical condition.

2. Replace spark plugs regularly. New spark plugs alone can increase fuel economy 3%.

3. Be sure the spark plugs are the correct type (heat range) for your vehicle. See the Tune-Up Specifications.

Heat range refers to the spark plug's ability to conduct heat away from the firing end. It must conduct the heat away in an even pattern to avoid becoming a source of pre-ignition, yet it must also operate hot enough to burn off conductive deposits that could cause misfiring.

The heat range is usually indicated by a number on the spark plug, part of the manufacturer's designation for each individual spark plug. The numbers in bold-face indicate the heat range in each manufacturer's identification system.

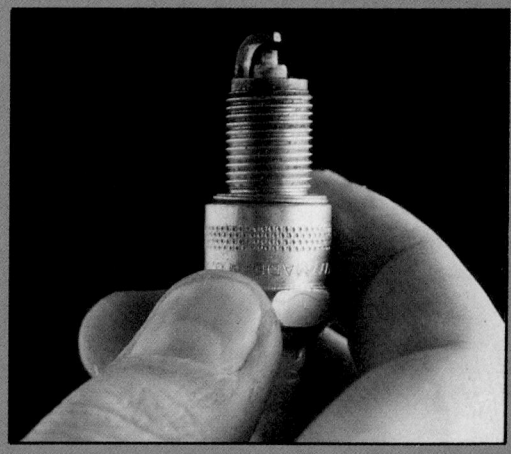

Periodically, check the spark plugs to be sure they are firing efficiently. They are excellent indicators of the internal condition of your engine.

Manufacturer	Typical Designation
AC	R **45** TS
Bosch (old)	WA **145** T30
Bosch (new)	HR **8** Y
Champion	RBL **15** Y
Fram/Autolite	4**15**
Mopar	P-**62** PR
Motorcraft	BRF-**42**
NGK	BP **5** ES-15
Nippondenso	W **16** EP
Prestolite	14GR **5** 2A

On AC, Bosch (new), Champion, Fram/Autolite, Mopar, Motorcraft and Prestolite, a higher number indicates a hotter plug. On Bosch (old), NGK and Nippondenso, a higher number indicates a colder plug.

4. Make sure the spark plugs are properly gapped. See the Tune-Up Specifications in this book.

5. Be sure the spark plugs are firing efficiently. The illustrations on the next 2 pages show you how to "read" the firing end of the spark plug.

6. Check the ignition timing and set it to specifications. Tests show that almost all cars have incorrect ignition timing by more than 2°.

veyed over 6,000 cars nationwide, they found that a tune-up, on cars that needed one, increased fuel economy over 11%. Replacing worn plugs alone, accounted for a 3% increase. The same test also revealed that 8 out of every 10 vehicles will have some maintenance deficiency that will directly affect fuel economy, emissions or performance. Most of this mileage-robbing neglect could be prevented with regular maintenance.

Modern engines require that all of the functioning systems operate properly for maximum efficiency. A malfunction anywhere wastes fuel. You can keep your vehicle running as efficiently and economically as possible, by being aware of your vehicle's operating and performance characteristics. If your vehicle suddenly develops performance or fuel economy problems it could be due to one or more of the following:

PROBLEM	POSSIBLE CAUSE
Engine Idles Rough	Ignition timing, idle mixture, vacuum leak or something amiss in the emission control system.
Hesitates on Acceleration	Dirty carburetor or fuel filter, improper accelerator pump setting, ignition timing or fouled spark plugs.
Starts Hard or Fails to Start	Worn spark plugs, improperly set automatic choke, ice (or water) in fuel system.
Stalls Frequently	Automatic choke improperly adjusted and possible dirty air filter or fuel filter.
Performs Sluggishly	Worn spark plugs, dirty fuel or air filter, ignition timing or automatic choke out of adjustment.

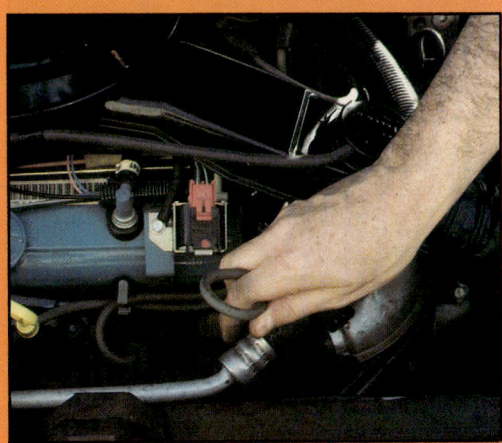

Check spark plug wires on conventional point type ignition for cracks by bending them in a loop around your finger.

Be sure that spark plug wires leading to adjacent cylinders do not run too close together. (Photo courtesy Champion Spark Plug Co.)

7. If your vehicle does not have electronic ignition, check the points, rotor and cap as specified.

8. Check the spark plug wires (used with conventional point-type ignitions) for cracks and burned or broken insulation by bending them in a loop around your finger. Cracked wires decrease fuel efficiency by failing to deliver full voltage to the spark plugs. One misfiring spark plug can cost you as much as 2 mpg.

9. Check the routing of the plug wires. Misfiring can be the result of spark plug leads to adjacent cylinders running parallel to each other and too close together. One wire tends to pick up voltage from the other causing it to fire "out of time".

10. Check all electrical and ignition circuits for voltage drop and resistance.

11. Check the distributor mechanical and/or vacuum advance mechanisms for proper functioning. The vacuum advance can be checked by twisting the distributor plate in the opposite direction of rotation. It should spring back when released.

12. Check and adjust the valve clearance on engines with mechanical lifters. The clearance should be slightly loose rather than too tight.

SPARK PLUG DIAGNOSIS

Normal

APPEARANCE: This plug is typical of one operating normally. The insulator nose varies from a light tan to grayish color with slight electrode wear. The presence of slight deposits is normal on used plugs and will have no adverse effect on engine performance. The spark plug heat range is correct for the engine and the engine is running normally.

CAUSE: Properly running engine.

RECOMMENDATION: Before reinstalling this plug, the electrodes should be cleaned and filed square. Set the gap to specifications. If the plug has been in service for more than 10-12,000 miles, the entire set should probably be replaced with a fresh set of the same heat range.

Oil Deposits

APPEARANCE: The firing end of the plug is covered with a wet, oily coating.

CAUSE: The problem is poor oil control. On high mileage engines, oil is leaking past the rings or valve guides into the combustion chamber. A common cause is also a plugged PCV valve, and a ruptured fuel pump diaphragm can also cause this condition. Oil fouled plugs such as these are often found in new or recently overhauled engines, before normal oil control is achieved, and can be cleaned and reinstalled.

RECOMMENDATION: A hotter spark plug may temporarily relieve the problem, but the engine is probably in need of work.

Incorrect Heat Range

APPEARANCE: The effects of high temperature on a spark plug are indicated by clean white, often blistered insulator. This can also be accompanied by excessive wear of the electrode, and the absence of deposits.

CAUSE: Check for the correct spark plug heat range. A plug which is too hot for the engine can result in overheating. A car operated mostly at high speeds can require a colder plug. Also check ignition timing, cooling system level, fuel mixture and leaking intake manifold.

RECOMMENDATION: If all ignition and engine adjustments are known to be correct, and no other malfunction exists, install spark plugs one heat range colder.

Photos Courtesy Fram Corporation

Carbon Deposits

APPEARANCE: Carbon fouling is easily identified by the presence of dry, soft, black, sooty deposits.

CAUSE: Changing the heat range can often lead to carbon fouling, as can prolonged slow, stop-and-start driving. If the heat range is correct, carbon fouling can be attributed to a rich fuel mixture, sticking choke, clogged air cleaner, worn breaker points, retarded timing or low compression. If only one or two plugs are carbon fouled, check for corroded or cracked wires on the affected plugs. Also look for cracks in the distributor cap between the towers of affected cylinders.

RECOMMENDATION: After the problem is corrected, these plugs can be cleaned and reinstalled if not worn severely.

MMT Fouled

APPEARANCE: Spark plugs fouled by MMT (Methycyclopentadienyl Maganese Tricarbonyl) have reddish, rusty appearance on the insulator and side electrode.

CAUSE: MMT is an anti-knock additive in gasoline used to replace lead. During the combustion process, the MMT leaves a reddish deposit on the insulator and side electrode.

RECOMMENDATION: No engine malfunction is indicated and the deposits will not affect plug performance any more than lead deposits (see Ash Deposits). MMT fouled plugs can be cleaned, regapped and reinstalled.

High Speed Glazing

APPEARANCE: Glazing appears as shiny coating on the plug, either yellow or tan in color.

CAUSE: During hard, fast acceleration, plug temperatures rise suddenly. Deposits from normal combustion have no chance to fluff-off; instead, they melt on the insulator forming an electrically conductive coating which causes misfiring.

RECOMMENDATION: Glazed plugs are not easily cleaned. They should be replaced with a fresh set of plugs of the correct heat range. If the condition recurs, using plugs with a heat range one step colder may cure the problem.

Ash (Lead) Deposits

APPEARANCE: Ash deposits are characterized by light brown or white colored deposits crusted on the side or center electrodes. In some cases it may give the plug a rusty appearance.

CAUSE: Ash deposits are normally derived from oil or fuel additives burned during normal combustion. Normally they are harmless, though excessive amounts can cause misfiring. If deposits are excessive in short mileage, the valve guides may be worn.

RECOMMENDATION: Ash-fouled plugs can be cleaned, gapped and reinstalled.

Detonation

APPEARANCE: Detonation is usually characterized by a broken plug insulator.

CAUSE: A portion of the fuel charge will begin to burn spontaneously, from the increased heat following ignition. The explosion that results applies extreme pressure to engine components, frequently damaging spark plugs and pistons.

Detonation can result by over-advanced ignition timing, inferior gasoline (low octane) lean air/fuel mixture, poor carburetion, engine lugging or an increase in compression ratio due to combustion chamber deposits or engine modification.

RECOMMENDATION: Replace the plugs after correcting the problem.

Photos Courtesy Champion Spark Plug Co.

EMISSION CONTROLS

13. Be aware of the general condition of the emission control system. It contributes to reduced pollution and should be serviced regularly to maintain efficient engine operation.

14. Check all vacuum lines for dried, cracked or brittle conditions. Something as simple as a leaking vacuum hose can cause poor performance and loss of economy.

15. Avoid tampering with the emission control system. Attempting to improve fuel econ-

FUEL SYSTEM

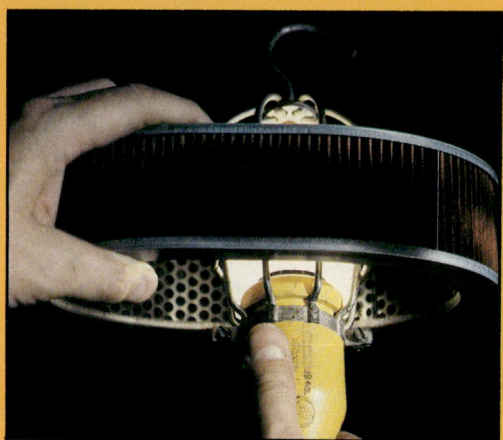

Check the air filter with a light behind it. If you can see light through the filter it can be reused.

Extremely clogged filters should be discarded and replaced with a new one.

18. Replace the air filter regularly. A dirty air filter richens the air/fuel mixture and can increase fuel consumption as much as 10%. Tests show that ⅓ of all vehicles have air filters in need of replacement.

19. Replace the fuel filter at least as often as recommended.

20. Set the idle speed and carburetor mixture to specifications.

21. Check the automatic choke. A sticking or malfunctioning choke wastes gas.

22. During the summer months, adjust the automatic choke for a leaner mixture which will produce faster engine warm-ups.

COOLING SYSTEM

29. Be sure all accessory drive belts are in good condition. Check for cracks or wear.

30. Adjust all accessory drive belts to proper tension.

31. Check all hoses for swollen areas, worn spots, or loose clamps.

32. Check coolant level in the radiator or expansion tank.

33. Be sure the thermostat is operating properly. A stuck thermostat delays engine warm-up and a cold engine uses nearly twice as much fuel as a warm engine.

34. Drain and replace the engine coolant at least as often as recommended. Rust and scale

TIRES & WHEELS

38. Check the tire pressure often with a pencil type gauge. Tests by a major tire manufacturer show that 90% of all vehicles have at least 1 tire improperly inflated. Better mileage can be achieved by over-inflating tires, but never exceed the maximum inflation pressure on the side of the tire.

39. If possible, install radial tires. Radial tires deliver as much as ½ mpg more than bias belted tires.

40. Avoid installing super-wide tires. They only create extra rolling resistance and decrease fuel mileage. Stick to the manufacturer's recommendations.

41. Have the wheels properly balanced.

omy by tampering with emission controls is more likely to worsen fuel economy than improve it. Emission control changes on modern engines are not readily reversible.

16. Clean (or replace) the EGR valve and lines as recommended.

17. Be sure that all vacuum lines and hoses are reconnected properly after working under the hood. An unconnected or misrouted vacuum line can wreak havoc with engine performance.

23. Check for fuel leaks at the carburetor, fuel pump, fuel lines and fuel tank. Be sure all lines and connections are tight.

24. Periodically check the tightness of the carburetor and intake manifold attaching nuts and bolts. These are a common place for vacuum leaks to occur.

25. Clean the carburetor periodically and lubricate the linkage.

26. The condition of the tailpipe can be an excellent indicator of proper engine combustion. After a long drive at highway speeds, the inside of the tailpipe should be a light grey in color. Black or soot on the insides indicates an overly rich mixture.

27. Check the fuel pump pressure. The fuel pump may be supplying more fuel than the engine needs.

28. Use the proper grade of gasoline for your engine. Don't try to compensate for knocking or "pinging" by advancing the ignition timing. This practice will only increase plug temperature and the chances of detonation or pre-ignition with relatively little performance gain.

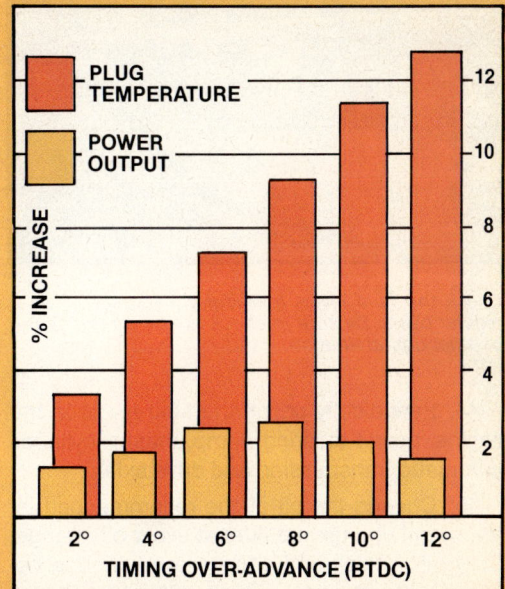

Increasing ignition timing past the specified setting results in a drastic increase in spark plug temperature with increased chance of detonation or preignition. Performance increase is considerably less. (Photo courtesy Champion Spark Plug Co.)

that form in the engine should be flushed out to allow the engine to operate at peak efficiency.

35. Clean the radiator of debris that can decrease cooling efficiency.

36. Install a flex-type or electric cooling fan, if you don't have a clutch type fan. Flex fans use curved plastic blades to push more air at low speeds when more cooling is needed; at high speeds the blades flatten out for less resistance. Electric fans only run when the engine temperature reaches a predetermined level.

37. Check the radiator cap for a worn or cracked gasket. If the cap does not seal properly, the cooling system will not function properly.

42. Be sure the front end is correctly aligned. A misaligned front end actually has wheels going in differed directions. The increased drag can reduce fuel economy by .3 mpg.

43. Correctly adjust the wheel bearings. Wheel bearings that are adjusted too tight increase rolling resistance.

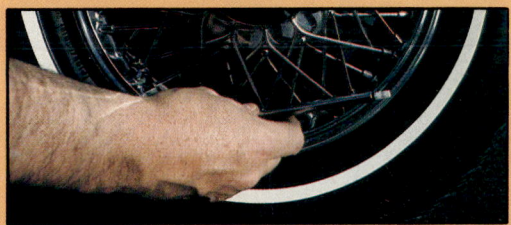

Check tire pressures regularly with a reliable pocket type gauge. Be sure to check the pressure on a cold tire.

GENERAL MAINTENANCE

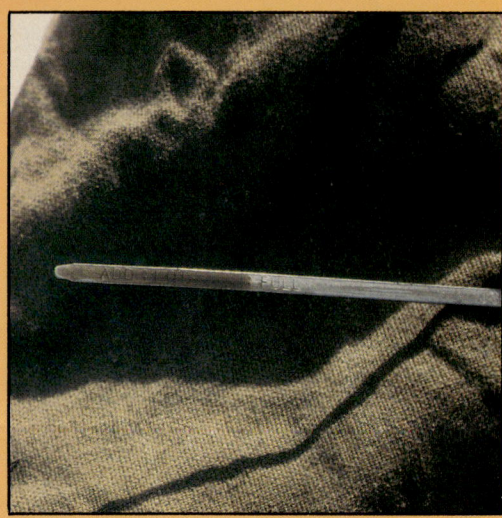

Check the fluid levels (particularly engine oil) on a regular basis. Be sure to check the oil for grit, water or other contamination.

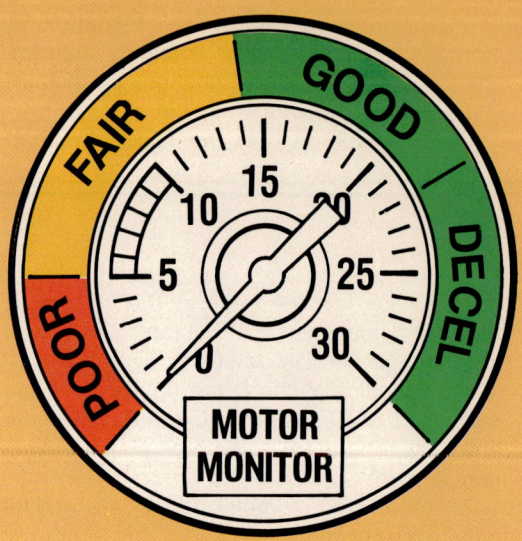

A vacuum gauge is another excellent indicator of internal engine condition and can also be installed in the dash as a mileage indicator.

44. Periodically check the fluid levels in the engine, power steering pump, master cylinder, automatic transmission and drive axle.

45. Change the oil at the recommended interval and change the filter at every oil change. Dirty oil is thick and causes extra friction between moving parts, cutting efficiency and increasing wear. A worn engine requires more frequent tune-ups and gets progressively worse fuel economy. In general, use the lightest viscosity oil for the driving conditions you will encounter.

46. Use the recommended viscosity fluids in the transmission and axle.

47. Be sure the battery is fully charged for fast starts. A slow starting engine wastes fuel.

48. Be sure battery terminals are clean and tight.

49. Check the battery electrolyte level and add distilled water if necessary.

50. Check the exhaust system for crushed pipes, blockages and leaks.

51. Adjust the brakes. Dragging brakes or brakes that are not releasing create increased drag on the engine.

52. Install a vacuum gauge or miles-per-gallon gauge. These gauges visually indicate engine vacuum in the intake manifold. High vacuum = good mileage and low vacuum = poorer mileage. The gauge can also be an excellent indicator of internal engine conditions.

53. Be sure the clutch is properly adjusted. A slipping clutch wastes fuel.

54. Check and periodically lubricate the heat control valve in the exhaust manifold. A sticking or inoperative valve prevents engine warm-up and wastes gas.

55. Keep accurate records to check fuel economy over a period of time. A sudden drop in fuel economy may signal a need for tune-up or other maintenance.

the ECM until the engine has been operated at part throttle for about 5 to 18 minutes. A trouble code will be stored until terminal "R" of the ECM has been disconnected from the battery for 10 seconds.

An easy way to erase the computer memory on the CCC System is to disconnect the battery terminals from the battery. If this method is used, don't forget to reset clocks and elec-tronic preprogramable radios. Another method is to remove the fuse marked ECM in the fuse panel. Not all models have such a fuse.

Air Injection Reactor

The system helps reduce HC and CO emissions in the same basic manner of a typical air pump type air injection system, except that the

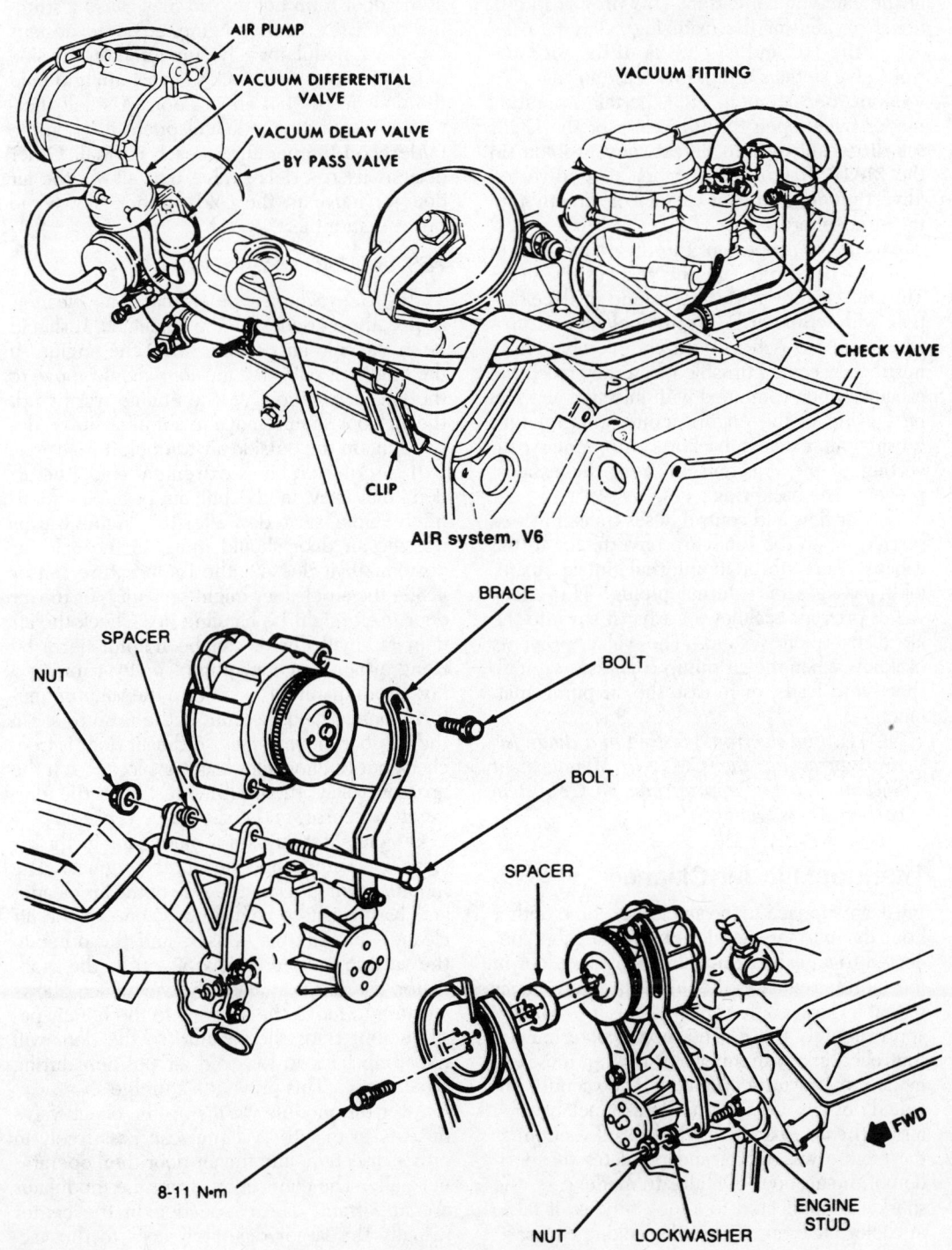

AIR system, V6

Removing the air pump

MAIR system is controlled by signals from the electronic control module (ECM). When the engine is cold, the ECM energizes an Air Control Solenoid. This allows air to flow to an Air Switching Valve, which is energized to direct air to the exhaust ports.

On a warm engine or when in "Closed Loop" operation, the ECM de-energizes the Air Switching Valve, directing air between the beds of the catalytic converter. This provides additional oxygen for the oxidizing catalyst to decrease the HC and CO levels. If the Air Control Valve detects a rapid increase in manifold vacuum (deceleration, etc.), certain operating modes (wide open throttle, etc.), or the ECM self-diagnostic system detects any problem in the MAIR system as a whole, air is diverted (divert mode) to the air cleaner or directly into the atmosphere.

A primary purpose of the divert mode is to prevent backfiring in the exhaust system. Throttle closure at the beginning of deceleration will temporarily create fuel/air mixtures which are too rich to burn completely. These mixtures become burnable when they reach the exhaust, when combined with injection air. The next firing of the engine ignites this mixture causing an exhaust backfire. Momentary diverting of the injection air from the exhaust prevents the backfiring.

The air flow and control hoses transmit pressurized air to the catalytic converter or to the exhaust ports through internal (intake manifold) passages or external piping. The check valves prevent backflow of exhaust gas into the air distribution system. The valve prevents backflow when the air pump "bypasses" at high speed and loads, or in case the air pump malfunctions.

NOTE: *For electronic testing and diagnosis procedures on the GM Air Management System, see the appropriate CCC System earlier in this section.*

Thermostatic Air Cleaner

Fresh air supplied to the air cleaner comes either from the normal snorkle, or from a tube connected to an exhaust manifold stove. A door in the snorkle regulates the source of incoming air so that a warm engine always takes in warm air, approximately 100°F. The snorkle door may be controlled in any number of ways, but most late models are vacuum operated. The vacuum operated designs use a thermostatic bimetal switch inside the air cleaner that bleeds off vacuum as the engine warms up, and regulates the position of the air door. On all late model cars, the snorkle is connected to a long tube so it takes in cooler air from outside the engine compart-

ment. In hot climates the cool air tube is necessary because underhood air can easily reach 200°F.

Vacuum operated air doors are all designed so that the air cleaner takes in cold air when there is no vacuum. This means that an air door in the hot air position will switch to the cold position at wide open throttle because of the loss of manifold vacuum. The sudden switching of the door from hot to cold may cause a stumble or misfire in the engine, so some designs include a modulator valve mounted on the side of the air cleaner to block the vacuum and hold the door in the hot air position. A small thermostat inside the modulator opens it when the underhood temperatures reach normal. Other designs used a delay valve that allows the air door to move to the cold position slowly, to prevent stumble.

TESTING TAC OPERATION

To test the vacuum type of heated air cleaner, inspect the air door with the engine off. It should be in the cold air position. Start the engine. If the engine is cold, the air door should move to the hot air position. As the engine warms up, the air door should move to a mid position, depending on the outside air temperature.

If the outside air is extremely cold, the air door may stay in the hot air position indefinitely. On a warm day, after the engine warms up, the air door should move to the cold air position. If it doesn't, the temperature sensor inside the air cleaner might be faulty, or the air door itself might be hanging up. Check the air door (a small mirror can be helpful here) by using a hand vacuum pump, or by running a hose from manifold vacuum to the vacuum motor. Connect and disconnect the hose to see if the air door moves freely. If the air door is free, check out the hoses for leaks or blockage. If the hoses are okay, the trouble must be in the temperature sensor, and it should be replaced.

General Motors uses a modulator in the air cleaner vacuum line on some engines. The modulator mounts on the side of the air cleaner and has two hose connections, one to the air cleaner temperature sensor, and the other to the vacuum motor. Below 50–80°F. the modulator is a one-way check valve, which allows vacuum to move the air door to the hot air position, but traps the vacuum so the door will not jump back to the cold air position during acceleration. This prevent a stumble.

After the module warms up the check valve unseats so that the vacuum can pass freely in either direction, and the air door then operates normally. The connections from the modulator are important. The connection in the center (usually the larger diameter) goes to the vacuum motor, and the connection on the edge goes to the vacuum source, which is the temperature sensor.

To test the modulator on a cold engine, apply enough vacuum to the edge port to move the air door to the hot position. Then remove the hose from the port, and the air door should stay in the hot position. Make the same test when the engine is warmed up, and the air door should move to the cold position when you pull off the hose.

GENERAL FUEL SYSTEM SERVICE

Fuel Pump

REMOVAL AND INSTALLATION

Gasoline Engines

1. Disconnect the negative battery terminal, open the fuel tank filler door and disconnect the sending unit feed wire.

2. Siphon the fuel from the fuel tank. If the rear of the car is raised one foot higher than the front, more fuel can be taken out.

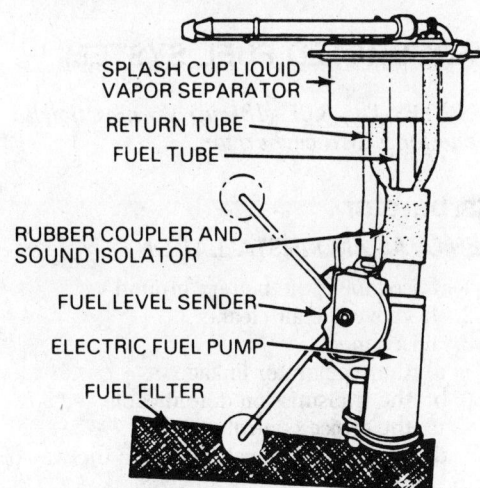

Fuel pump used on the 3.0L V6 & 4.1L V8

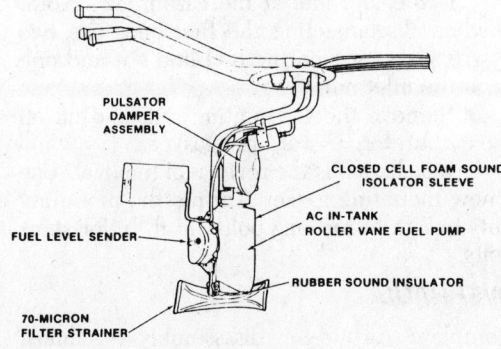

Fuel pump used on the 3.8L V6

3. Raise the rear of the car and remove the screw securing the ground wire to the cross member.

4. Disconnect the fuel line, evaporative emission lines and the fuel return lines at the front of the tank.

5. Support the tank with a jack and wooden block and remove one screw on each side securing the fuel tank support straps to the body at the front of the tank.

6. Lower the jack and tank enough so that the fuel pump electrical lead can be disconnected. Disconnect the wire.

7. Remove the fuel tank from the car.

8. Remove the locknuts securing the fuel gauge tank unit and fuel pump feed wires to the tank unit.

9. Turn the cam locking ring counterclockwise with a soft non-ferrous punch and hammer. When the lock ring is disengaged, remove it and lift the gauge/pump unit from the tank.

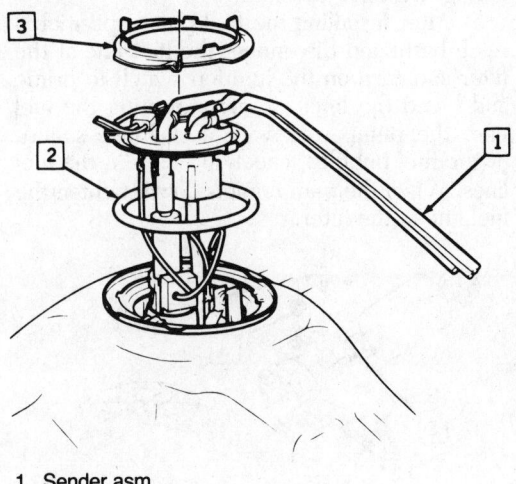

1. Sender asm.
2. Gasket
3. Cam

Fuel pump removal

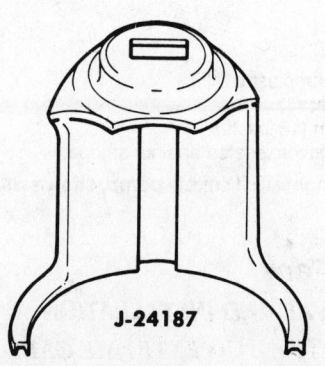

J-24187

Fuel tank lockring removal and installation tool

10. Install in the reverse order of removal. Tighten the fuel tank retaining strap screws to 25 ft. lbs.

Diesel Engine Fuel Supply Pump

NOTE: *The fuel pump used on the V6 diesel engine is located at the front of the engine, next to the fuel heater.*

1. Disconnect negative battery cable, remove the air cleaner, and unplug all electrical connectors from the pump.

2. Place a rag under the pump inlet and outlet fittings, and carefully unscrew the inlet and the outlet fittings. Cap all fittings to keep dirt out.

3. Remove the pump mounting bracket nut, then the fuel pump.

4. To install fuel pump, reverse above procedure, and tighten the nut of the pump mounting bracket to 18 ft. lbs. Then torque inlet and outlet line fittings to 19 ft. lbs.

NOTE: *In some cases you may have to adjust pump position slightly to align pump fittings with the fuel lines.*

5. After installing the fuel pump, position a catch basin and disconnect the fuel line at the filter and turn on the ignition switch to prime and bleed the lines. If after torquing the fuel line, the pump runs with a click-like sound, or the fuel bubbles, check for leaks in the fuel lines. When the pump quiets down, tighten the fuel line at the filter.

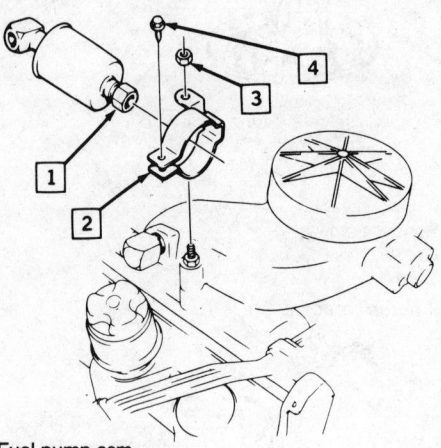

1. Fuel pump asm.
2. Bracket
3. 24 N·m (18 lbs. ft.)
4. Fully driven, seated and not stripped

Diesel engine fuel supply pump, shown with external EGR

Fuel Tank

REMOVAL AND INSTALLATION

CAUTION: *Use EXTREME CARE when removing the fuel tank. Gasoline, and to a lesser extent, diesel fuel, is VERY FLAMMABLE and HIGHLY EXPLOSIVE! Never smoke around the fuel tank. Never allow any sparks to come in contact with the tank or fuel vapors. Never operate any kind of electrical equipment near an open fuel tank. It's a good idea to have a class B fire extinguisher handy when working on the fuel tank.*

1. Disconnect the battery ground. Drain the tank by siphoning off as much fuel as possible. Don't use your mouth as a suction-starting device. There are many good, cheap devices available.

2. Raise and support the rear of the car on jackstands.

3. Disconnect the sending unit harness connector.

4. Disconnect the filler pipe and vent hoses.

5. Disconnect the tank sending unit hoses.

6. Remove the lower intermediate exhaust pipe heat shield.

7. Support the tank with a floor jack. Remove the tank support straps and lower the tank from the car.

8. Installation is the reverse of removal. Torque the strap bolts to 25 ft.lb.

CARBURETED FUEL SYSTEM

NOTE: *The 3.0L (181cid) V6 uses the Rochester E2ME carburetor.*

Carburetor

REMOVAL AND INSTALLATION

1. Disconnect the battery ground.
2. Remove the air cleaner.
3. Disconnect:
   a. the accelerator linkage
   b. the transmission detent cable
   c. the cruise control
   d. all wiring connected to or in the way of the carburetor
   e. all vacuum lines connected to the carburetor
   f. the fuel line at the carburetor. Note, when disconnecting the fuel line, use two wrenches, one on the fuel line nut and one on the inlet nut.
4. Remove the 4 mounting nuts and lift off the carburetor. Discard the gasket.
5. Installation is the reverse of removal. Use a new mounting gasket. Torque the mounting nuts to 7 ft.lb. for long bolts; 11 ft.lb. for short bolts.

OVERHAUL

Complete carburetor disassembly is almost never necessary. Also, the expertise required is usually beyond the scope of the do-it-your-selfer. Therefore, we suggest that you limit your carburetor overhaul work to those procedures required by the purchasing of a carburetor overhaul kit. Complete instructions for overhaul related to the parts in the kit will befound with the kit itself.

ADJUSTMENTS

NOTE: *Adjustments to this carburetor are difficult and require special tools and a relatively high level of skill. The adjustments listed below are limited to those practical to the user. Adjustments requiring too high a level of expertise and/or expensive special tools are not given in this book.*

Mixture Control Solenoid Plunger

1. Remove the air horn.
2. Remove the solenoid adjusting screw.
3. Remove the rich limit stop.
4. Remove the mixture control solenoid plunger.
5. Remove the primary metering rods and spring.
6. Remove the plastic filler block.
7. Remove the mixture control solenoid.
8. Check that the solenoid bore or plunger are not worn. Check the metering rods for sticking or improper installation.
9. Check for foreign material in the jets.
10. Install mixture control solenoid gauging tool J-33815-1/BT-8253-A, or equivalent, over the throttle side metering jet rod guide and temporarily install the solenoid adjustment screw spring, mixture control solenoid plunger, rich limit stop and solenoid adjusting screw.
11. Using just enough light finger pressure to hold the solenoid plunger in the DOWN position, use tool J-28696-10/BT-7928, or equivalent, to turn the solenoid adjusting screw clockwise until the plunger contacts the gauging tool. The plunger should, at this point, be in contact with both the solenoid stop and the gauging tool.

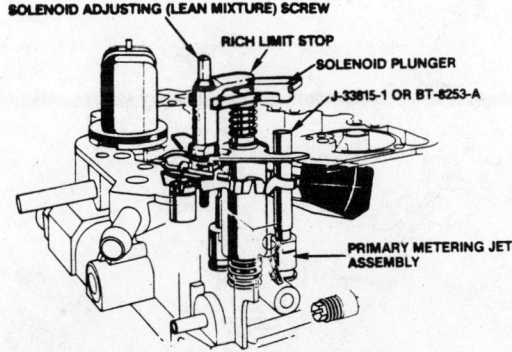

Installing the mixture control solenoid gauging tool

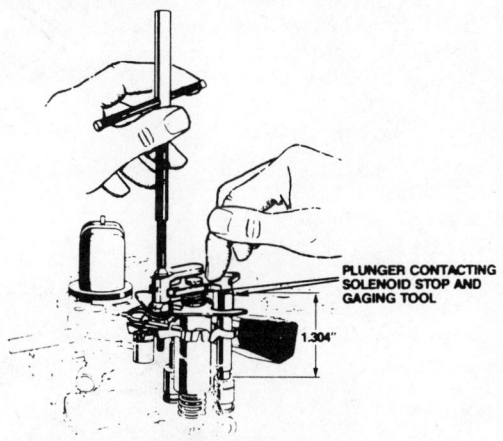

Adjusting the solenoid lean mixture screw

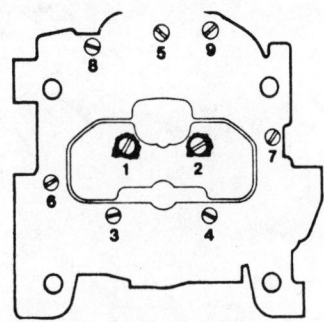

Air horn bolt tightening sequence

12. In this position, note the exact position of the adjustment screw tool tee handle and record the number of turns clockwise necessary to make the mixture control solenoid bottom out against the float bowl.
13. Remove the temporarily installed items and the gauging tools.
14. Reassemble all carburetor parts properly. Tighten the air horn bolts in a criss-cross pattern.
15. Disconnect the vacuum line from the canister purge valve and plug the line.
16. Ground the diagnostic "test" terminal and connect a dwell meter at the dwell meter test connector.
17. Run the engine to normal operating temperature. This should put the system in the closed loop mode.
18. Run the engine to 3,000 rpm and hold it there. Take a dwell reading. If the dwell is 10° to 50°, the mixture control solenoid adjustment is correct. A reading higher than 50° indicates a too rich mixture; lower than 10° indicates a too lean mixture.

Idle Speed Control

NOTE: *This adjustment should be necessary only when replacing the ISC unit, to estab-*

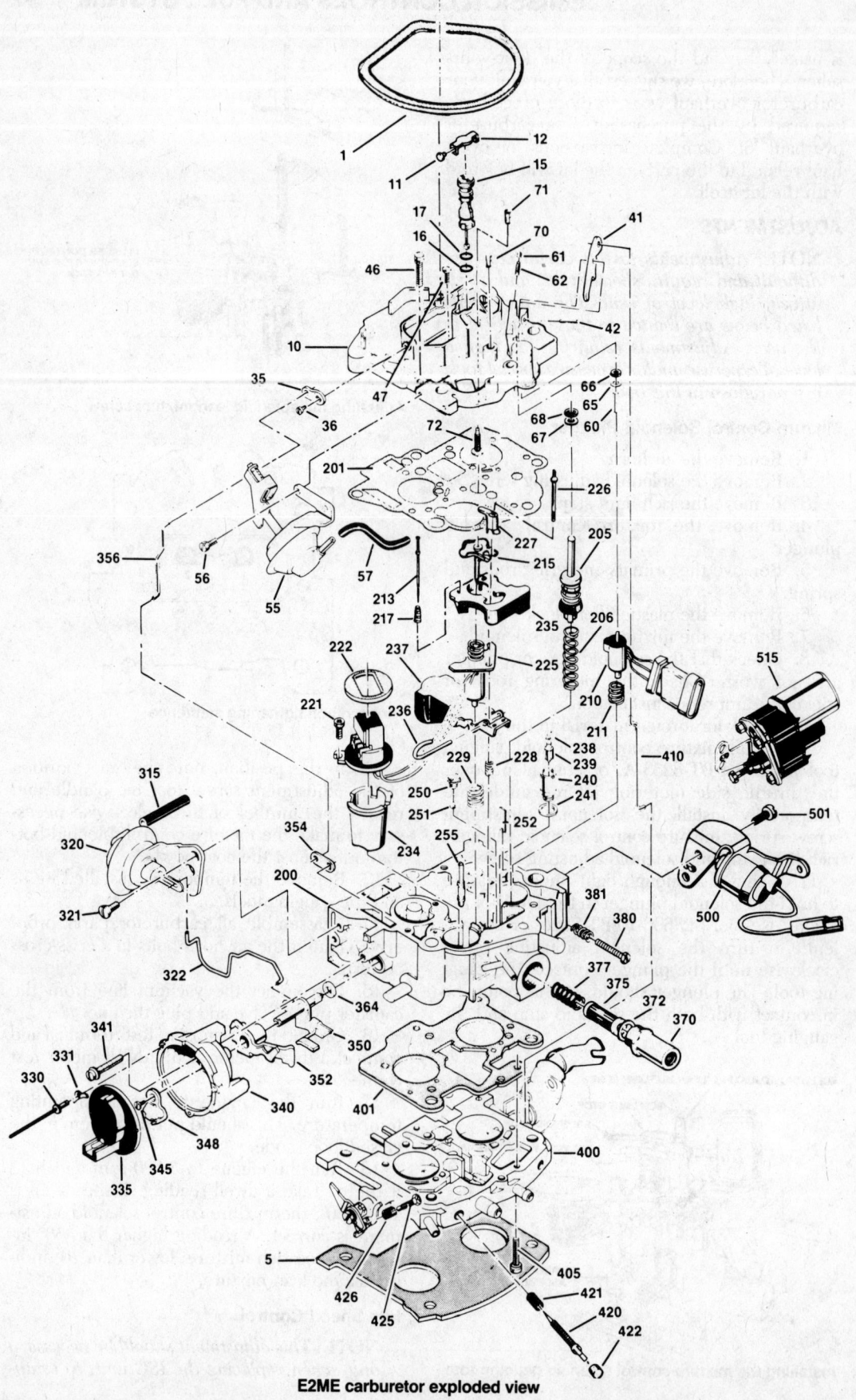

E2ME carburetor exploded view

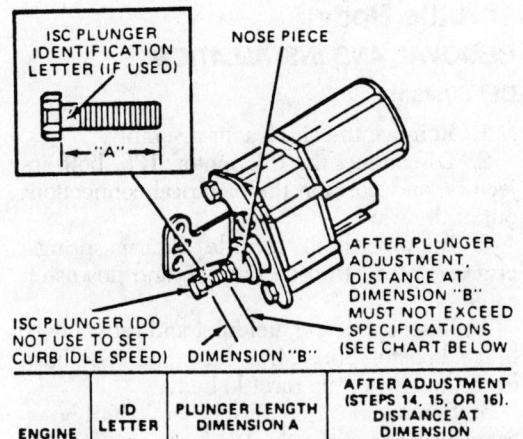

ISC PLUNGER IDENTIFICATION LETTER (IF USED)

NOSE PIECE

ISC PLUNGER (DO NOT USE TO SET CURB IDLE SPEED)

AFTER PLUNGER ADJUSTMENT, DISTANCE AT DIMENSION "B" MUST NOT EXCEED SPECIFICATIONS (SEE CHART BELOW)

DIMENSION "B"

ENGINE	ID LETTER	PLUNGER LENGTH DIMENSION A	AFTER ADJUSTMENT (STEPS 14, 15, OR 16). DISTANCE AT DIMENSION "B" MUST NOT EXCEED
3.0L AT-4 SPD	L	27.5 mm (1-3/32")	19.0 mm (3/4")
3.0L AT-3 SPD	J	30.0 mm (1-3/16")	21.5 mm (27/32")

ISC adjustment information

PLUNGER IN — BATT

PLUNGER OUT — BATT

A* B* C D

*NEVER CONNECT VOLTAGE SOURCE ACROSS TERMINALS "A" AND "B"

1. ISC plunger (do not use to set curb idle speed)
2. Electrical connector
3. Motor

ISC connector and plunger

1. Gasket–air cleaner
5. Gasket–flange
10. Air horn assembly
11. Rivet–cover attaching
12. Cover–air bleed valve
15. Air bleed valve assembly
16. O-ring–air bleed valve–lower
17. O-ring–air bleed valve–upper
35. Lever–choke
36. Screw–choke lever attaching
41. Lever–pump
42. Pin–pump lever hinge
46. Screw assembly–air horn to float bowl
47. Screw–air horn to float bowl (countersunk)
55. Vacuum break assembly–primary side (front)
56. Screw–primary side (front) vacuum break assembly attaching
57. Hose–primary side (front) vacuum break
60. Plunger–sensor actuator
61. Plug–TPS adjusting screw
62. Screw–TPS adjusting
65. Retainer–TPS seal
66. Seal–TPS plunger
67. Retainer–pump stem seal
68. Seal–pump stem
70. Plug–solenoid adjusting screw
71. Plug–solenoid stop screw
72. Screw–solenoid stop (rich mixture)
200. Float bowl assembly
201. Gasket–air horn to float bowl
205. Pump assembly
206. Spring–pump return
210. Sensor–throttle position (TPS)
211. Spring–sensor adjusting
213. Rod–primary metering
215. Plunger–solenoid
217. Spring–primary metering rod (E2M, E4M only)
221. Screw–solenoid connector attaching
222. Gasket–solenoid connector to air horn
225. Mixture control solenoid assembly
226. Screw–solenoid adjusting (lean mixture)
227. Stop–rich limit
228. Spring–solenoid adjusting screw
229. Spring–solenoid return
234. Insert–aneroid cavity
235. Insert–float bowl

236. Hinge pin–float description–E4ME
237. Float
238. Pull clip–float needle
239. Needle–float
240. Seat–float needle
241. Gasket–float needle seat
250. Plug–pump discharge (retainer)
251. Ball–pump discharge
252. Baffle–pump well
255. Primary metering jet assembly
315. Hose–secondary side (rear) vacuum break
320. Vacuum break assembly–secondary side (rear)
321. Screw–secondary side (rear) vacuum break assembly attaching
322. Link–secondary side (rear) vacuum break to choke
330. Rivet–choke cover attaching
331. Retainer–choke cover
335. Electric choke cover and stat assembly
340. Choke housing assembly
341. Screw and washer assembly–choke housing to float bowl
345. Screw–choke stat lever attaching
348. Lever–choke stat
350. Intermediate choke shaft, lever and link assembly
352. Fast idle cam assembly
354. Lever–intermediate choke
356. Link–choke
364. Seal–intermediate choke shaft
370. Nut–fuel inlet
372. Gasket–fuel inlet nut
375. Filter–fuel inlet
377. Spring–fuel filter
380. Screw–throttle stop
381. Spring–throttle stop screw
400. Throttle body assembly
401. Gasket–float bowl to throttle body
405. Screw assembly–float bowl to throttle body
410. Link–pump
420. Needle–idle mixture
421. Spring–idle mixture needle
422. Plug–idle mixture needle
425. Screw–fast idle adjusting
426. Spring–fast idle adjusting screw
500. Solenoid and bracket assembly
501. Screw–bracket attaching
515. Idle speed control assembly

lish a minimum and maximum speed range for the new unit. *Never disconnect or connect the ISC motor connector with the engine running.*

1. Place the transaxle in PARK, set the parking brake and block the drive wheels.

2. Check the ISC plunger, near the bolt head end, for an identification letter. Using this letter, check, and if necessary, adjust the distance between the bolt head and the locknut to 1³⁄₃₂″ for letter L, or 1³⁄₁₆″ for letter J, by loosening the locknut and turning the bolt.

3. Connect a tachometer.

4. Connect a dwell meter to the mixture control solenoid green (dwell) lead. Set the dwell meter on the 6-cylinder scale.

5. Turn the air conditioning OFF.

6. Start the engine and run it to normal operating temperature. When the dwell meter reading begins to waver noticably, the system has entered the closed loop mode. Turn the ignition OFF.

7. Unplug the ISC motor wire.

8. Apply a 12v source to terminal C (second from bottom) of the motor connector, and ground terminal D (bottom). This will retract the plunger. As soon as the plunger retracts, disconnect the power source and ground leads. Prolonged contact with a power source will damage the ISC motor.

9. Start the engine and watch the dwell meter, noting when the system enters the closed loop mode. Place the transaxle in DRIVE.

10. With the plunger retracted, adjust the carburetor base idle stop screw to obtain an idle speed of 500 rpm.

11. Apply a 12v source to terminal D of the connector, and ground terminal C, to fully extend the plunger. Remove the wires as soon as the plunger is fully extended.

NOTE: *The fully extended plunger should be in contact with the throttle lever to prevent damage to the unit.*

12. Adjust, if necessary, the maximum idle speed to 1300 rpm, by loosening the locknut and turning the plunger bolt head.

13. Place the transaxle in PARK, shut off the engine, remove all test equipment and reconnect the ISC motor connector.

GASOLINE FUEL INJECTION SYSTEMS

NOTE: *Two different injection systems are described below. They are as follows:*
- Digital Fuel Injection (DFI) used on the 4.1L V8
- Multi Port Fuel Injection (MPFI) usedon the 3.8L V6

Throttle Body
REMOVAL AND INSTALLATION
DFI System

1. Remove the air cleaner assembly.

2. Disconnect the ISC motor, IPS, both injectors, and position the electrical connections out of the way.

3. Remove both throttle return springs, cruise control, throttle linkage, and downshift cable.

4. Disconnect the fuel inlet and return line, brake booster line, MAP hose and AIR hose from the rear of the throttle body.

5. Remove the PCV, EVAP, and EGR hoses from the front of the throttle body.

6. Remove the three throttle body mounting bolts and remove the throttle body and gasket.

7. Installation is the revere of removal.

After installation, check and adjust the throttle position sensor (TPS) and the idle speed control (ISC) motor as necessary.

MPFI

NOTE: *The throttle body is actually part of the idle air control valve unit. Also incorporated in the unit is the throttle position sensor.*

1. Remove the air cleaner.

2. Disconnect, and tag, all hoses from throttle body.

3. Disconnect the idle air control valve connector and the throttle plosition sensor connector.

4. Disconnect the throttle linkage and cruise control linkage.

5. Remove the two attaching bolts and lift off the throttle body. Discard the gasket.

6. Installation is the reverse of removal. Use a new gasket, and tighten the attaching bolts to 13 ft.lb.

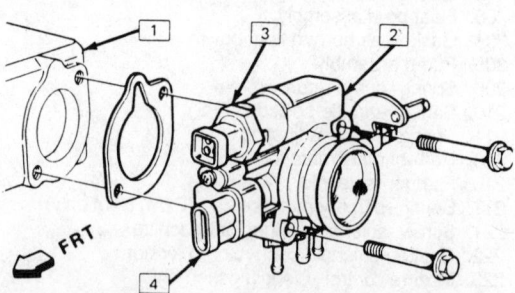

1. Intake manifold
2. Throttle body asm.
3. Idle air control valve
4. Throttle position sensor

Throttle body and air control valve removal

Throttle Position Switch
REMOVAL AND INSTALLATION

DFI

TPS replacement requires a complex series of adjustments using expensive special tools. This procedure is best refered to a qaulified service technician.

MPFI

See Throttle Body Removal and Installation, above.

Fuel Injector
REMOVAL AND INSTALLATION

DFI Systems

NOTE: *Never immerse any injection system component in any type of cleaner!*

1. Disconnect the negative battery terminal.

2. Remove the air cleaner assembly.

3. Remove the 8 screws securing the fuel meter cover, noting the placement of the 4 short screws, and remove the cover.

4. With the fuel meter cover still in place to prevent damage to the casting, carefully pry the injector from the fuel meter body.

5. Discard the upper and lower O-rings.

6. Lubricate a new smaller O-ring with automatic transmission fluid prior to installing it on the injector.

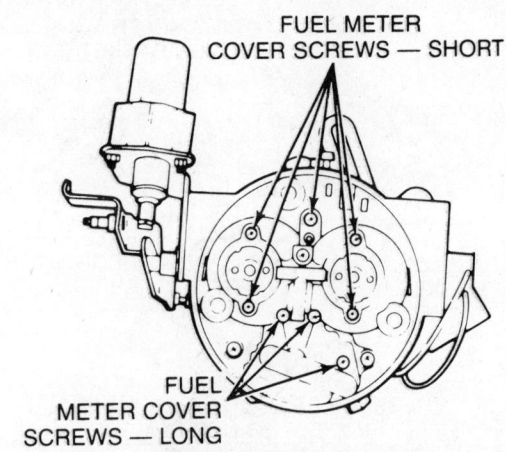

Removing the DFI fuel meter cover

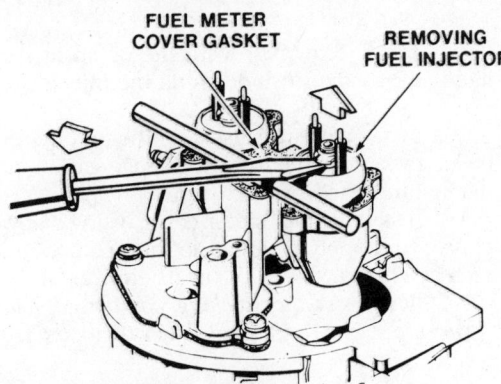

Removing the DFI fuel injector

7. Install the steel back-up washer in the recess in the fuel meter body.

8. Lubricate the new larger O-ring with ATF and instal it directly above the steel back-up ring, pressing it down into the fuel meter body. The O-ring should be flush with the casing surface.

9. Install the injector with a pushing/twisting motion to center the O-ring in the bottom of the injector cavity and align the raised

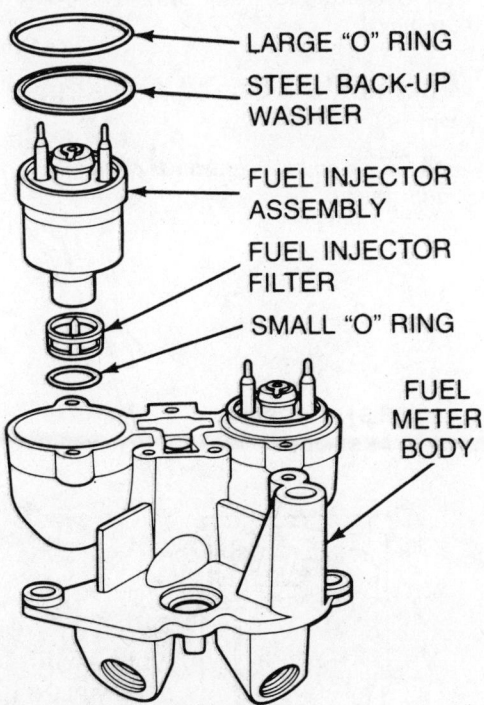

DFI fuel injector components

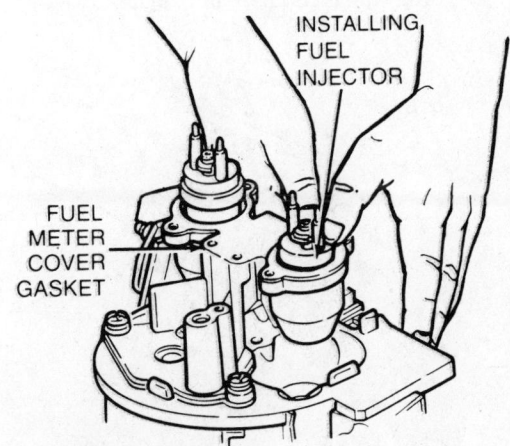

Installing the DFI injector

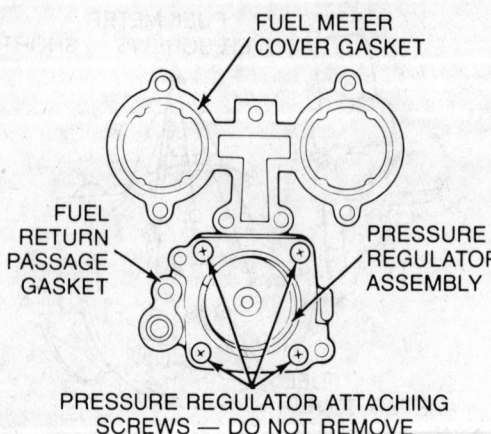

FUEL METER
COVER GASKET

FUEL
RETURN
PASSAGE
GASKET

PRESSURE
REGULATOR
ASSEMBLY

PRESSURE REGULATOR ATTACHING
SCREWS — DO NOT REMOVE

Installing the meter cover

lug on the injector base with the notch in the fuel meter body. Push down on the injector to make sure it is fully seated.

10. Install a new dust seal on the fuel meter body, and a new fuel return passage gasket on the fuel meter cover.

11. Install the fuel meter cover, using a new gasket, and apply a thread locking compound such as Threadlocker®262, or its equivalent.

NOTE: *A stronger locking compound will cause damage to the unit when the bolts are next removed.*

12. Install the electrical connector, the air cleaner and start the engine. Check for leaks.

MPFI

NOTE: *Use great care in handling the injectors. They are delicate electronic components! Also, they should NEVER be immersed in any type of cleaner!*

1. Connect a fuel pressure gauge to the fuel pressure valve. Wrap a shop towel around the fitting while connecting the gauge to avoid fuel spillage.

2. Run the bleed hose into an approved gas-oline container, open the valve and bleed the system.

3. With the ignition OFF, disconnect the electrical connectors at the injectors.

4. Unbolt and remove the fuel rail.

5. Remove the injectors. Discard the O-rings.

6. Installation is the reverse of removal. Lubricate the new O-rings with clean engine oil.

Fuel Pressure Regulator

NOTE: *Do not attempt to make any adjustments to the pressure regulator during this procedure as all adjustments are preset at the factory.*

REMOVAL AND INSTALLATION
MPFI

CAUTION: *Relieve the fuel pressure as described in injector removal for this fuel system.*

1. Remove the vacuum hose from the top of the pressure regulator.

2. Bleed off the pressure in the fuel delivery system as outlined in Steps 1–4 of the "Chassis-Mounted Fuel Pump" removal procedure and disconnect the flexible fuel hose between the fuel rail and the regulator. Disconnect the fuel return line.

3. Remove the one nut securing the pressure regulator to the bracket. This nut has metric threads.

4. Remove the regulator.

5. Install the regulator in the reverse order of removal.

Minimum Idle Speed Adjustment
MPFI

NOTE: *A digital voltmeter is essential for this adjustment*

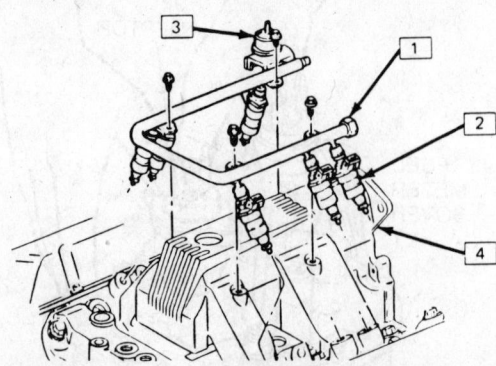

1. Fuel rail assembly
2. Injector
3. Pressure regulator
4. Intake manifold

Removing the MPFI fuel rail and injectors

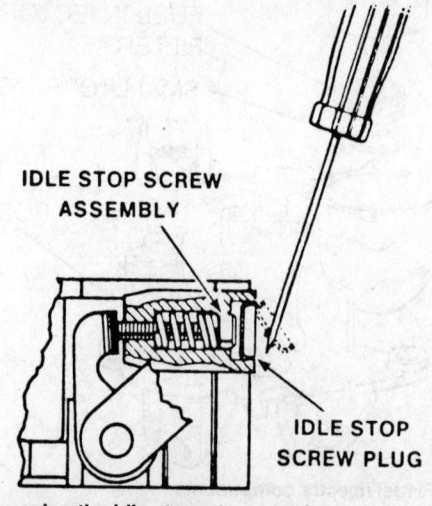

IDLE STOP SCREW
ASSEMBLY

IDLE STOP
SCREW PLUG

Removing the idle stopset screw plug

This adjustment is necessary only when throttle body parts have been replaced.

1. Run the engine to normal operating temperature. Shut off the engine.
2. Connect a tachometer.
3. Ground the diagnostic terminal.
4. Turn the ignition switch to ON, but don't start the engine. Wait at least 30 seconds.
5. Disconnect the idle air control connector.
6. Remove the ground from the diagnostic connector and start the engine.
7. Place the transaxle in DRIVE. Adjust the idle stopset screw to 500 ± 50 rpm.
8. Turn the ignition OFF and reconnect the IAC.
9. Using a digital voltmeter, adjust the TPS voltage to 0.40 ± 0.05v.
10. Start the engine and check for normal operation.

DIESEL FUEL SYSTEM

The fuel system is the heart of the diesel engine. The main components are the injection pump, injection lines and fuel injectors. The fuel injection pump is a small, high pressure rotary pump which delivers a small, metered amount of fuel to the injection nozzles at the proper time. The high pressure lines are all of equal length to avoid differences in timing. The nozzles project into the combustion chambers and spray/atomize the fuel entering the chambers. A small, low pressure transfer pump is employed in the inlet line to the injection pump to keep the injection pump supplied. Engine rpm is controlled by a rotary fuel metering valve operated by the accelerator linkage. A fuel filter is located between the transfer pump and the injection pump.

On all engines, the fuel pump is of the mechanical diaphragm type, mounted on the engine.

Tachometer Hook-Up

A magnetic pickup tachometer is necessary because of the lack of an ignition system. The tachometer probe is inserted into the hole in the timing indicator.

Fuel Injection Pump and Lines
REMOVAL AND INSTALLATION

NOTE: *This procedure contains throttle rod and transmission cable adjustments.*

1. Remove the air cleaner.
2. Remove the filters and pipes from the valve covers and air crossover.
3. Remove the air crossover and cap the

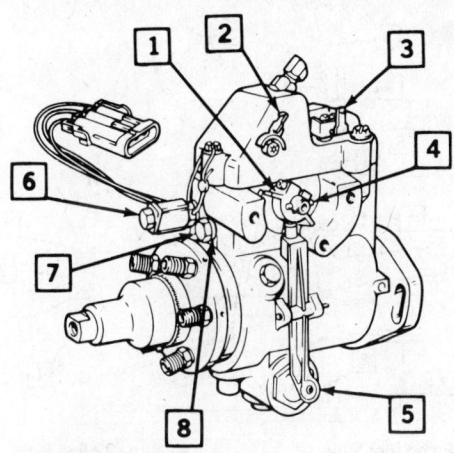

1. Throttle shaft advance cam
2. Housing pressure cold advance
3. Fuel solenoid terminal
4. Throttle shaft pin
5. Advance cam actuating arm
6. Metering valve sensor
7. Guide stud
8. Washer

Diesel injection pump, with internal EGR

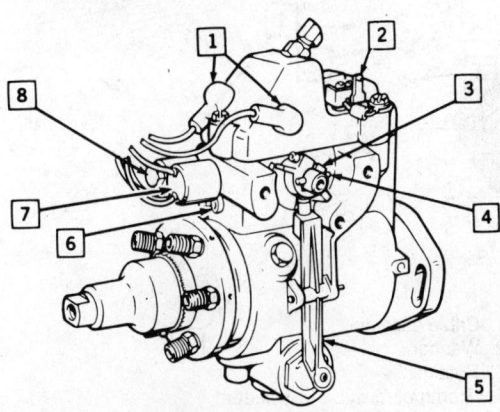

1. Housing pressure cold advance leads
2. Fuel shut-off solenoid
3. Throttle shaft advance cam
4. Throttle shaft pin
5. Advance cam actuating arm
6. Guide stud
7. Altitude fuel limiter
8. Metering valve sensor

Diesel injection pump, with external EGR

intake manifold with screened covers (J-29657), or tape.

4. Disconnect the throttle rod and return spring.
5. Remove the bellcrank.
6. Remove the throttle and transmission cables from the intake manifold brackets.
7. Disconnect the fuel lines from the filter and remove the filter.
8. Disconnect the fuel inlet line at the pump.

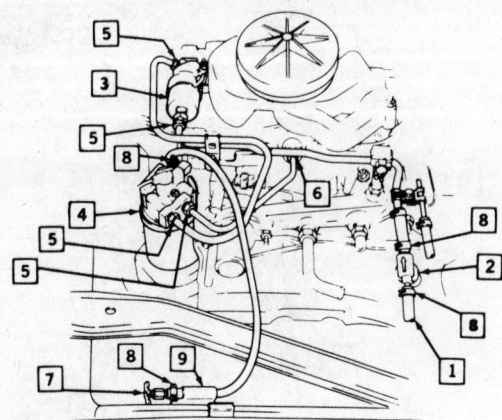

1. From fuel tank
2. In-line filter assembly
3. Fuel pump assembly
4. Fuel filter assembly
5. 25 N·m (18 lbs. ft.)
6. 30 N·m (22 lbs. ft.)
7. Drain valve
8. Clamps
9. Bracket

Fuel supply lines on engines with internal EGR

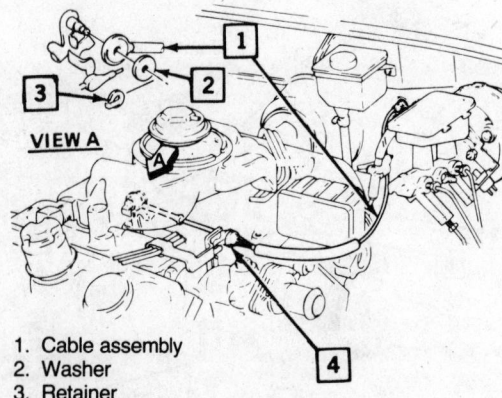

VIEW A

1. Cable assembly
2. Washer
3. Retainer
4. Grommet must be fully seated

Throttle cable

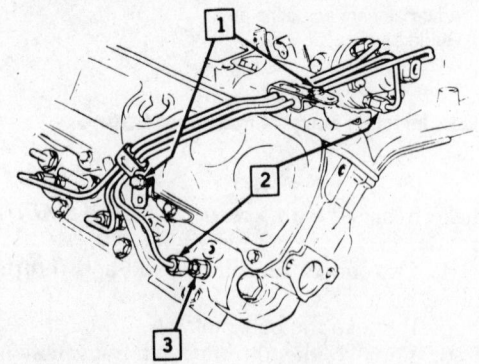

1. Fully driven, seated and not stripped
2. 34 N·m (25 lbs. ft.)
3. Use back-up wrench on upper (inlet fitting) hex of nozzle

Injection lines

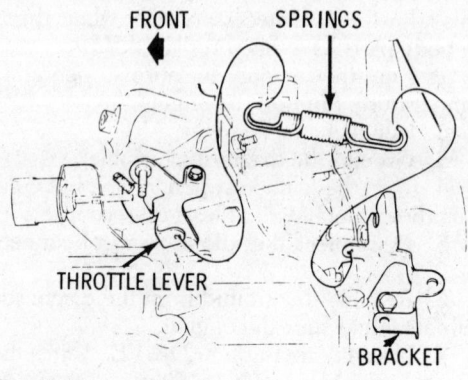

Throttle return spring

9. Remove the rear A/C compressor brace and remove the fuel line.

10. Disconnect the fuel return line from the injection pump.

11. Remove the clamps and pull the fuel return lines from each injection nozzle.

12. Using two wrenches, disconnect the high pressure lines at the nozzles.

13. Remove the three injection pump retaining nuts with tool J-26987 or its equivalent.

14. Remove the pump and cap all lines and nozzles.

To install:

15. Remove the protective caps from all lines and nozzles. Place the engine on TDC for the No. 1 cylinder. The mark on the harmonic balancer on the crankshaft will be aligned with the zero mark on the timing tab, and both valves for No. 1 cylinder will be closed. The index mark on the injection pump driven gear should be offset to the right when No. 1 is at TDC. Check that all of these conditions are met before continuing.

16. Line up the offset tang on the pump driveshaft with the pump driven gear and install the pump.

17. Install, but do not tighten the pump retaining nuts.

18. Connect the high pressure lines at the nozzles.

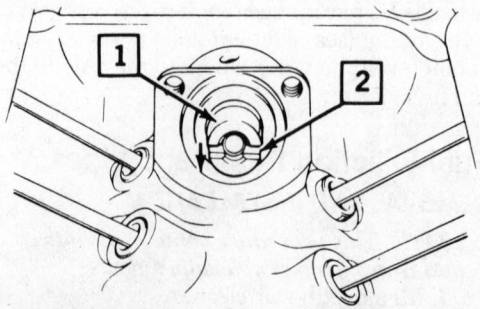

1. Pump driven gear
2. Offset

Injection pump offset

19. Using two wrenches, torque the high pressure line nuts to 25 ft. lbs.

20. Connect the fuel return lines to the nozzles and pump.

21. Align the timing mark on the injection pump with the line on the timing mark adapter and torque the mounting nuts to 35 ft. lbs.

NOTE: *A ¾ in. open end wrench on the boss at the front of the injection pump will aid in rotating the pump to align the marks.*

22. Adjust the throttle rod:

a. remove the clip from the cruise control rod and remove the rod from the bellcrank.

b. loosen the locknut on the throttle rod a few turns, then shorten the rod several turns.

c. rotate the bellcrank to the full throttle stop, then lengthen the throttle rod until the injection pump lever contacts the injection pump full throttle stop, then release the bellcrank.

d. tighten the throttle rod locknut.

23. Install the fuel inlet line between the transfer pump and the filter.

24. Install the rear A/C compressor brace.

25. Install the bellcrank and clip.

26. Connect the throttle rod and return spring.

27. Adjust the transmission cable:

a. push the snap-lock to the disengaged position.

b. rotate the injection pump lever to the full throttle stop and hold it there.

c. push in the snap-lock until it is flush.

d. release the injection pump lever.

28. Start the engine and check for fuel leaks.

29. Remove the screened covers or tape and install the air crossover.

30. Install the tubes in the air flow control valve in the air crossover and install the ventilation filters in the valve covers.

31. Install the air cleaner.

32. Start the engine and allow it to run for two minutes. Stop the engine, let it stand for two minutes, then restart. This permits the air to bleed off within the pump.

Slow Idle Speed

ADJUSTMENT

1. Run the engine to normal operating temperature.

2. Insert the probe of a magnetic pickup tachometer into the timing indicator hole.

3. Set the parking brake and block the drive wheels.

4. Place the transmission in Drive and turn the A/C Off.

5. Turn the slow idle screw on the injection pump to obtain the idle specification on the emission control label.

Fast Idle Solenoid

ADJUSTMENT

1. With the ignition off, disconnect the single green wire from the fast idle relay located on the front of the firewall.

2. Set the parking brake and block the drive wheels.

3. Start the engine and adjust the solenoid (energized) to the specifications on the underhood emission control label.

4. Turn off the engine and reconnect the green wire.

Cruise Control Servo Relay Rod

ADJUSTMENT

1. Turn the engine Off.

2. Adjust the rod to minimum slack then put the clip in the first free hole closest to the bellcrank, but within the servo ball.

Injection Timing

CHECKING AND/OR ADJUSTING TIMING USING J-33075 TIMING METER

The timing meter picks up the engine speed and crankshaft position from the crankshaft balancer. It uses a luminosity signal through a glow plug probe to determine combustion timing. Certain engine malfunctions may cause incorrect timing readings. Engine malfunctions should be corrected before timing adjustment is made. The marks on the pump and adapter flange will normally be aligned within 1.27mm (.050").

NOTE: *Alignment of timing marks may be used in emergency situations (i.e. timing meter not available). However for optimum engine operation, the timing should be adjusted with the timing meter as soon as possible.*

1. Place the transmission selector lever in park, apply the parking brake and block the drive wheels.

2. Start the engine and let it run at idle until fully warmed up. Then shut off the engine.

NOTE: *Failure to have the engine fully warmed up will result in incorrect timing reading and adjustments.*

3. Remove the air cleaner assembly and install cover J-26996-1. The EGR valve hose must be disconnected.

4. Clean any dirt from the engine probe holder (RPM counter) and crankshaft balancer rim.

5. Clean the lens on both ends of the glow plug probe and clean the lens in the photoelectric pick-up. Use a dulled tooth pick to scrape the carbon from the combustion cham-

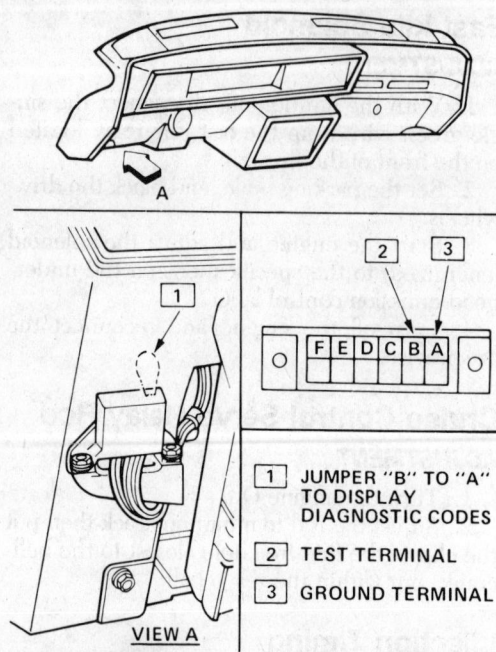

1	JUMPER "B" TO "A" TO DISPLAY DIAGNOSTIC CODES
2	TEST TERMINAL
3	GROUND TERMINAL

VIEW A

Typical ALCL

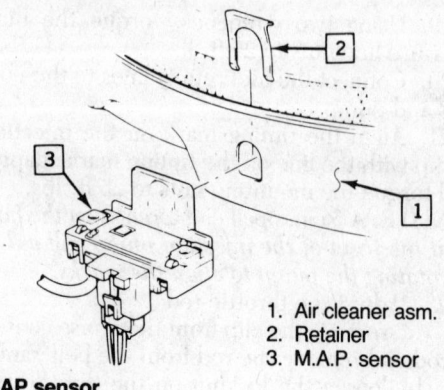

1. Air cleaner asm.
2. Retainer
3. M.A.P. sensor

MAP sensor

ber side of the glow plug probe. Look through the probe to be sure it's clean. Retarded readings will result if the probe is not clean.

6. Install the RPM probe into the crankshaft RPM counter (probe holder).

7. Remove the glow plug from No. 1 cylinder. Install the glow plug from No. 1 cylinder. Install the glow plug probe in the glow opening. Torque the probe to 9 ft. lbs.

8. Set the timing meter offset selector to A (20).

9. Connect the battery leads; red to positive, black to negative.

10. Start the engine and adjust the RPM (speed) to the speed specified on the "Vehicle Emission Control Information Label".

11. Observe the timing reading then at 2 minute intervals, again observe the reading. When the readings stabilize over the 2 minute interval, compare that reading to the one specified on the "Vehicle Emission Control Information Label". The timing reading, when set to specification will be "Negative" (after top dead center).

12. Disconnect the timing meter.

13. Lubricate only the threads of the removed glow plug with lubricant 9985462 or equivalent.

NOTE: *Failure to apply the correct lubricant can cause engine damage.*

14. Install the removed glow plug. Torque the glow plug to 15 ft. lbs.

15. Install the air cleaner being certain to reconnect the EGR valve hose.

ADJUSTMENT

1. Shut off the engine.

2. Note the relative position of the marks on the pump flange and pump intermediate adapter.

3. Loosen the bolts holding the pump to the adapter to a point where the pump can be rotated. Use a 1" open end wrench. (Tool J-25304 has the proper offset on the handle to clear the fuel return line.

4. Rotate the pump to the left to advance the timing and to the right to retard the timing. The width of the mark on the intermediate adapter is about ⅔ degree. Move the pump the amount that is needed and tighten the pump retaining bolts to 35 ft. lbs.

5. Start the engine and recheck the timing reading as outlined previously. Reset and recheck the timing if needed.

6. Reset the idle speed.

Please note the following:

a. Sooty or dirty probes will result in retarded readings.

b. The luminosity probe will soot up very fast when used in a cold engine.

c. Wild needle fluctuations on the timing meter indicate a cylinder not firing properly. Correction of this condition must be made prior to adjusting the timing.

Injection Nozzle
REMOVAL AND INSTALLATION

The injection nozzles on these engines are simply unbolted from the cylinder head, after the fuel lines are removed, in similar fashion to a spark plug. Be careful not to damage the nozzle end and make sure you remove the copper nozzle gasket from the cylinder head if it does not come off with the nozzle.

Clean the carbon off the tip of the nozzle with a soft brass wire brush and install the nozzles, with gaskets.

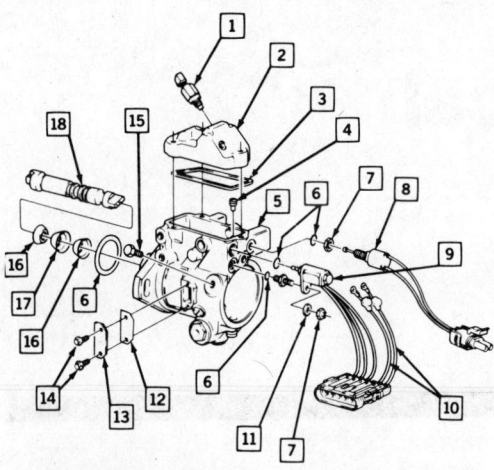

1. Housing pressure regulator and return line connector
2. Injection pump cover
3. Cover gasket
4. Vent screw assembly
5. Housing
6. "O" ring seal
7. Lock nut
8. Altitude fuel limiter solenoid
9. Metering valve sensor
10. HPCA leads
11. Washer
12. Timing line cover gasket
13. Timing line cover
14. Screw
15. Head locking screw
16. Drive shaft seal (black)
17. Drive shaft seal (red)
18. Drive shaft

Pump housing and drive group with internal EGR

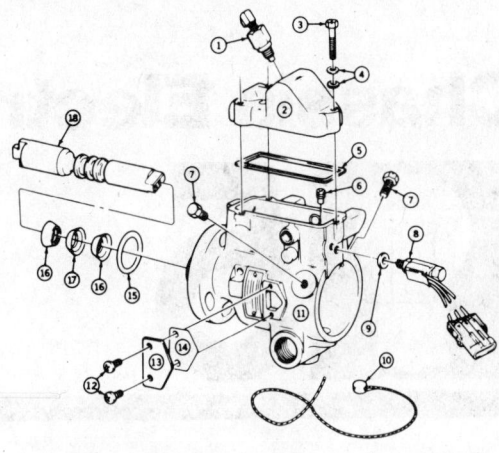

1. Return line connector
2. Cover, governor control
3. Screw
4. Lock washer or plain washer
5. Governor cover gasket
6. Vent screw assembly
7. Head locking screw
8. MVS
9. Seal
10. Seal wire
11. Housing
12. Screw
13. Timing line cover timing line
14. Cover gasket
15. "O" ring
16. Drive shaft seal (black)
17. Drive shaft seal (red)
18. Drive shaft

Pump housing and drive group with external EGR

NOTE: *These engines use two types of injectors, CAV Lucas and Diesel Equipment. When installing the inlet fittings, torque the Diesel Equipment injector fitting to 45 ft. lbs. and the CAV Lucas to 25 ft. lbs.*

Injection Pump Adapter, Seal, Timing Mark

REMOVAL AND INSTALLATION

NOTE: *Skip Steps 4 and 9 if a new adapter is not being installed.*

1. Remove injection pump and lines as described earlier.
2. Remove the injection pump adapter.
3. Remove the seal from the adapter.
4. File the timing mark from the adapter. Do not file the mark off the pump.
5. Position the engine at TDC of No. 1 cylinder. Align the mark on the balancer with the zero mark on the indicator. The index is offset to the right when No. 1 is at TDC.
6. Apply chassis lube to the seal areas. Install, but do not tighten the injection pump.

7. Install the new seal on the adapter using tool J-28425, or its equivalent.
8. Torque the adapter bolts to 25 ft. lbs.
9. Install timing tool J-26896 into the injection pump adapter. Torque the tool, toward No. 1 cylinder, to 50 ft. lbs. Mark the injection pump adapter. Remove the tool.
10. Install the injection pump.

GLOW PLUGS

There are two types of glow plugs used on General Motors Corp. diesels; the "fast glow" type and the "slow glow" type. The fast glow type use pulsing current applied to 6 volt glow plugs while the slow glow type use continuous current applied to 12 volt glow plugs.

An easy way to tell the plugs apart is that the fast glow (6 volt) plugs have a 5/16 in. wide electrical connector plug while the slow glow (12 volt) connector plug is 1/4 in. wide. Do not attempt to interchange any parts of these two glow plug systems.

Chassis Electrical

5

UNDERSTANDING AND TROUBLESHOOTING ELECTRICAL SYSTEMS

For any electrical system to operate, it must make a complete circuit. This simply means that the power flow from the battery must make a complete circle. When an electrical component is operating, power flows from the battery to the component, passes through the component causing it to perform its function (lighting a light built), and then returns to the battery through the ground of the circuit. This ground is usually (but not always) the metal part of the car or truck on which the electrical component is mounted.

Perhaps the easiest was to visualize this is to think of connecting a light bulb with two wires attached to it to the battery. If one of the two wires attached to it to the battery. If one of the two wires attached to the light built were attached to the negative post of the battery and the other were attached to the positive post of the battery, you would have a complete circuit. Current from the battery would flow to the light bulb, causing it to light, and return to the negative post of the battery.

The normal automotive circuit differs from this simple example in two ways. First, instead of having a return wire from the bulb to the battery, the light bulb returns the current to the battery through the chassis of the vehicle. Since the negative battery cable is attached to the chassis and the chassis is made of electrically conductive metal, the chassis of the vehicle can serve as ground wire to complete the circuit. Secondly, most automotive circuits contain switches to turn components on and off as required.

Every complete circuit from a power source must include a component which is using the power from the power source. If you were to disconnect the light bulb from the wires and touch the two wires together (don't do this) the power supply wire to the component would be grounded before the normal ground connection for the circuit.

Because grounding a wire from a power source makes a complete circuit, less the required component to use the power, this phenomenon is called a short circuit. Common causes are: broken insulation (exposing the metal wire to a metal part of the car or truck), or a shorted switch.

Some electrical components which require a large amount of current to operate also have a relay in their circuit. Since these circuits carry a large amount of current, the thickness of the wire in the circuit (gauge size) is also greater. If this large wire were connected from the component to the control switch on the instrument panel, and then back to the component, a voltage drop would occur in the circuit. To prevent this potential drop in voltage, an electromagnetic switch (relay) is used. The large wires in the circuit are connected from the battery to one side of the relay, and from the opposite side of the relay to the component. The relay is normally open, preventing current from passing through the circuit. An additional, smaller, wire is connected from the relay to the control switch for the circuit. When the control switch is turned on, it grounds the smaller wire from the relay and completes the circuit. This closes the relay and allows current to flow from the battery to the component. The horn, headlight, and starter circuits are three which use relays.

It is possible for larger surges of current to pass through the electrical system of your car or truck. If this surge of current were to reach an electrical component, it could burn it out. To prevent this, fuses, circuit breakers or fusible links are connected into the current supply wires of most of the major electrical systems. When an electrical current of excessive power passes throughout the component's fuse, the fuse blows out and breaks the circuit, saving the component from destruction.

A circuit breaker is basically a self-repairing fuse. The circuit breaker opens the circuit the same way a fuse does. However, when either the short is removed from the circuit or the surge subsides, the circuit breaker resets itself and does not have to be replaced as a fuse does.

A fuse link is a wire that acts as a fuse. It is normally connected between the starter relay and the main wiring harness. This connection is usually under the hood. The fuse link (if installed) protects all the chassis electrical components, and is the probable cause of trouble when none of the electrical components function, unless the battery is disconnected or dead.

Electrical problems generally fall into one of three areas:

1. The component that is not functioning is not receiving current.

2. The component itself is not functioning.

3. The component is not properly grounded. The electrical system can be checked with a test light and a jumper wire. A test light is a device that looks like a pointed screwdriver with a wire attached to it and has a light bulb in its handle. A jumper wire is a piece of insulated wire with an alligator clip attached to each end.

If a component is not working, you must follow a systematic plan to determine which of the three causes is the villain.

1. Turn on the switch that controls the inoperable component.

2. Disconnect the power supply wire from the component.

3. Attach the ground wire on the test light to a good metal ground.

4. Touch the probe end of the test light to the end of the power supply wire that was disconnected from the component. If the component is receiving current, the test light will go on.

NOTE: *Some components work only when the ignition switch is turned on.*

If the test light does not go on, then the problem is in the circuit between the battery and the component. This includes all the switches, fuses and relays in the system. Follow the wire that runs back to the battery. The problem is an open circuit between the battery and the component. If the fuse is blown and, when replaced, immediately blows again, there is a short circuit in the system which must be located and repaired. If there is a switch in the system, bypass it with a jumper wire. This is done by connecting one end of the jumper wire to the power supply wire into the switch and the other end of the jumper wire to the wire coming out of the switch. If the test light lights with the jumper wire installed, the switch or whatever was bypassed is defective.

NOTE: *Never substitute the jumper wire for the component, since it is required to use the power from the power source.*

5. If the bulb in the test light goes on, then the current is getting to the component that is not working. This eliminates the first of the three possible causes. Connect the power supply wire and connect a jumper wire from the component to a good metal ground. Do this with the switch which controls the component turned on, and also the ignition switch turned on if it is required for the component to work. If the component works with the jumper wire installed, then it has a bad ground. This is usually caused by the metal area on which the component mounts to the chassis being coated with some type of foreign matter.

6. If neither test located the source of the trouble, then the component itself is defective. Remember that for any electrical system to work, all connections must be clean and tight.

HEATER

Blower Motor
REMOVAL AND INSTALLATION

1. Disconnect the wiring from the motor.

2. Disconnect the cooling line.

3. Remove the mounting screws and lift the blower from the case.

4. Installation is the reverse of removal. Make sure that the sealer around the blower motor flange is intact.

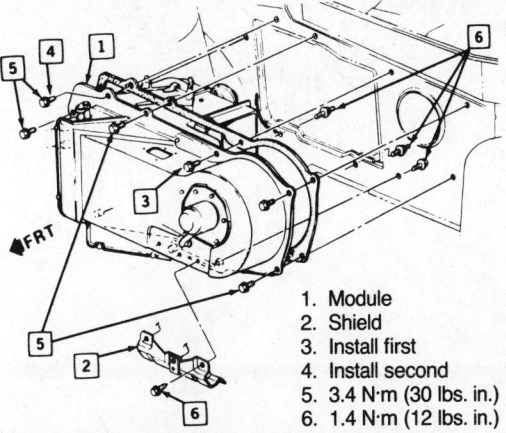

1. Module
2. Shield
3. Install first
4. Install second
5. 3.4 N·m (30 lbs. in.)
6. 1.4 N·m (12 lbs. in.)

Evaporator/heatercore/blower motor housing

Heater Core
REMOVAL AND INSTALLATION
Oldsmobile and Buick

1. Disconnect the battery ground.

2. Drain the cooling system.

3. Remove the right sound insulator.

4. Remove the center instrument panel trim plate.

5. Remove the lower instrument panel trim plate.

6. Remove the right speaker grille and the speaker, giving access to the A/C programmer attaching bolt.

7. Remove the wires and hoses from the programmer. Tag each for installation.

8. Remove the programmer linakge cover and disconnect the linkage.

9. Remove the programmer.

10. Remove the heater core cover.

11. Remove the splash shield and disconnect the hoses from the core.

12. Remove the core.

13. Installation is the reverse of removal. Make sure that all insulating material is replaced intact.

Cadillac

1. Remove the left and right sound insulators.

2. Remove the upper instrument panel trim pad.

3. Remove the lower steering column trim filler.

4. Remove the steering column support bracket, and carefully lower the column, supporting it in the lowered position.

5. Disconnect any wiring in the way of access to the heater/evaporator module.

6. Unbolt and remove the instrument panel pad. An assistant will be necessary for this job.

7. Remove the glove box and lower sound insulator.

8. Remove the programmer and body control module.

9. Remove the ECC module and heater core cover from the evaporator housing, in the engine compartment.

10. Drain the cooling system.

11. Disconnect the heater hoses from the core tubes.

12. Remove the two heater core attaching screws and lift out the core.

13. Installation is the reverse of removal.

INSTRUMENT PANEL

Headlight Switch

REMOVAL AND INSTALLATION

1. Remove the instrument panel trim plate.

2. Remove the 2 switch mounting screws.

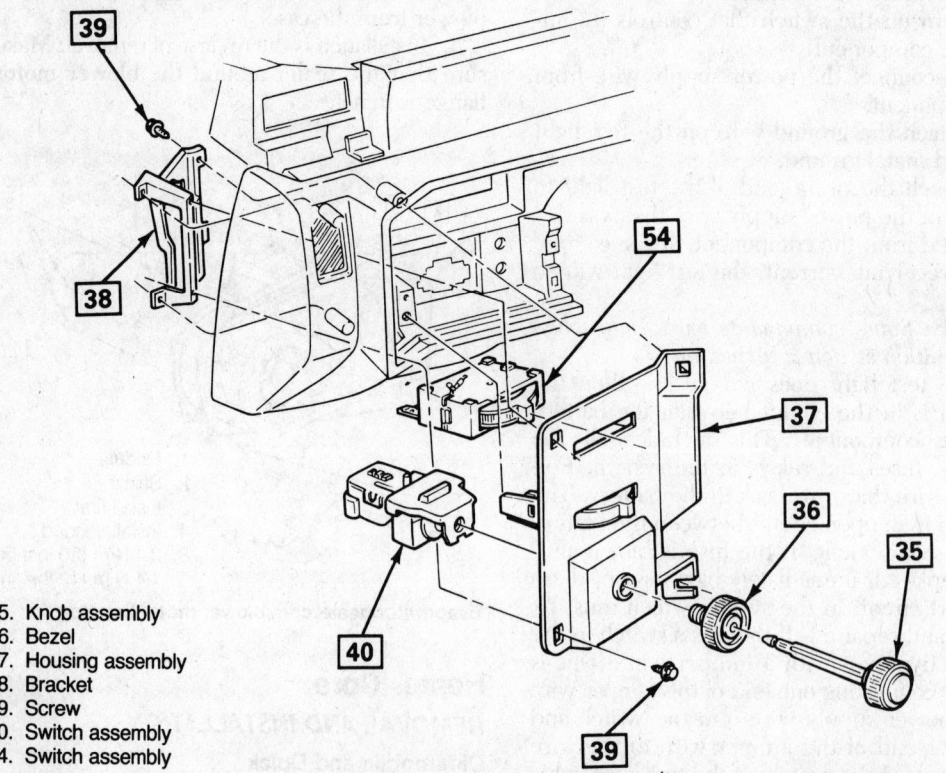

35. Knob assembly
36. Bezel
37. Housing assembly
38. Bracket
39. Screw
40. Switch assembly
54. Switch assembly

Cadillac headlight switch

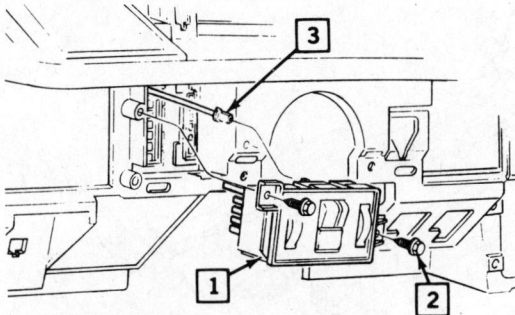

1. Headlamp switch
2. Fully driven, seated and not stripped
3. Plug fiber optic into rear of switch asm.

Oldsmobile and Buick headlight switch

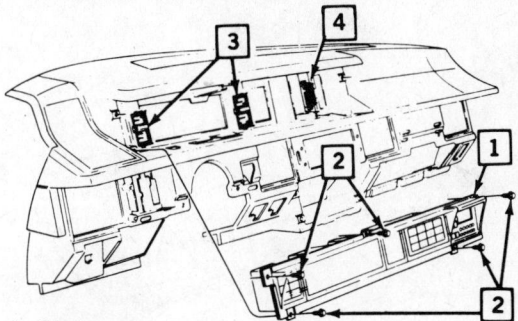

1. Cluster asm.
2. Fully driven, seated and not stripped
3. I.P. wire connectors must be installed before the cluster asm.
4. Auto-calculator connector

Oldsmobile instrument cluster removal

3. Pull the swithc rearward and disconnect the wiring.
4. Installation is the reverse of removal.

Instrument Cluster
REMOVAL AND INSTALLATION
Oldsmobile and Buick

1. Remove the accessory trim plate.
2. Remove the cluster trim plate.
3. Disconnect the speedometer cable at the transaxle or cruise control transducer.
4. Remove the screws attaching the cluster to the instrument panel.
5. Pull the cluster out far enough to reach behind and disconnect the speedometer cable. Then, remove the cluster.
6. Installation is the reverse of removal.

Cadillac

The gauges are removed individually after the cluster trim panel is removed.

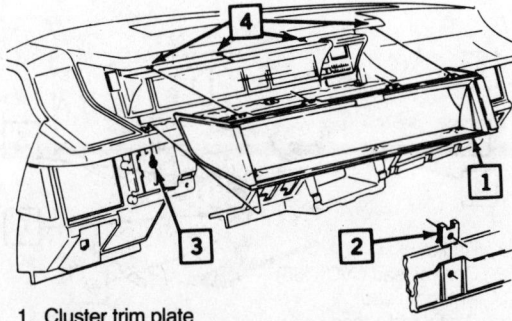

1. Cluster trim plate
2. Nut (3)
3. Fully driven, seated and not stripped
4. Wedge trim plate existing clips and I.P. pad assembly

Oldsmobile cluster trim plate removal

Emissions Indicator

An emissions indicator flag may appear in the odometer window of the speedometer. The flag could say "Sensor", "Emissions" or "Catalyst" depending on the part or assembly that is scheduled for regular emissions maintenance replacement. The word "Sensor" indicates a need for oxygen sensor replacement and the words "Emissions" or "Catalyst" indicate the need for catalytic converter replacement.

RESET PROCEDURE

1. Remove the instrument panel trim plate.
2. Remove the instrument cluster lens.
3. Locate the flag indicator reset notches at the drivers side of the odometer.
4. Use a pointed tool to apply light downward pressure on the notches, until the indicator is reset.

Speedometer Cable
REMOVAL AND INSTALLATION

See the instrument cluster removal procedures above. When the cable is disconnected at both ends it may be removed from the car.

WINDSHIELD WIPERS

Motor and Transmission
REMOVAL AND INSTALLATION

1. Remove the wiper arm assemblies.
2. Loosen, but do not remove, the nuts attaching the wiper drive links to the motor crank arm.
3. Remove the cowl panel.
4. Remove the transmission drive links from the motor crank arm.

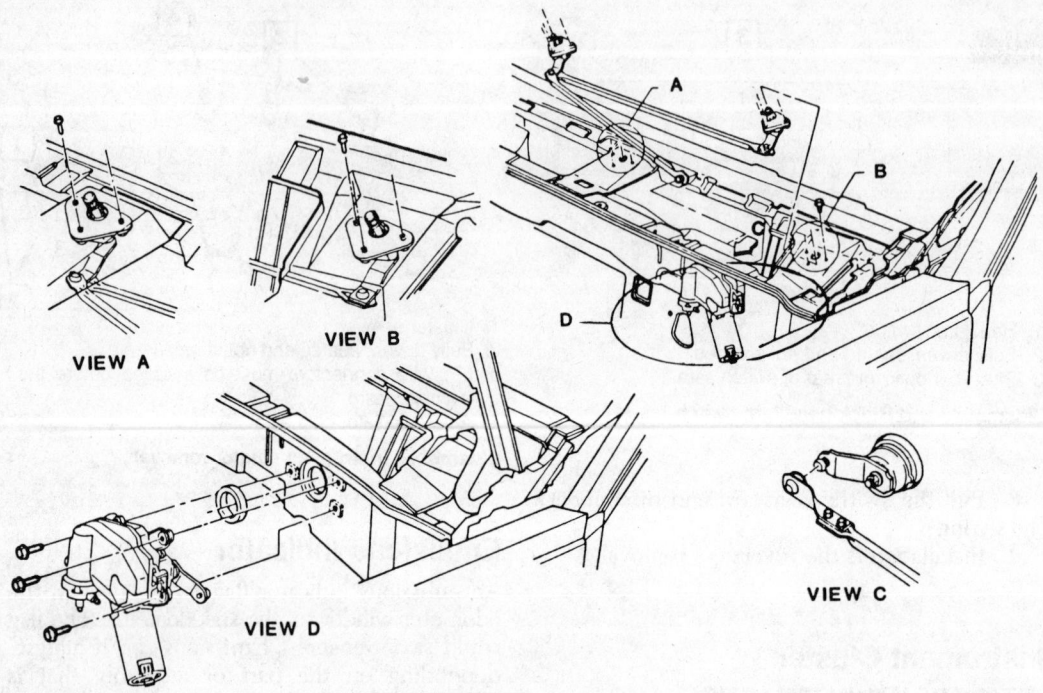

Wiper motor and transmission

5. Remove the transmission-to-body screws.

6. Guide the transmission through the hole in the upper shroud panel.

7. Disconnect the wiring harness at the motor, unbolt and remove the motor.

8. Installation is the reverse of removal.

RADIO

REMOVAL AND INSTALLATION

Oldsmobile and Buick

1. Remove the right instrument panel trim cover.

2. Remove the 4 radio-to-bracket mounting screws.

3. Pull the radio out far enough to reach behind and disconnect the antenna lead and wiring.

4. Remove the radio.

5. Installation is the reverse of removal.

Cadillac

1. Remove the radio trim plate.

2. Open the ashtray and remove the 3 radio mounting screws.

3. Remove the light bulb and socket.

4. Disconnect the wiring and antenna lead, and pull out the radio.

5. Installation is the reverse of removal.

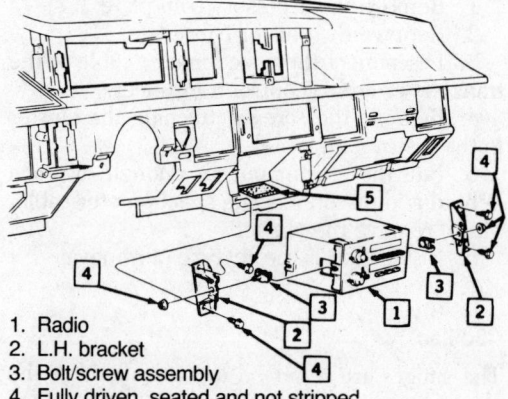

1. Radio
2. L.H. bracket
3. Bolt/screw assembly
4. Fully driven, seated and not stripped
5. Insulator

Oldsmobile or Buick radio removal

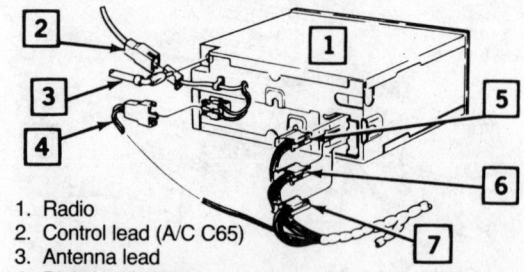

1. Radio
2. Control lead (A/C C65)
3. Antenna lead
4. Digital radio lead
5. Rear speaker (blue connector)
6. Front speaker (white connector)
7. Ignition and ground (black connector)

Radio wiring

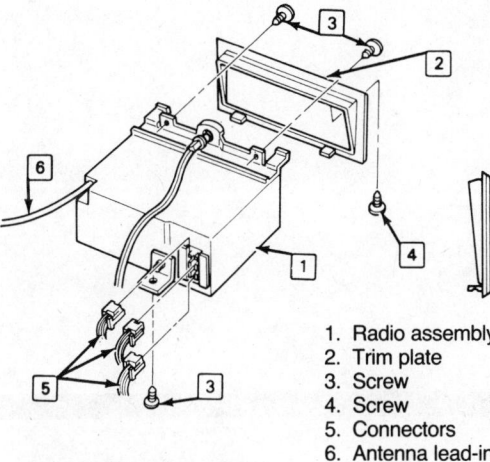

1. Radio assembly
2. Trim plate
3. Screw
4. Screw
5. Connectors
6. Antenna lead-in

Cadillac radio installation

FUSES

The fuse block is located beneath the instrument panel above the headlight dimmer floor switch. Fuse holders are labeled as to their service and the correct amperage. Always replace blown fuses with new ones of the correct amperage. Otherwise electrical overloads and possible wiring damage will result.

Electrical Component Location

FUSE PANEL

The fuse panel is located on the left side of the vehicle. It is under the instrument panel assembly. In order to gain access to the fuse panel, it may be necessary to first remove the under dash padding.

ELECTRONIC CONTROL MODULE

The electronic control module is located on the right side of the vehicle. It is positioned in front of the right hand kick panel. In order to gain access to the assembly you must first remove the trim panel.

TURN SIGNAL FLASHER

The turn signal flasher is located directly under the steering column of the vehicle. It is secured in place with a plastic retainer. In order to gain access to the component, it may first be necessary to remove the under dash padding.

CONVENIENCE CENTER

The convenience center is located on the right side of the vehicle. It is positioned under the dash panel. In order to gain access to the convenience center it may be necessary to remove instrument panel sound absorber.

CIRCUIT BREAKER

A circuit breaker is an electrical switch which breaks the circuit during an electrical overload. The circuit breaker will remain open until the short or overload condition in the circuit is corrected.

FUSIBLE LINKS

Fusible links are sections of wire, with special insulation, designed to melt under electrical overload. Replacements are simply spliced into the wire. There may be as many as five of these in the engine compartment wiring harnesses. These are:

1. Horn relay to fuse panel circuit—one link.
2. Charging circuit, from the starter solenoid to the horn relay—two links.
3. Starter solenoid to ammeter circuit—one link.
4. Horn relay to rear window defroster circuit—one link.

The fusible links are all two wire gauge sizes smaller than the wires they protect.

NOTE: *Most models have fusible links at these locations.*

REPLACEMENT

1. Disconnect the battery ground cable.
2. Disconnect the fusible link from the junction block or starter solenoid.
3. Cut the harness directly behind the connector to remove the damaged fusible link.
4. Strip the harness wire approximately ½ in.
5. Connect the new fusible link to the harness wire using a crimp on connector. Solder the connection using resin core solder.
6. Tape all exposed wires with plastic electrical tape.
7. Connect the fusible link to the junction block or starter solenoid and reconnect the battery ground cable.

WIRING DIAGRAMS

Wiring diagrams have been left out of this book. As cars have become more complex and available with longer and longer option lists, wiring diagrams have grown in size and complexity also. It has become virtually impossible to provide a readable reproduction in a reasonable number of pages.

Transaxle

6

AUTOMATIC TRANSAXLE

NOTE: *GM C-Body cars use the Turbo Hydra-Matic 440–T4 transaxle*

Adjustments

CAUTION: *Any inaccuracies in shift linkage adjustments may result in premature failure of the transmission due to operation without the controls in full detent. Such operation results in reduced fluid pressure and in turn, partial engagement of the affected clutches. Partial engagement of the clutches, with sufficient pressure to permit apparently normal vehicle operation will result in failure of the clutches and/or other internal parts after only a few miles of operation.*

THROTTLE VALVE CABLE

NOTE: *On the diesel, the throttle link assembly must be correctly adjusted before making this adjustment.*

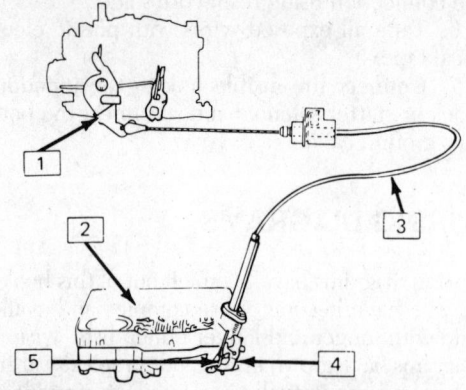

1. Throttle lever
2. Control valve asm.
3. T.V. cable
4. T.V. lever and bracket asm.
5. T.V. link

Throttle valve cable

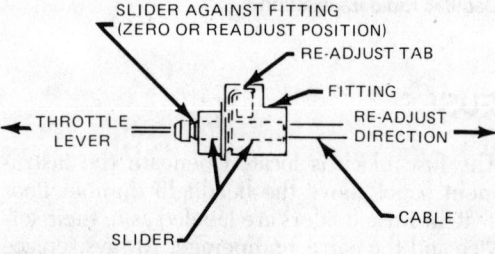

Throttle valve cable adjuster

1. With the engine OFF, depress and hold down the metal readjusting tab at the engine end of the T.V. cable.
2. Move the slider until it stops against the fitting.
3. Release the adjustment tab.
4. Rotate the throttle lever to its full travel position. The slider must move toward the lever when the lever is moved.
5. Make sure that the cable moves freely, especially when the engine is hot. Road test the car.

SHIFT CONTROL CABLE

1. Place the shift lever in NEUTRAL.
2. Loosen the nut on the transaxle shift lever pin.
3. Hold the transaxle lever in the NEUTRAL detent and tighten the nut to 20 ft.lb.

NEUTRAL START AND BACK-UP LIGHT SWITCH

1. Place the trasaxle control shifter in NEUTRAL.
2. Loosen the switch attaching screws.
3. Rotate the switch until the holes in the switch and bracket align.
4. Insert a 3/32" drill bit in the hole to a depth of 5/8"(15mm).
5. Tighten the attaching screws to 20 ft.lb. and remove the drill bit.

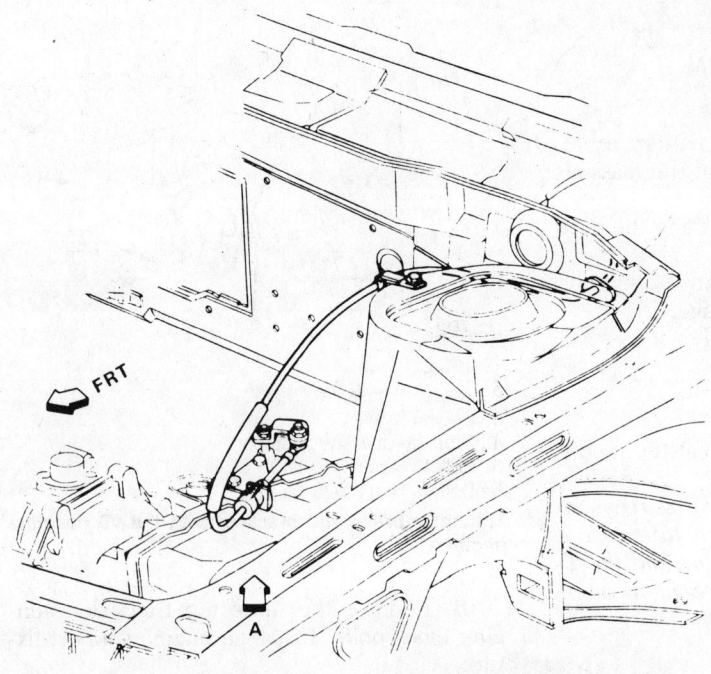

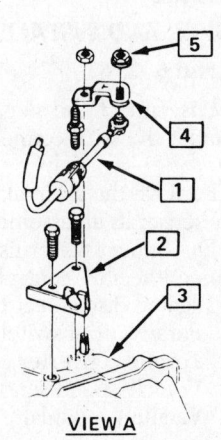

VIEW A

1	SHIFT CONTROL CABLE
2	CABLE BRACKET
3	TRANSMISSION
4	SHIFT LEVER
5	NUT (27 N·m/20 FT. LBS.)

Shift control cable adjustment

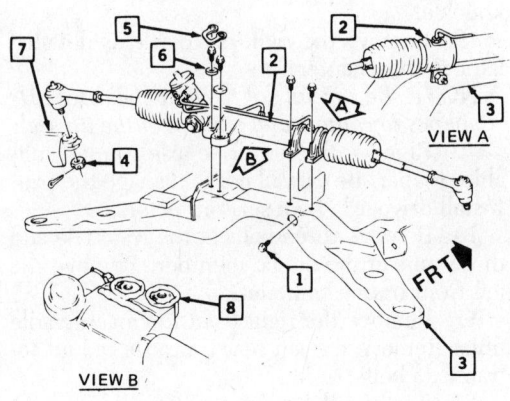

VIEW A

VIEW B

3. Transaxle assm.
6. Neutral start & back-up lamp switch
7. Bolt (27 N·m/20 ft. lbs.)

Neutral start and back-up light switch

Fluid and Filter
REPLACEMENT

1. Raise and support the car on jackstands.
2. Place a drain pan under the transaxle.
3. Remove the pan bolts from the front and sides of the pan and loosen the rear pan bolts about half way. This will allow the fluid to drain relatively neatly.

NOTE: *If you have to pry the pan loose, be very carefull to avoid damaging the mating surface of the pan!*

4. When the fluid has drained, remove the remaing bolts and the pan. Discard the gasket.
5. Remove the screen and filter. Discard the filter and wash the screen in a safe solvent. The filter uses a lip ring seal which should be removed only if replacement is necessary.
6. Install the screen and new filter.
7. Install the pan and new gasket. Torque the pan bolts to 5-7 ft.lb.

Transaxle

REMOVAL AND INSTALLATION

6–181 and 6–231

1. Disconnect the negative battery cable. Disconnect the wire connector at the mass air flow sensor (6–231 only).

2. Remove the air intake duct and the mass air flow sensor as an assembly.

3. Disconnect the cruise control assembly. Disconnect the shift control linkage.

4. Tag and disconnect the following:
   a. Park/Neutral switch
   b. Torque converter clutch
   c. Vehicle speed sensor
   d. Vacuum modulator hose at the modulator.

NOTE: *Care must be exercised on reassembly of the Park/Neutral switch to ensure a proper fit of both the connector and the T-latch. Failure to do so may result in intermittent loss of switch functions.*

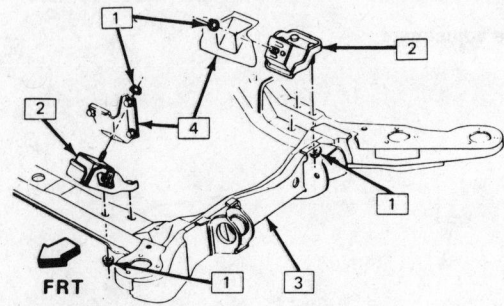

1. Nut 41 N·m (30 ft. lbs.)
2. Mount asm.
3. Frame asm.
4. Transaxle mounting bracket

Left side transaxle mounts

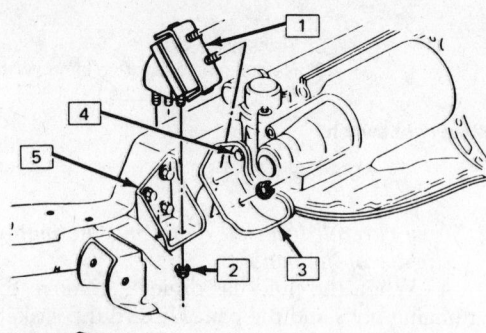

1. Mount asm.
2. Nut 30 N·m (22 ft. lbs.)
3. Transaxle mount bracket
4. Transaxle to bracket bolt 55 N·m (40 ft. lbs.)
5. Transaxle mount bracket

Right side transaxle mounts

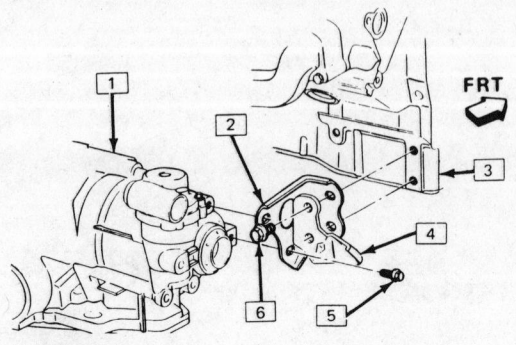

1. Transaxle
2. Brace–Transaxle
3. Engine asm.
4. Bracket–driveline absorber
5. Bolt 45 N·m (33 ft. lbs.)
6. Bolt 95 N·m (70 ft. lbs.)

Transaxle brace and brackets with the V6 gasoline engine

5. Remove the three top transaxle-to-engine block bolts. Install an engine support fixture.

6. Remove both front wheels and then turn the steering wheel to the full left position.

7. Remove the right front ball joint nut and separate the control arm from the steering knuckle.

8. Remove the right drive axle as detailed later in this chapter.

NOTE: *Be careful not to allow the drive axle splines to contact any portion of the lip seal.*

9. Remove the left drive axle using a suitable pry bar. Be careful not to damage the pan. Install drive axle boot seal protectors.

10. Remove three bolts at the transaxle and three nuts at the cradle member. Remove the left front transaxle mount.

11. Remove the right front mount-to-cradle nuts. Remove the left rear transaxle mount-to-transaxle bolts.

12. Remove the right rear transaxle mount as in Step 10. Remove the engine support bracket-to-transaxle case bolts.

13. Remove the flywheel cover. Remove the flywheel-to-converter bolts.

NOTE: *Be sure to matchmark the flywheel-to-converter relationship for proper alignment upon reassembly.*

14. Remove the bolts attaching the rear cradle member to the front cradle dog leg.

15. Remove the front left cradle-to-body bolt. Remove the front cradle dog leg-to-right cradle member bolts.

16. Install a transaxle support fixture into position.

17. Remove the cradle assembly by swinging it aside and supporting it with a suitable stand.

18. Disconnect and cap the oil cooler lines at the transaxle.

NOTE: *One bolt is located between the transaxle and the engine block and is installed in the opposite direction.*

19. Remove the remaining lower transaxle-to-engine bolts. And then lower the transaxle assembly away from the car.

20. Installation is in the reverse order or removal. Check the fluid level and all adjustments.

8–250

1. Disconnect the negative battery cable, the air cleaner and the TV cable.

2. Disconnect the shift linkage at the transaxle. Install a suitable engine support fixture.

3. Tag and disconnect the following:
   a. Converter clutch
   b. Vehicle speed sensor
   c. Neutral start/back-up light switch
   d. Vacuum line at the modulator

4. Remove the upper bolts and studs securing the bell housing to the block.

5. Raise and support the car and remove both front wheels.

6. Disconnect the lower ball joint from the left steering knuckle. Remove both drive axles from the transaxle.

7. Remove the stabilizer bar mounting bolt from the left control arm.

8. Remove the left front cradle assembly.

9. Remove the extension housing-to-engine block support bracket.

10. Disconnect and cap the oil cooler lines at the transaxle case.

11. Remove the right and left transaxle mount attachments.

12. Remove the flexplate splash shield. Remove the converter-to-flexplate bolts.

13. Remove all the lower bell housing bolts except the lower rear on (No. 6).

14. Position a jack under the transaxle and then remove the last bell housing bolt.

NOTE: *To reach the last bell housing bolt, you will need a 3 in. socket wrench extension and you must come through the right wheel arch opening.*

15. Remove the transaxle assembly.

16. Installation is in the reverse order of removal. Check the fluid level and all adjustments.

6–263 Diesel

1. Disconnect the negative battery cable. Disconnect the TV cable at the injection pump and transaxle.

2. Remove the crossover pipe shield and disconnect the shift control linkage.

3. Tag and disconnect the following:

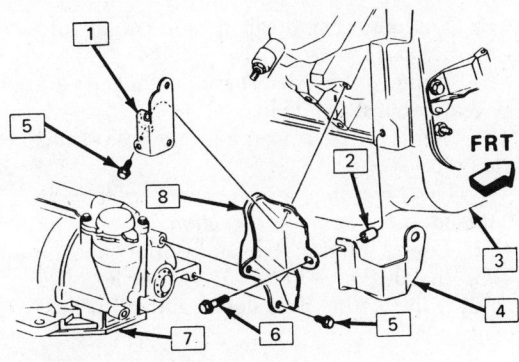

1. Power steering pump bracket
2. Spacer
3. Engine
4. Shock absorber bracket
5. Bolt 55 N·m (40 ft. lbs.)
6. Bolt 45 N·m (33 ft. lbs.)
7. Transaxle
8. Brace

Transaxle brace and brackets with the V6 diesel engine

   a. Park/Neutral switch
   b. Torque converter clutch
   c. Vehicle speed sensor
   d. Vacuum hose at the modulator.

4. Remove the three upper engine-to-transaxle bolts.

5. Loosen, but do not remove, the engine-to-transaxle bolt at the starter.

6. Install a suitable engine support fixture.

7. Raise and support the car. Remove both front wheels and then turn the steering wheel to the full left position.

8. Disconnect the right front ball joint from the steering knuckle. Remove the right drive axle from the transaxle.

9. Remove the left front and rear transaxle-to-cradle mounts.

10. Remove the transaxle brace and its bracket.

11. Disconnect the speedometer cable.

12. Remove the right rear transaxle mount and disconnect the left stabilizer link.

13. Remove the flywheel cover and then remove the flywheel-to-torque converter bolts.

14. Remove the bolts attaching the rear cradle member to the front cradle dog leg. Remove one stabilizer brace and loosen the other.

15. Remove the front cradle-to-body bolt and the right front motor mount.

16. Remove the wiring harness cover on the cradle and position it out of the way.

17. Install a suitable transmission support fixture.

18. Slide the cradle assembly to one side and support it.

19. Disconnect and cap the oil cooler lines at the transaxle.

20. Remove the exhaust connector pipe and the rear exhaust manifold.

21. Remove the remaining engine-to-transaxle bolts.

NOTE: *One engine-to-transaxle bolt is installed in the opposite direction.*

22. Lower the transaxle and remove it.

23. Installation is in the reverse order of removal. Check the fluid level and all adjustments.

DRIVESHAFTS, DRIVE AXLES AND U-JOINTS

Drive Axle

REMOVAL AND INSTALLATION

CAUTION: *Use care when removing the drive axle. Tri-pots can be damaged if the drive axle is over-extended.*

1. Remove the hub nut.

2. Raise the front of the car. Remove the wheel and tire.

3. Install an axle shaft boot seal protector, G.M. special tool no. J-28712 or the equivalent, onto the seal.

4. Disconnect the brake hose clip from the MacPherson strut, but do not disconnect the hose from the caliper. Remove the brake caliper from the spindle, and hang the caliper out

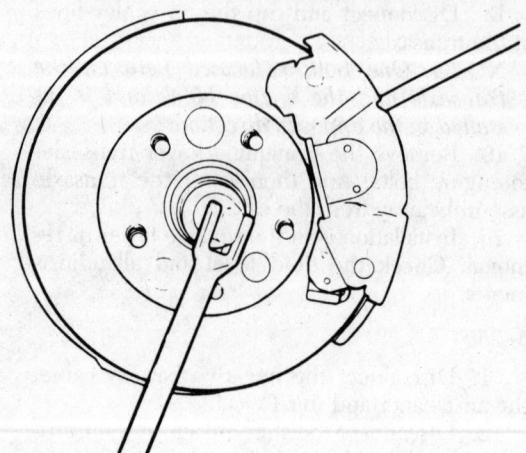

Removing hub nut

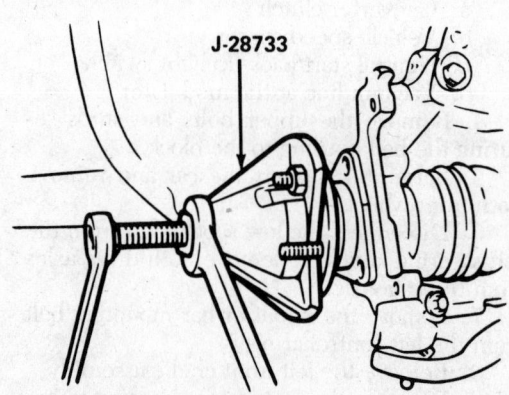

Loosening the splines between the drive axle and hub

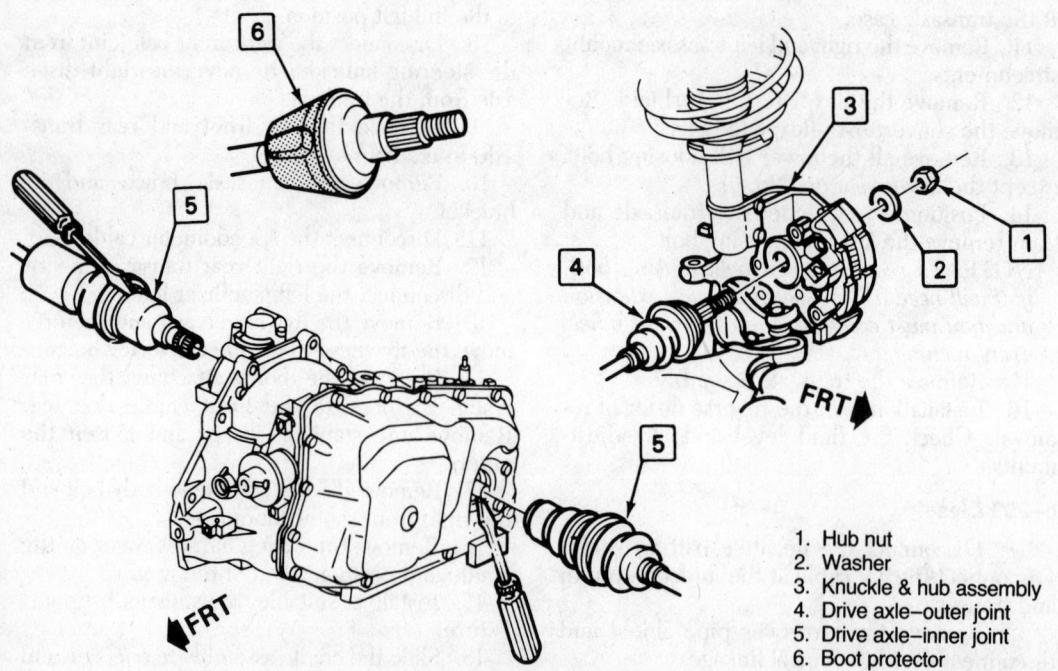

1. Hub nut
2. Washer
3. Knuckle & hub assembly
4. Drive axle—outer joint
5. Drive axle—inner joint
6. Boot protector

Removing the drive axle

of the way by a length of wire. Do not allow the caliper to hang by the brake hose.

5. Mark the camber alignment cam bolt for reassembly. Remove the cam bolt and the upper attaching bolt from the strut and spindle.

6. Pull the steering knuckle assembly from the strut bracket.

7. Using B.M. special tool J-28733 or the equivalent spindle remover, remove the axle shaft from the hub and bearing assembly.

To install:

1. If a new drive axle is to be installed, a new knuckle seal should be installed first.

2. Loosely install the drive axle into the transaxle and steering knuckle.

3. Loosely attach the steering knuckle to the suspension strut.

4. The drive axle is an interference fit in the steering knuckle. Press the axle into place, then install the hub nut. When the shaft begins to turn with the hub, insert a drift through the caliper into one of the cooling slots in the rotor to keep it from turning. Insert a long bolt in the hub flange to prevent the shaft from turning. Tighten the hub nut to 70 ft. lbs. to completely seat the shaft.

5. Install the brake caliper. Tighten the bolts to 30 ft. lbs.

6. Load the hub assembly by lowering it onto a jackstand. Align the camber cam bolt marks made during removal, install the bolt and tighten to 140 ft. lbs. Tighten the upper nut to the same value.

7. Install the axle shaft all the way into the transaxle using a screwdriver inserted into the groove provided on the inner retainer. Tap the screwdriver until the shaft seats in the transaxle. Remove the boot seal protector.

8. Connect the brake hose clip to the strut. Install the tire and wheel, lower the car, and tighten the hub nut to 225 ft. lbs. (1980–82); 185 ft. lbs. (1983 and later).

Constant Velocity Joints

Front wheel drive vehicles present several unique problems to engineers because the driveshaft must do three things, simultaneously. It must allow the wheels to turn for steering, telescope to compensate for road surface vibrations, and it must transmit torque continuously without vibration.

To compensate for these three factors a two-joint driveshaft allows the front wheels to perform these functions. This driveshaft mates disc type straight groove ball joint design with the bell type Rzeppa CV universal joint. The Rzeppa joint on the outboard end of each driveshaft provides steering ability by allowing drive wheels to steer up to 43° while transmitting all

available torque to the wheels. The inboard joint allows telescoping (up to 1 ½") through the rolling actions of balls in straight grooves and operates at angles up to 20°. The combined action of these two ball type U-joints eliminates vibration.

The typical front wheel drive vehicle uses two driveshaft assemblies; one to each driving wheel. Each assembly has a CV-joint at the wheel end is called the inboard joint. This joint may be either the ball or tripode type. It allows the slip motion required when the driveshaft must shorten or lengthen in response to suspension action when traveling over an irregular surface.

Constant velocity joints are precision machined parts that have difficult jobs to perform in a hostile enviornment. They are exposed to heat, shock, torque, and many thousands of miles of service. For this reason, the lubricants used are specially formulated to be compatible with the rubber boot and give proper lubrication. Most CV-joint repair kits have this special lubricant included.

NOTE: *Wear pattern in a used ball or tripode CV-joint are impossible to match during reassembly. If there are any signs of wear, abnormal operating noise, corrosion, heat discoloration, the joint must be replaced.*

TROUBLESHOOTING

Noises from the engine, drive axles, suspension and steering in the front drive cars can be misleading to the untrained ear. Ideally a smooth road serves best for detecting operating condition(s) that cause noise.

- A humming noise could indicate that early stage of insufficient or incorrect lubricant.
- Worn driveshaft joints will cause a continuous knock at low speeds.
- A popping or clicking sound on sharp turns indicates trouble in the outer or wheel end joint.
- The click noise at acceleration from coasting or deceleration from a load pull indicated two possibilities: damaged inner or transaxle joint or differential problem(s).
- An inner joint will create a vibration during acceleration due to plunging action hanging up and releasing repeatedly. Probable cause would be foreign particles or lack of lubrication, or improper assembly.
- Remember that tires, suspension, engine, and exhaust system are all up front to add their noises.
- Make a check with front wheels elevated off ground. Spin the wheels by hand to determine if wheel bearing could be noisy or if out of round tires are causing vibration.

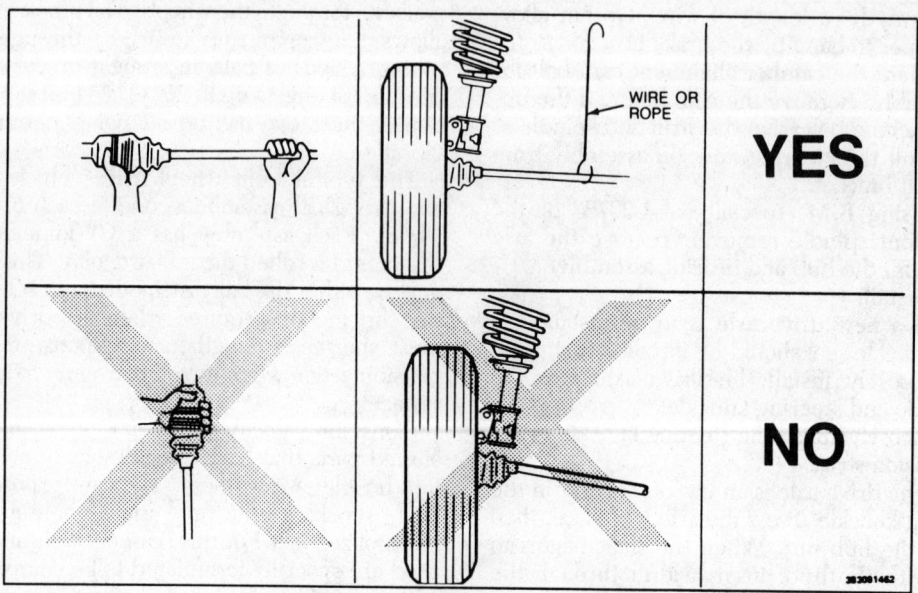

Precautions for handling the Tri-pot joints

1. Race, C.V. joint outer
2. Cage, C.V. joint
3. Race, C.V. joint inner
4. Ring, shaft retaining
5. Ball (6)
6. Clamp, seal retaining
7. Seal, C.V. joint
8. Clamp, seal retaining
9. Shaft, axle (L.H.)
10. Seal, tri-pot joint
11. Spider, tri-pot joint
12. Roller, needle
13. Ball, tri-pot joint (3)
14. This no. not used
15. Housing assy., tri-pot (L.H.)
16. Housing assy., tri-pot (R.H.)
17. Shaft, damper & axle (R.H.)
18. Ring, spacer
19. Ring, race retaining
20. Clamp, seal retaining
21. Retainer, needle
22. Ring, needle retainer
23. Ring, joint retaining
24. Ring, deflector

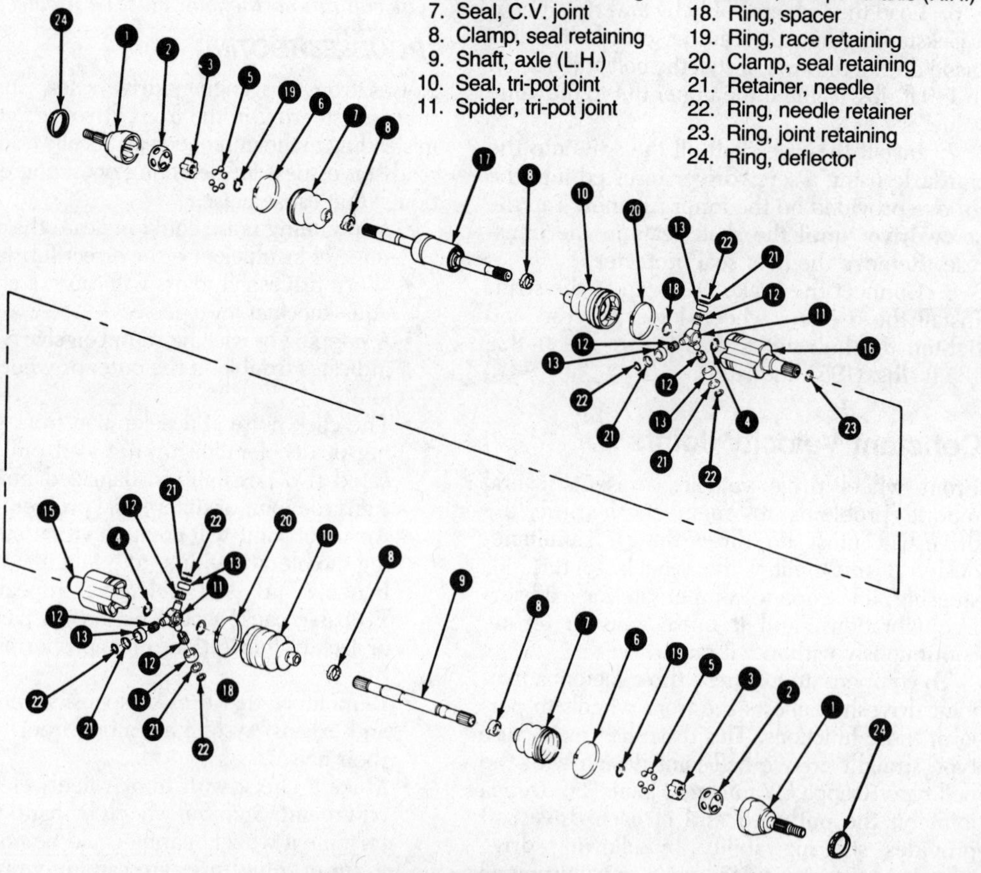

Tri-pot joint exploded view

Many wheel bearings are prelubed and sealed at the factory.

CAUTION: *Personal can occur from spinning wheels by engine power. Spinning a wheel at excess speed may cause damage to CV-joints that could be operating at angles too steep when wheels are allowed to hang. Over speeding might also cause damage to tires and the differential.*

TRI-POT JOINT OVERHAUL

For overhaul of these joints, follow the picture sequence below:

Hub and Bearing
REMOVAL AND INSTALLATION

NOTE: *Several special tools are required for this procedure.*

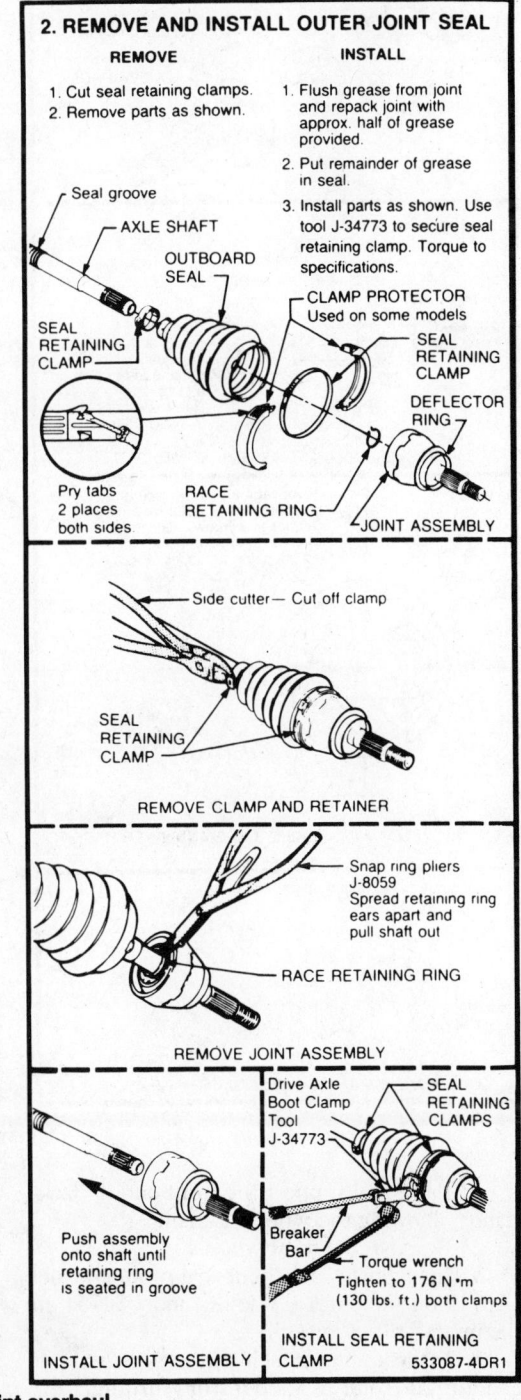

1. REMOVE AND INSTALL DEFLECTOR RING

REMOVE

1. For damaged deflector ring, remove parts as shown.

INSTALL

1. Install part as shown.

— DEFLECTOR RING

AXLE ASSEMBLY WITH STEEL DEFLECTOR RING

— DEFLECTOR RING

AXLE ASSEMBLY WITH RUBBER DEFLECTOR RING

— DEFLECTOR RING —
To install, stretch ring and seat properly in groove

REMOVE AND INSTALL DEFLECTOR RING (RUBBER)

Use brass drift to tap off deflector ring

DEFLECTOR RING

REMOVE DEFLECTOR RING (STEEL)

SHEET STEEL (3mm MIN THICKNESS) WITH 24mm DRILLED HOLE

DEFLECTOR RING — 2½" PIPE COUPLING — M20 x 1.5 NUT

INSTALL DEFLECTOR RING (STEEL)

2. REMOVE AND INSTALL OUTER JOINT SEAL

REMOVE

1. Cut seal retaining clamps.
2. Remove parts as shown.

INSTALL

1. Flush grease from joint and repack joint with approx. half of grease provided.
2. Put remainder of grease in seal.
3. Install parts as shown. Use tool J-34773 to secure seal retaining clamp. Torque to specifications.

Seal groove
AXLE SHAFT
OUTBOARD SEAL

SEAL RETAINING CLAMP

CLAMP PROTECTOR Used on some models

SEAL RETAINING CLAMP

DEFLECTOR RING

Pry tabs 2 places both sides

RACE RETAINING RING

JOINT ASSEMBLY

Side cutter — Cut off clamp

SEAL RETAINING CLAMP

REMOVE CLAMP AND RETAINER

Snap ring pliers J-8059 Spread retaining ring ears apart and pull shaft out

RACE RETAINING RING

REMOVE JOINT ASSEMBLY

Push assembly onto shaft until retaining ring is seated in groove

Drive Axle Boot Clamp Tool J-34773

SEAL RETAINING CLAMPS

Breaker Bar

Torque wrench Tighten to 176 N·m (130 lbs. ft.) both clamps

INSTALL JOINT ASSEMBLY

INSTALL SEAL RETAINING CLAMP 533087-4DR1

Tri-pot joint overhaul

3. DISASSEMBLE AND ASSEMBLE OUTER JOINT ASSEMBLY

REMOVE

1. Remove parts as shown.

INSTALL

1. Put a light coat of recommended grease on ball grooves of inner and outer races.
2. Install parts as shown.

   NOTICE: Be sure retaining ring side of inner race faces axle shaft.
3. Pack joint with recommended grease.

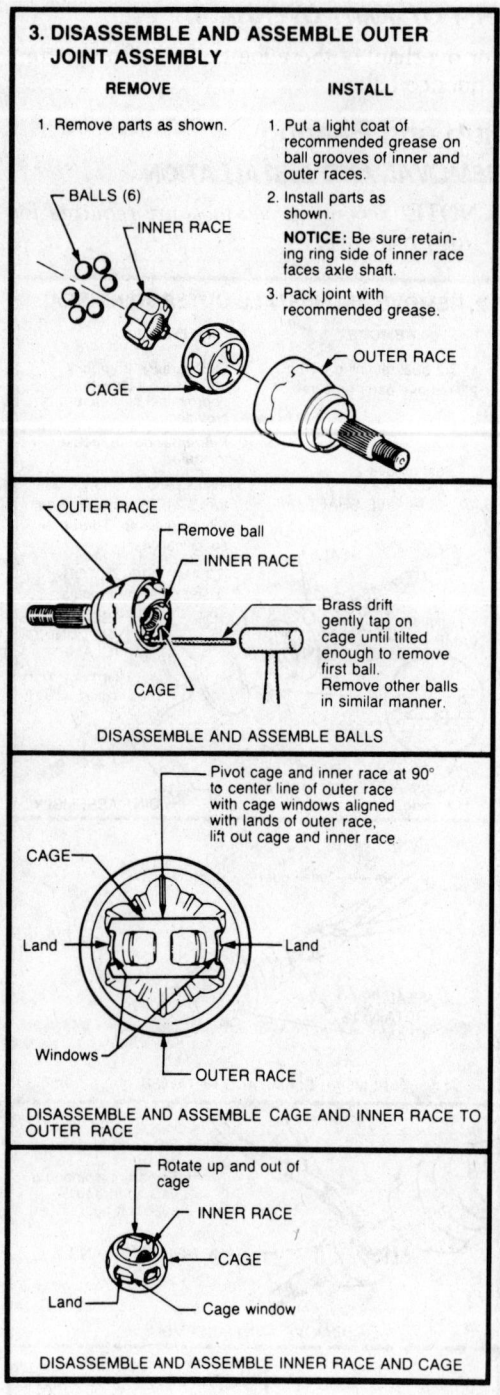

BALLS (6)

INNER RACE

CAGE

OUTER RACE

OUTER RACE

Remove ball

INNER RACE

Brass drift gently tap on cage until tilted enough to remove first ball. Remove other balls in similar manner.

CAGE

DISASSEMBLE AND ASSEMBLE BALLS

Pivot cage and inner race at 90° to center line of outer race with cage windows aligned with lands of outer race, lift out cage and inner race.

CAGE

Land

Land

Windows

OUTER RACE

DISASSEMBLE AND ASSEMBLE CAGE AND INNER RACE TO OUTER RACE

Rotate up and out of cage

INNER RACE

CAGE

Land

Cage window

DISASSEMBLE AND ASSEMBLE INNER RACE AND CAGE

4. REMOVE AND INSTALL INNER TRI-POT SEAL

REMOVE

1. Remove parts as shown.

INSTALL

1. Flush grease from housing and repack housing with approx. half of grease furnished with new seal.
2. Put remainder of grease in seal.
3. Install parts as shown. Use tool J-22610 to secure seal retaining clamps.

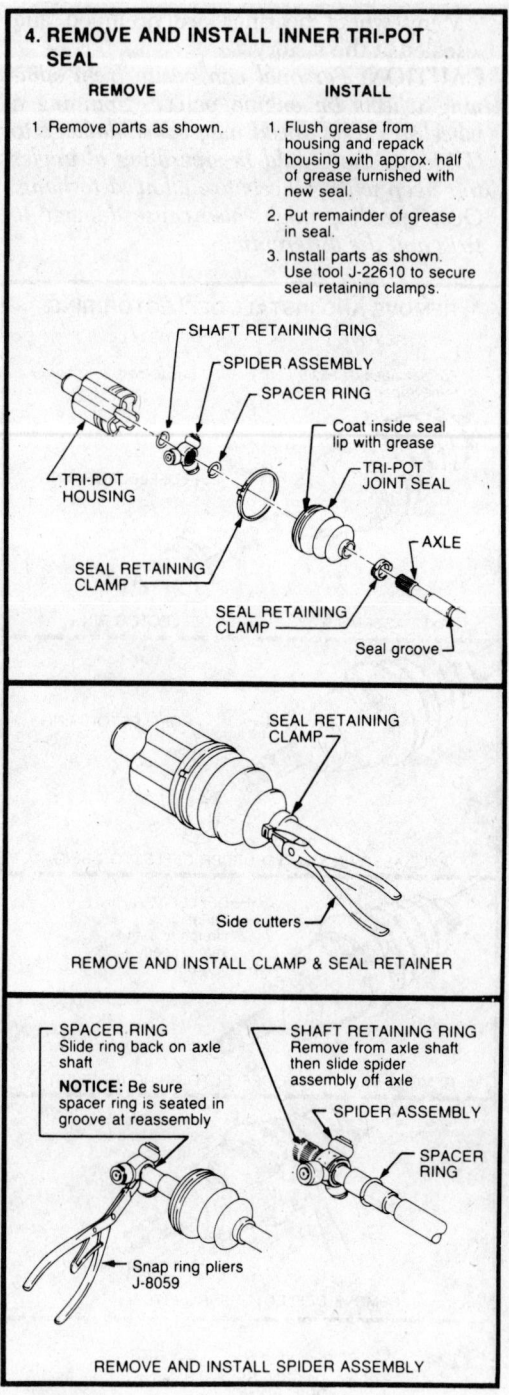

SHAFT RETAINING RING

SPIDER ASSEMBLY

SPACER RING

Coat inside seal lip with grease

TRI-POT JOINT SEAL

TRI-POT HOUSING

AXLE

SEAL RETAINING CLAMP

SEAL RETAINING CLAMP

Seal groove

SEAL RETAINING CLAMP

Side cutters

REMOVE AND INSTALL CLAMP & SEAL RETAINER

SPACER RING
Slide ring back on axle shaft
NOTICE: Be sure spacer ring is seated in groove at reassembly

SHAFT RETAINING RING
Remove from axle shaft then slide spider assembly off axle

SPIDER ASSEMBLY

SPACER RING

Snap ring pliers J-8059

REMOVE AND INSTALL SPIDER ASSEMBLY

Tri-Pot Joint Overhaul

1. Raise and support the front end on jackstands, allowing the wheels to hang.
2. Remove the front wheels.
3. Install drive axle boot seal protector tool J-28712 on the outer CV joints and J-34754 on the inner Tri-pot joints.
4. Insert a long punch through the caliper and into the rotor to keep it from turning.
5. Clean the axle threads and lubricate them with a thread lubricant.
6. Remove the hub nut and washer.
7. Remove and support the caliper out of the way.
8. Remove the rotor.
9. Using puller J-28733, loosen the splined fit between the hub and shaft.

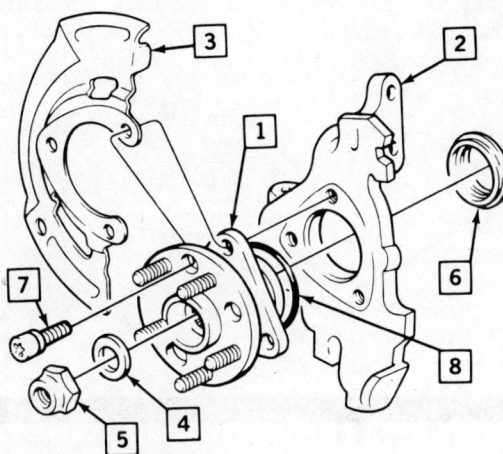

1. Hub and bearing assembly
2. Steering knuckle
3. Shield
4. Washer
5. Hub nut 245 N·m (180 ft. lbs.)
6. Seal
7. Hub and bearing retaining bolt (55 torx) 95 N·m (70 ft. lbs.)
8. "O" ring

Front wheel bearing

10. Remove the three hub attaching bolts, shield, hub and bearing assembly, and O-ring.

NOTE: *The hub and bearing are serviced as an assembly only.*

11. Remove the bearing seal from the knuckle.

12. Installation is the reverse of removal. Use a new O-ring and bearing seal. Lubricate the new bearing seal and the bearing with wheel bearing grease. Tighten the hub bolts to 70 ft.lb.; the caliper bolts to 28 ft.lb.; the hub nut to 181 ft.lb.

FRONT SUSPENSION

MacPherson Strut
REMOVAL AND INSTALLATION

1. Remove the three nuts attaching the top of the strut assembly to the body.

2. Raise the car and support it with jack stands under the engine cradle.

3. Lower the car slightly so that the weight rests on the jackstands.

4. Remove the wheels and tires.

NOTE: *Always install drive axle boot seal protectors. Care must be taken to prevent overextension of the inner Tri-Pot joints.*

5. Remove the brake line bracket bolt from the strut assembly. Do not disconnect the brake line from the caliper.

6. Remove the strut-to-steering knuckle bolts and then carefully remove the strut assembly.

7. Installation is in the reverse order of removal. Please note the following:

   a. Check wheel alignment

   b. Tighten the strut-to-body bolts to 18 ft. lbs.

   c. Tighten the strut-to-steering knuckle bolts to 144 ft. lbs.

Front Wheel Bearings
ADJUSTMENT

All models covered in this utilize a permanently sealed and lubricated front wheel bearing assembly. No adjustments are either necessary or possible.

Ball Joints
INSPECTION

1. Raise the front of the car with a lift placed under the engine cradle. The front wheel should be clear of the ground.

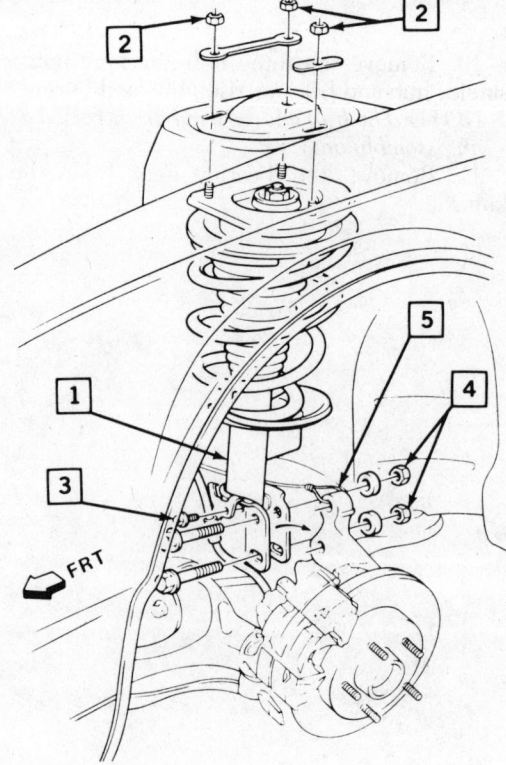

1. Strut assembly
2. Strut to body nuts 24 N·m (18 ft. lbs.)
3. Brake line bracket bolt 17 N·m (13 ft. lbs.)
4. Strut to steering knuckle nuts 195 N·m (144 ft. lbs.)
5. Retain steering knuckle with wire once strut assembly is removed

Front strut removal or installation

2. Grasp the wheel at the top and bottom and shake the wheel in and out.

3. If any movement is seen of the steering knuckle relative to the control arm, the ball joints are defective and must be replaced. Note that movement elsewhere may be due to loose

wheel bearings or other troubles; watch the knuckle-to-control arm connection.

4. If the ball stud is disconnected from the steering knuckle and any looseness is noted, often the ball joint stud can be twisted in its socket with your fingers, replace the ball joints.

REMOVAL AND INSTALLATION

1. Raise the front of the car and support it with jackstands underneath the engine cradle. Lower the car slightly so that the weight rests primarily on the jack stands.

2. Remove the wheel and tire assemblies.

3. Install drive axle covers to protect the drive axle boot seals.

4. Pull the cotter pin from the ball joint and install a ball joint separator tool. Turn the castellated nut counterclockwise to separate the ball joint from the steering knuckle.

5. Use a ⅛ in. drill bit to drill a hole approximately ¼ in. deep in the center of each of the three ball joint rivets.

6. Use a ½ in. drill bit to drill off the rivet heads. Drill only enough to remove the rivet head.

7. Use a hammer and punch to remove the rivets. Drive them out from the bottom.

8. Loosen the stabilizer bar bushing assembly nut.

9. Pull down on the control arm and remove the ball joint from the steering knuckle and control arm.

10. Install the new ball joint in the steering

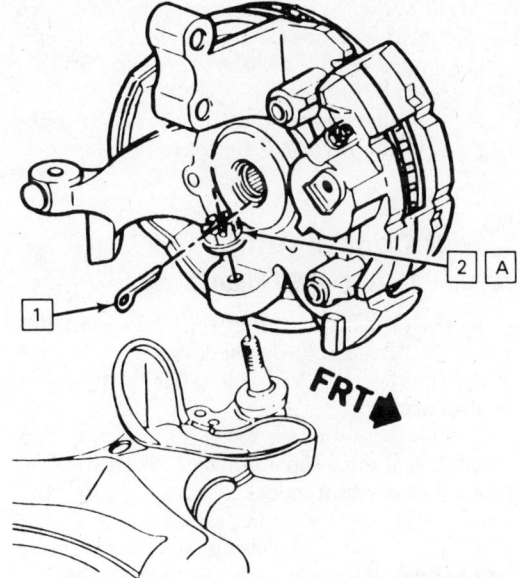

1. Cotter pin
2. Nut to be tightened to 10 N·m (88 lb. in.) then tightened an additional 120° (2 flats on nut) during which a torque of 50 N·m (37 ft. lbs.) must be attained

Ball joint removal

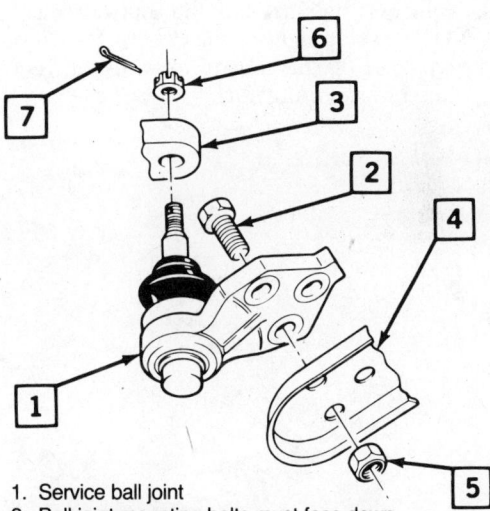

1. Service ball joint
2. Ball joint mounting bolts must face down
3. Steering knuckle
4. Control arm
5. Ball joint mounting nuts 68 N·m (50 ft. lbs.)
6. Ball joint to steering knuckle nut 110 N·m (81 ft. lbs.) before cotter pin installation
7. Cotter pin

Ball joint replacement

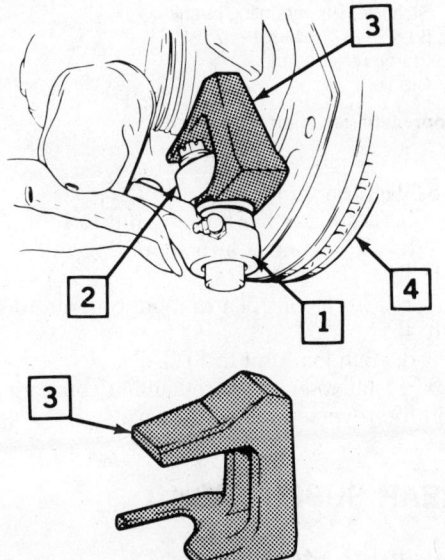

1. Ball joint
2. Steering knuckle
3. Ball joint seperator J-34505
4. Rotor

Separating ball joint from knuckle

knuckle and line up the holes with those in the control arm.

11. Install the three ball joint nuts facing down and tighten the nuts to 50 ft. lbs.

12. Install the castellated nut and tighten to 81 ft. lbs.

NOTE: *Tightening the nut for cotter pin alignment is allowed, but do not loosen it once the torque value has been reached.*

13. Install the cotter pin.

14. Installation of the remaining components is in the reverse order of removal.

Lower Control Arm

REMOVAL AND INSTALLATION

1. Perform Steps 1–3 of the "Ball Joint Removal and Installation" procedure.

2. Remove the stabilizer bar bushing-to-control arm bolt.

3. Pull the cotter pin from the ball joint and install a ball joint separator tool. Turn the castellated nut counterclockwise to separate the ball joint from the steering knuckle.

4. Remove the remaining control arm bolts and remove the control arm from the vehicle.

5. Position the control arm and install the mounting bolts, but DO NOT tighten.

6. Install the stabilizer bar bushing assembly. Reconnect the ball joint to the steering knuckle.

7. Hoist the vehicle slightly so the weight of the vehicle is supported by the control arms.

NOTE: *The weight of the vehicle MUST be supported by the control arms when tightening the mounting nuts.*

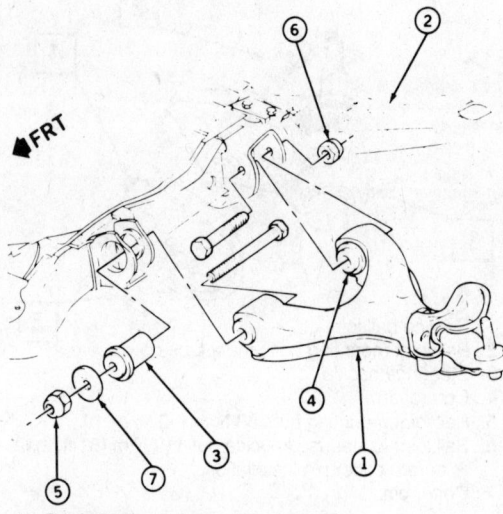

1. Control arm
2. Cradle
3. Cradle mounted bushing
4. Control arm mounted bushing
5. Cradle mounted bushing nut 190 N·m (140 ft. lbs.)
6. Control arm mounted bushing nut 123 N·m (90 ft. lbs.)
7. Washer

Control arm mounting

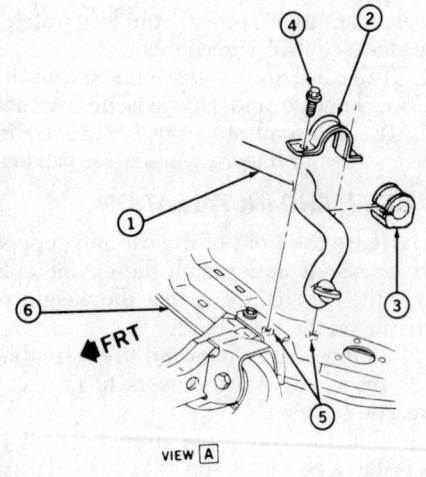

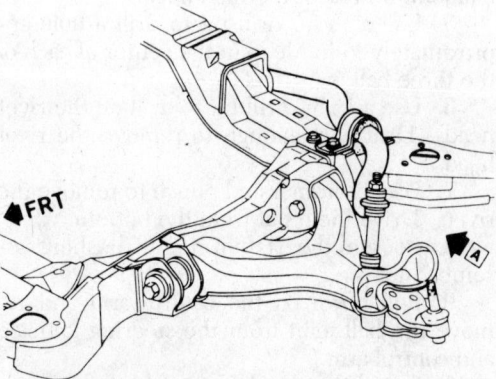

1. Stabilizer bar
2. Stabilizer bar mounting bracket
3. Stabilizer bar mounting bushing
4. 50 N·m (37 ft. lbs.)
5. Frame welded nuts
6. Cradle

Front stabilizer shaft

8. Tighten the:

  a. Stabilizer bar bushing nut to 13 ft. lbs.

  b. Rear control arm mounting nut to 90 ft. lbs.

  c. Front control arm mounting nut to 140 ft. lbs.

  d. Ball joint nut to 81 ft. lbs.

9. Installation of the remaining components is in the reverse order of removal.

REAR SUSPENSION

Superlift Strut

REMOVAL AND INSTALLATION

1. Remove the inner trunk side cover.

2. Raise and support the rear of the vehicle. Remove the wheels and tires.

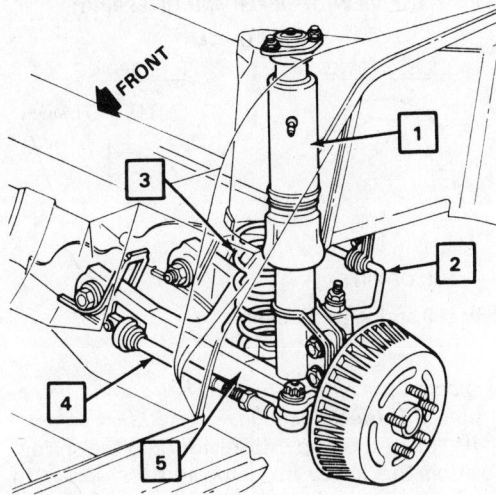

1. Superlift® strut
2. Stabilizer bar
3. Coil spring
4. Suspension adjustment link
5. Lower control arm

Rear suspension components

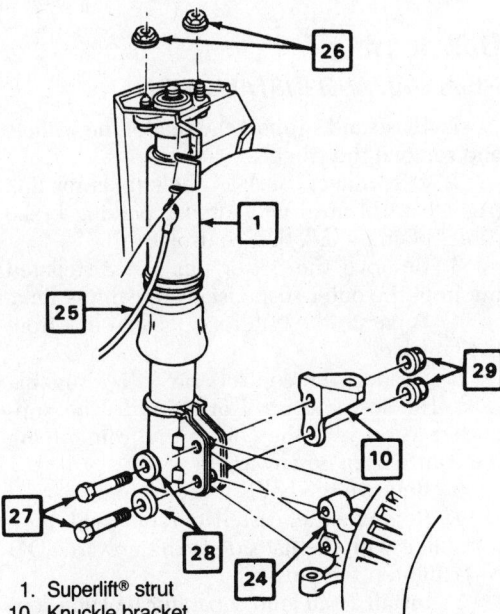

1. Superlift® strut
10. Knuckle bracket
24. Knuckle
25. ELC air line
26. Strut mounting nuts (25 N·m/19 ft. lbs.)
27. Strut anchor bolts
28. Strut anchor washers
29. Strut anchor nuts (195 N·m/144 ft. lbs.)

Superlift strut removal

3. Disconnect and plus the ELC air line.
4. Remove the strut tower mounting nuts from inside the trunk.
5. Remove the strut anchor bolts, washers

and nuts from the rear knuckle and knuckle bracket.

6. Remove the strut.
7. Installation is in the reverse order of removal. Please note the following:

a. Tighten the strut tower mounting nuts to 19 ft. lbs.

b. Tighten the strut anchor nuts to 144 ft. lbs.

c. Lightly pressurize the ELC system by momentarily grounding the compressor test lead in the engine compartment.

d. Check rear wheel alignment.

Coil Springs

REMOVAL AND INSTALLATION

1. Raise the rear of the vehicle and support it so that the control arms hang free. Remove the rear wheels.

2. Separate the rear stabilizer bar from the knuckle bracket and remove it.

3. Disconnect the ELC height sensor link (right control arm) and/ or the parking brake cable retaining clip (left control arm).

4. Position the special tool J-23028-01 or its equivalent, so as to cradle the control arm bushings.

NOTE: *Special tool J-23028-01 should be secured to a suitable jack.*

5. Raise the jack to remove the tension from the control arm pivot bolts.

CAUTION: *Secure a chain around the spring and through the control arm as a safety precaution.*

6. Remove the rear control arm pivot bolt and nut.

7. Slowly maneuver the jack so as to relieve any tension in the front control arm pivot bolt and then remove the bolt and nut.

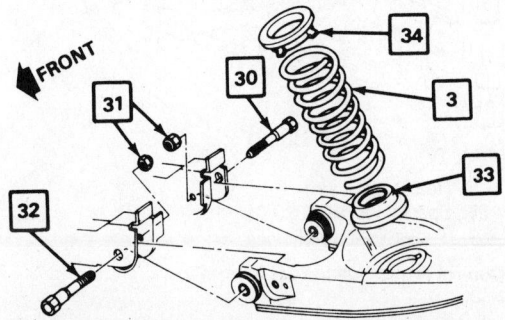

3. Coil spring
30. Control arm pivot bolt–rear (170 N·m/125 ft. lbs.)
31. Control arm pivot nuts (115 N·m/85 ft. lbs.)
32. Control arm pivot bolt–front (170 N·m/125 ft. lbs.)
33. Lower coil spring insulator
34. Upper coil spring insulator

Rear coil spring

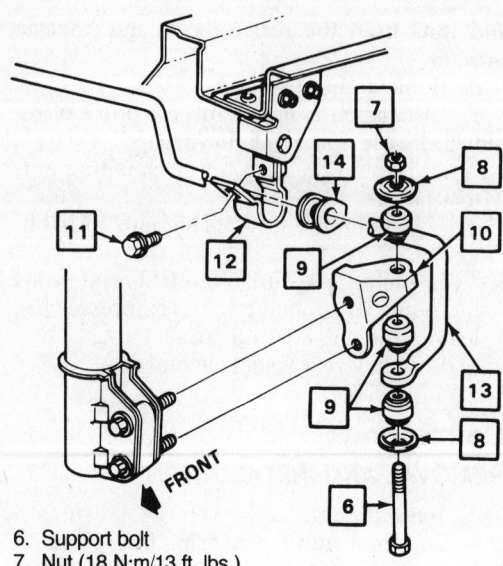

6. Support bolt
7. Nut (18 N·m/13 ft. lbs.)
8. Retainer
9. Insulators
10. Knuckle bracket
11. Bushing clip bolt (50 N·m/37 ft. lbs.)
12. Support assembly
13. Stabilizer bar
14. Bushing

Rear stabilizer bar

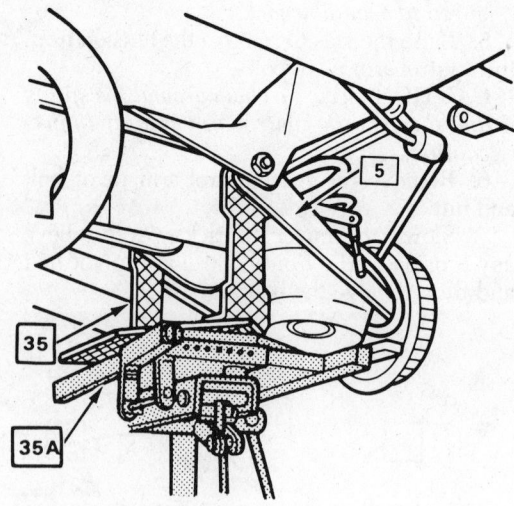

5. Rear control arm
35. Special tool J-23028-01
35A. Transmission jack

Control arm cradling tool in position

8. Lower the jack to allow the control arm to pivot downward.

9. When all pressure is removed from the coil spring, remove the safety chain, spring and insulators.

NOTE: *The spring insulators should be in-*

TOP VIEW OF UPPER END OF SPRING

FRONT OF CAR

L.H. SPRING   R.H. SPRING

END OF PIGTAIL

END OF PIGTAIL

Rear coil spring positioning

spected for cuts or tears. They should be replaced if the vehicle has over 50,000 miles.

10. Snap the upper insulator onto the spring. Position the lower insulator and the spring in the control arm. Install the coil springs so that the upper ends are positioned as shown in the illustration.

11. Installation of the remaining components is in the reverse order of removal. Control arm mounting nuts should not be tightened until the vehicle is unsupported and resting on its wheels at normal trim height.

Ball Joint

REMOVAL AND INSTALLATION

1. Raise and support the rear of the vehicle and remove the wheels.

2. Disconnect the ELC height sensor link (right control arm) and/ or the parking brake cable retaining link (left control arm).

3. Remove the cotter pin and castellated nut from the outer suspension adjustment link.

4. Separate the outer suspension link from the knuckle.

5. Support the control arm with a suitable jack. The lower control arm MUST be supported to prevent the coil spring from forcing the control arm downward.

6. Remove the ball stud cotter pin.

7. Remove the castellated nut and then reinstall it with the flat side facing upward. DO NOT tighten the nut.

8. Install a ball joint separator tool and separate the knuckle from the ball stud by backing off the inverted nut against the tool.

9. Separate the ball joint from the control arm.

10. Installation is in the reverse order of removal. Please note following:

   a. Tighten a NEW castellated nut to 7.5 ft. lbs. Tighten the nut an additional ⅔ of a turn.

   b. Align the slot in the nut to the cotter pin hole by tightening only. Do not loosen the nut to align the holes.

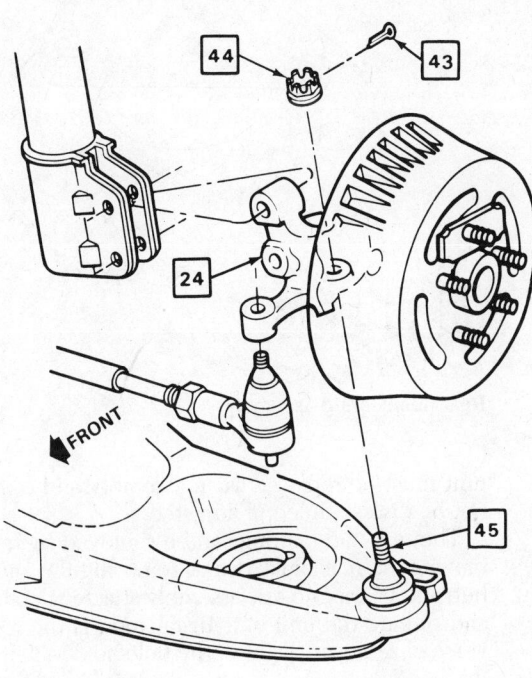

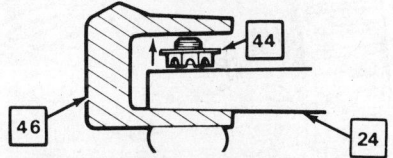

REMOVE CASTELLATED NUT AND REINSTALL WITH FLAT SIDE FACING UPWARD.

PLACE J-34505 INTO POSITION AS SHOWN. LOOSEN NUT AND BACK OFF UNTIL . . .

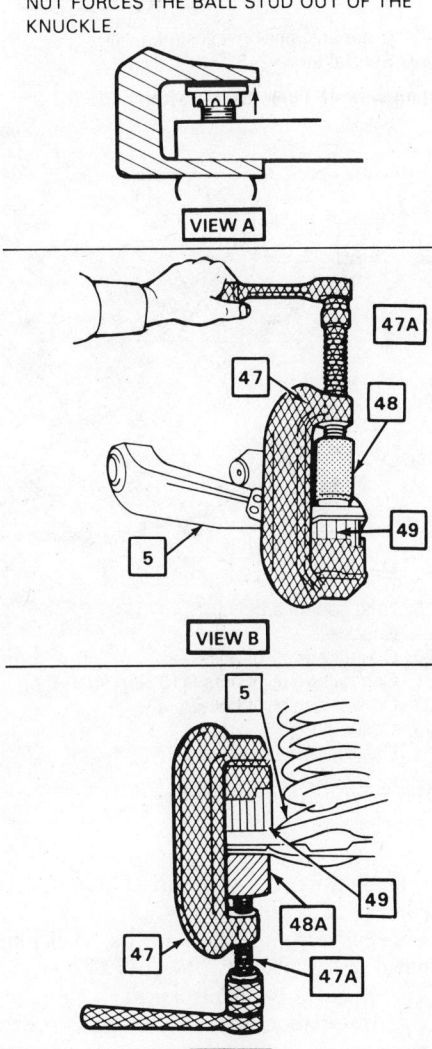

. . . THE NUT CONTACTS THE TOOL. CONTINUE BACKING OFF THE NUT UNTIL THE NUT FORCES THE BALL STUD OUT OF THE KNUCKLE.

VIEW A

VIEW B

VIEW C

5. Rear control arm
24. Knuckle
43. Cotter pin
44. Castellated nut
45. Ball joint
46. Special tool J-34505
47. Special tool J-9519-23 (clamp)
47A. Special tool J-9519-18 (screw)
48. Special tool J-9519-7
48A. Special tool J-9519-16
49. Special tool J-9519-17

Removing or installing the rear suspension ball joint

Control Arm
REMOVAL AND INSTALLATION

1. Perform Steps 1–2 of the "Ball Joint Removal & Installation" procedures.

2. Remove the suspension adjustment link retaining nut and retainer.

3. Separate the link assembly from the control arm.

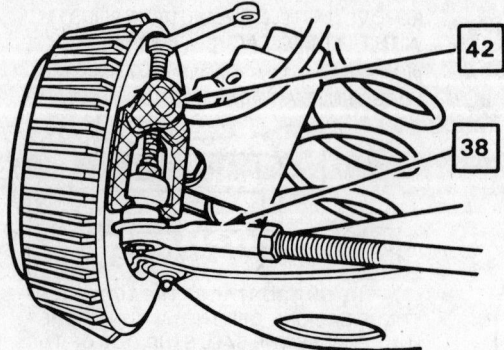

38. Outer suspension adjustment link
42. Special tool J-24319-01

Removing the link from the knuckle

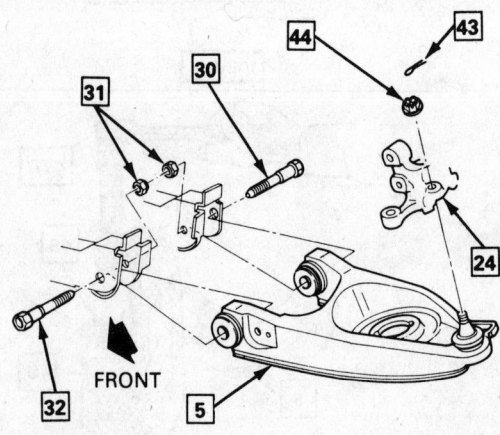

5. Rear control arm
24. Knuckle
30. Control arm pivot bolt—rear
31. Control arm pivot nuts (115 N·m/85 ft. lbs.)
32. Control arm pivot bolt—front
43. Cotter pin
44. Castellated nut

Rear control arm attachment

4. Remove the coil spring as detailed previously.

5. Perform Steps 6–9 of the "Ball Joint Removal & Installation" procedure.

6. Remove the control arm.

7. Installation is in the reverse order of removal.

Rear Axle Hub
REMOVAL AND INSTALLATION

A single unit hub and bearing assembly is bolted to both ends of the rear axle assembly. These take the place of "rear axles" used on rear wheel drive cars. The hub and bearing assembly is a sealed unit which requires no maintenance. The

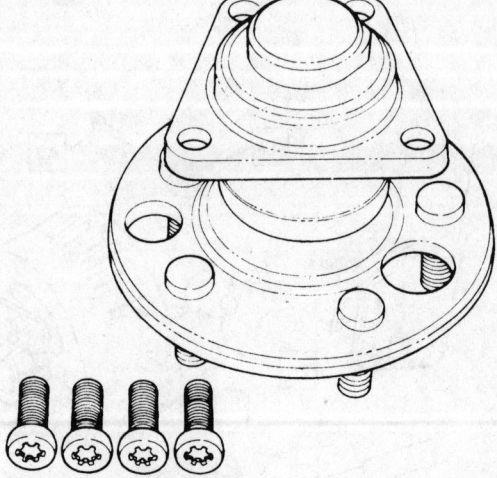

Rear hub/bearing assembly

unit must be replaced as an assembly and cannot be disassembled or adjusted.

The hub and bearing can be removed by removing the rear brake drum, removing the four hub and bearing-to-axle assembly attaching bolts and pulling the unit out. Installation is the reverse of removal. Tighten the bolts to 35–39 ft. lbs.

STEERING

Tie Rod
REMOVAL AND INSTALLATION

1. Loosen the jam nut on the steering rack (inner tie rod).

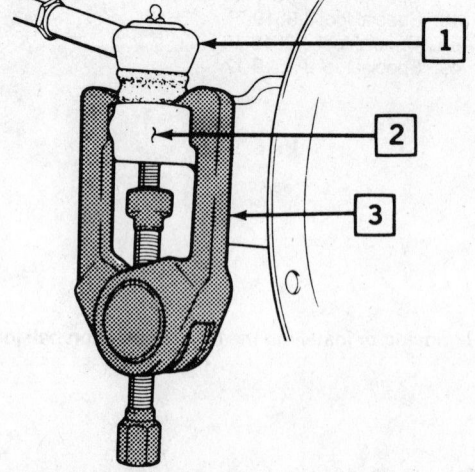

1. Tie rod end
2. Steering knuckle
3. Tie rod end puller J-24319-01

Separating tie rod end from the knuckle

2. Remove the tie rod end nut. Separate the tie rod end from the steering knuckle using a puller.

3. Unscrew the tie rod end, counting the number of turns.

4. To install, screw the tie rod end onto the steering rack (inner tie rod) the same number of turns as counted for removal. This will give approximately correct toe.

5. Install the tie rod end into the knuckle. Install the nut and tighten to 40 ft. lbs.

6. If the toe must be adjusted, use pliers to expand the boot clamp. Turn the inner tie rod to adjust. Replace the clamp.

7. Tighten the jam nut to 50 ft. lbs.

Rack and Pinion
REMOVAL AND INSTALLATION

1. Raise and support the front end of the car with jackstands under the frame members. Allow the front suspension to hang freely. Disconnect the power steering hoses from the gear, where equipped.

2. Move the intermediate shaft seal upward and remove the intermediate shaft-to-stub shaft pinch bolt.

3. Remove both front wheels.

4. Remove the cotter pins and nut from both tie rod ends. Disconnect the tie rod ends from the steering knuckles.

5. Remove the line retainer.

6. Remove the outlet and pressure hose.

7. Remove the five rack and pinion assembly mounting bolts.

8. Loosen the front engine cradle mounting bolts. Install jack stands and the lower the rear of the cradle about 3 in. (76mm).

CAUTION: *Do not lower the rear of the engine cradle too far.*

9. Remove the rack and pinion assembly.

10. Installation is in the reverse order of removal. Tighten the rack mounting bolts to 50 ft. lbs. Tighten the tie rod end nut to 35–52 ft. lbs. Bleed the power steering system and check for leaks.

Power Steering Pump
REMOVAL AND INSTALLATION
3.0L V6 and 3.8L V6

1. Disconnect the negative battery cable.

2. Remove the air cleaner assembly on the 3.0L.

3. Remove the drive belt and then the alternator itself.

4. Raise the front of the vehicle and support it on jack stands.

5. Disconnect and plug the pressure and return lines at the pump.

6. Remove the rear pump adjustment bracket-to-pump nut. Remove the power steering belt and lower the vehicle.

7. Remove the alternator adjustment bracket and support brace.

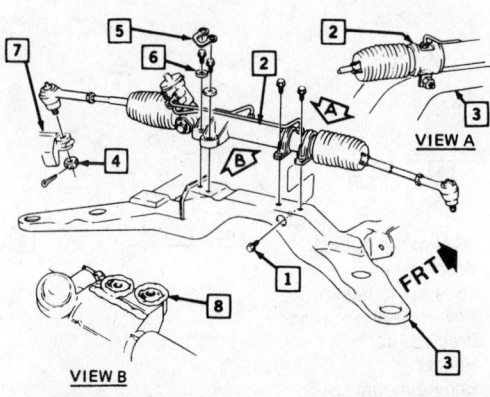

1. Bolt 68 N·m (50 ft. lbs.) after second reuse of bolt, locktite thread locking kit no. 1052624 must be used
2. Steering gear
3. Frame
4. 40 N·m (30 ft. lbs., 70 N·m (52 ft. lbs.) maximum permissible torque to align cotter pin slot. (⅙ turn maximum) Do not back off for cotter pin insertion
5. Retainer
6. Washer
7. Steering knuckle
8. RTV sealer around inserts

Rack and pinion removal or installation

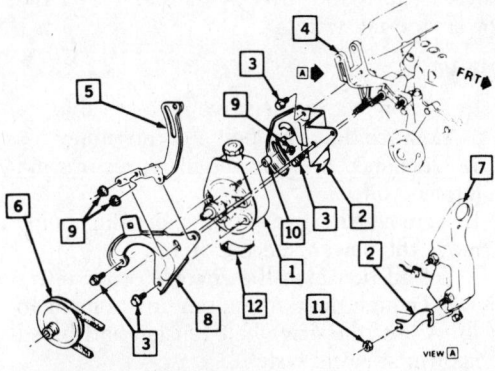

1. Power steering pump
2. Rear adj. bracket
3. Bolt–50 N·m (38 ft. lbs.)
4. Generator mounting bracket
5. Generator adjusting bracket
6. Pulley
7. Eng. lift bracket and shield
8. Front adj. bracket
9. Nut–50 N·m (38 ft. lbs.)
10. Rear bracket spacer
11. Nut–27 N·m (20 ft. lbs.)
12. Protector

Power steering pump mounting on the 3.0L or 3.8L V6

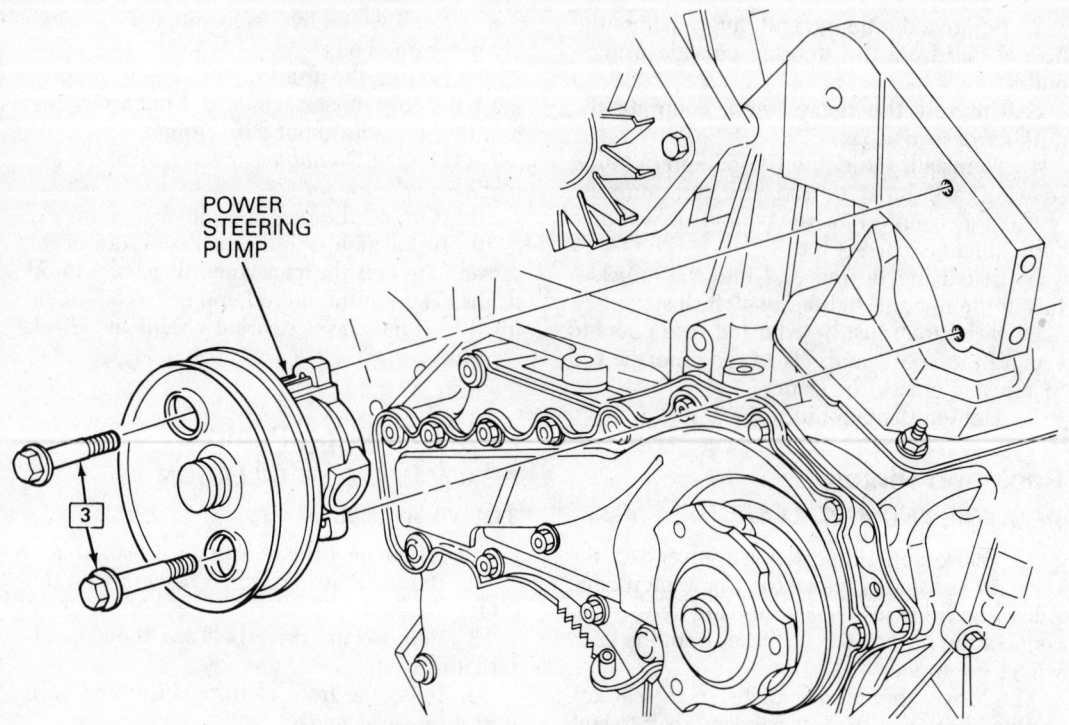

Power steering pump mounting on the 4.1L V8

8. Remove the rear pump adjustment bracket and then remove the pump assembly.

9. Remove the front pump adjustment bracket and then remove the pulley.

10. Installation is in the reverse order of removal. Adjust the drive belts and bleed the power steering system.

4.1L V8

1. Disconnect the negative battery cable.

2. Remove the drive belt and the pulley.

3. Disconnect and plus the high pressure and pump feed lines.

4. Remove the two pump mounting bolts. Remove the power steering pump.

5. Installation is in the reverse order of removal. Tighten the pump mounting bolts to 18 ft. lb. Adjust the drive belt tension and bleed the power steering system.

4.3L V6 Diesel

1. Disconnect the negative battery cable. Raise and support the front of the vehicle on jack stands.

2. Remove the engine splash shield. Remove the crankshaft pulley and the engine shock absorber.

3. Disconnect the reservoir hose from the power steering pump and drain the reservoir.

4. Remove the high pressure hose support. Disconnect and cap the high pressure hose.

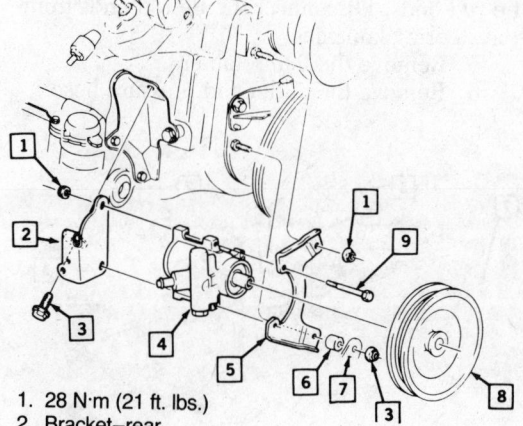

1. 28 N·m (21 ft. lbs.)
2. Bracket–rear
3. 55 N·m (40 ft. lbs.)
4. Power steering pump
5. Bracket–front
6. Spacer
7. Engine mount bracket
8. Pulley
9. Bolt (torque nut)

Power steering pump mounting on the 4.3L diesel with internal EGR

5. Remove the three bracket bolts. Remove the pump assembly along with its brackets.

6. Installation is in the reverse order of removal. Adjust the drive belt tension. Bleed the power steering system and check for leaks.

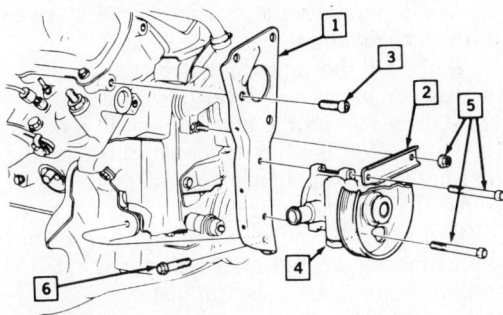

1. Support
2. Brace
3. 55 N·m (40 ft. lbs.)
4. Power steering pump
5. 28 N·m (21 ft. lbs.)
6. 45 N·m (34 ft. lbs.)

Power steering pump mounting on the 4.3L diesel with external EGR

BLEEDING THE POWER STEERING SYSTEM

1. Fill the fluid reservoir.

2. Let the fluid stand undisturbed for two minutes, then crank the engine for about two seconds. Refill reservoir if necessary.

3. Repeat Steps 1 and 2 above until the fluid level remains constant after cranking the engine.

4. Raise the front of the car until the wheels are off the ground, then start the engine. Increase the engine speed to about 1500 rpm.

5. Turn the wheels lightly against the stops to the left and right, checking the fluid level and refilling if necessary.

Steering Wheel

REMOVAL AND INSTALLATION

CAUTION: *Disconnect the battery ground cable before removing the steering wheel. When installing a steering wheel, always make sure that the turn signal lever is in the neutral position.*

1. Remove the trim retaining screws from behind the wheel. On wheels with a center cap, pull off the cap.

2. Lift the trim off and pull the horn wires from the turn signal cancelling cam.

3. Remove the retainer and the steering wheel nut.

4. Mark the wheel-to-shaft relationship, and then remove the wheel with a puller.

5. Install the wheel on the shaft aligning the previously made marks. Tighten the nut.

6. Insert the horn wires into the cancelling cam.

7. Install the center trim and reconnect the battery cable.

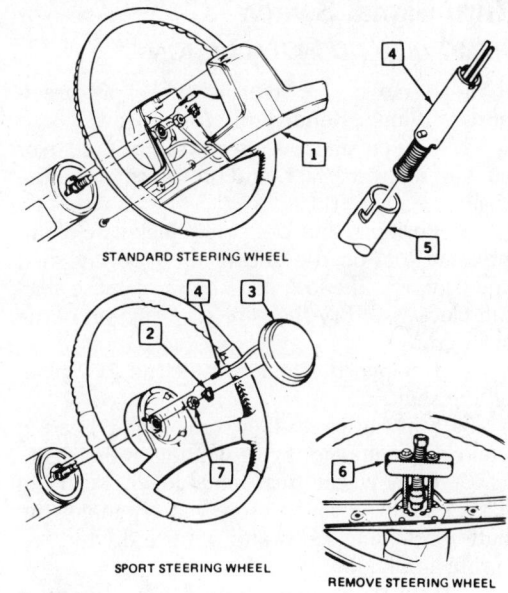

1. Pad
2. Retainer
3. Cap
4. Horn lead
5. Cam tower
6. J-1859-03 or BT-61-9
7. Nut–41 N·m (30 ft. lbs.)

Steering wheel mounting

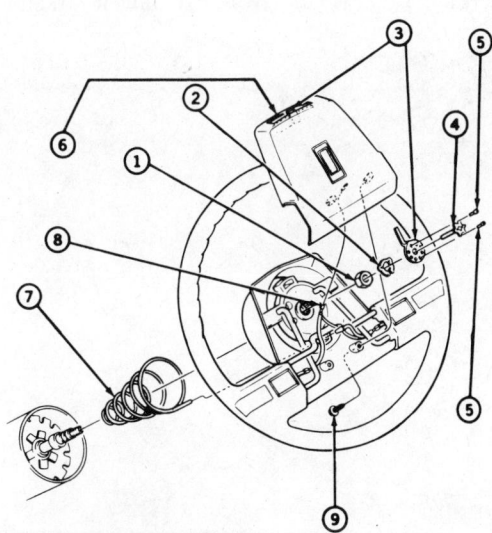

1. Steering wheel nut 41 N·m (30 ft. lbs.)
2. Steering wheel nut retainer
3. Telescoping adjuster lever
4. Steering shaft lock knob bolt
5. Steering shaft lock knob bolt positioning screw (2)
6. Steering wheel pad
7. Horn contact spring
8. Horn lead
9. Fully driven, seated and not stripped

Tilt wheel mounting

Turn Signal Switch
REMOVAL AND INSTALLATION

1. Remove the steering wheel as previously outlined. Remove the trim cover.

2. Loosen the cover screws. Pry the cover off with a screwdriver, and lift the cover off the shaft.

3. Position the U-shaped lockplate compressing tool on the end of the steering shaft and compress the lock plate by turning the shaft nut clockwise. Pry the wire snapring out of the shaft groove.

4. Remove the tool and lift the lock plate off the shaft.

5. Slip the cancelling cam, upper bearing preload spring, and thrust washer off the shaft.

6. Remove the turn signal lever. Push the flasher knob in and unscrew it. Remove the button retaining screw and remove the button, spring and knob.

7. Pull the switch connector out the mast jacket and tape the upper part to facilitate switch removal. Attach a long piece of wire to the turn signal switch connector. When installing the turn signal switch, feed this wire through the column first, and then use this wire to pull the switch connector into position. On tilt wheels, place the turn signal and shifter housing in low position and remove the harness cover.

8. Remove the three switch mounting screws. Remove the switch pulling it straight

up while guiding the wiring harness cover through the column.

9. Install the replacement switch by working the connector and cover down through the housing and under the bracket. On tilt models, the connector is worked down through the housing, under the bracket, and then the cover is installed on the harness.

10. Install the switch mounting screws and the connector on the mast jacket bracket. Install the column-to-dash trim plate.

11. Install the flasher know and the turn signal lever.

12. With the turn signal lever in neutral and the flasher knob out, slide the thrust washer, upper bearing preload spring, and cancelling cam onto the shaft.

13. Position the lock plate on the shaft and press it down until a new snapring can be inserted in the shaft groove. Always use a new snapring when assembling.

14. Install the cover and the steering wheel.

Ignition Switch
REMOVAL AND INSTALLATION

The switch is located on the upper side of the lower steering column area and is completely inaccessible without first lowering the steering column. The switch is actuated by a rod and rack assembly. A gear on the end of the lock cylinder engages the toothed upper end of the rod.

1. Lower the steering column; be sure to properly support it.

2. Disconnect the wiring from the switch.

3. Remove the two switch screws and remove the switch assembly.

4. Before installing, place the slider on the new switch in one of the following positions, depending on your steering column and accessories:

• Standard column with Key Release—extreme left detent

1. Turn signal lever
2. Insulator
3. Housing
4. Switch notch
5. Tang
6. Cruise control wiring

Multi-function switch removal

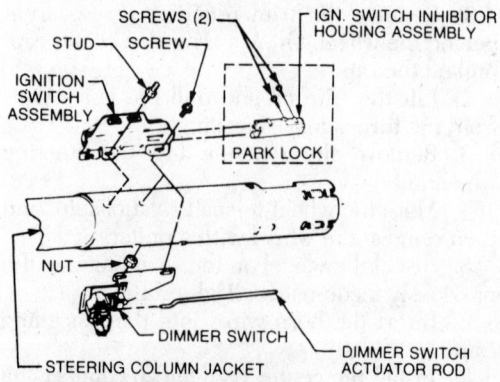

Typical ignition switch location

Wheel Alignment Specifications

| Model | Caster (deg.) | | Camber (deg.) | | Toe-in (in.) | Steering Axis Inclination (dg.) |
	Range	Pref.	Range	Pref.		
All	1° 48″P to 2° 48′P		2° 18″P	①	②	0

① Left wheel: 45″N to 15″N
 Right wheel: 15″P to 45″P
② Left wheel: 30″N
 Right wheel: 30″P

- Standard column with Park Lock—one dentent from extreme left
- All other standard columns—two detents from extreme left
- Adjustable column with ket release—extreme right detent
- Adjustable column with Park Lock—one detent from extreme right
- All other adjustable columns—two detents from extreme right

5. Install the activating rod into the switch and assemble the switch on the column. Tighten the mounting screws. Use only the specified screws since overlength screws could impair the collapsibility of the column.

6. Reinstall the steering column.

Ignition Lock Cylinder
REMOVAL AND INSTALLATION

1. Place the lock in the RUN position.
2. Remove the lock plate, turn signal switch and buzzer switch.
3. Remove the screw and lock cylinder.
If the screw is dropped on removal, it could fall into the column, requiring complete disassembly to retrieve the screw.
4. Rotate the cylinder clockwise to align cylinder key with the keyway in the housing.
5. Push the lock all the way in.
6. Install the screw. Tighten the screw to 14 inch lbs. for adjustable columns and 25 inch lbs. for standard columns.

Brakes

BRAKES

Master Cylinder

REMOVAL AND INSTALLATION

1. Disconnect and plug hydraulic lines, and drain the cylinder.
2. Remove the attaching nuts and remove the master cylinder from the power booster unit.
3. Reverse to install. Bleed the system.

OVERHAUL

1. Remove the unit from the car
2. Empty any fluid from the reservoir.
3. Secure the unit in a vise by clamping on the mounting flange.
4. With a small prybar, lever the reservoir from the master cylinder body.
5. Remove the lockring while depressing the primary piston.

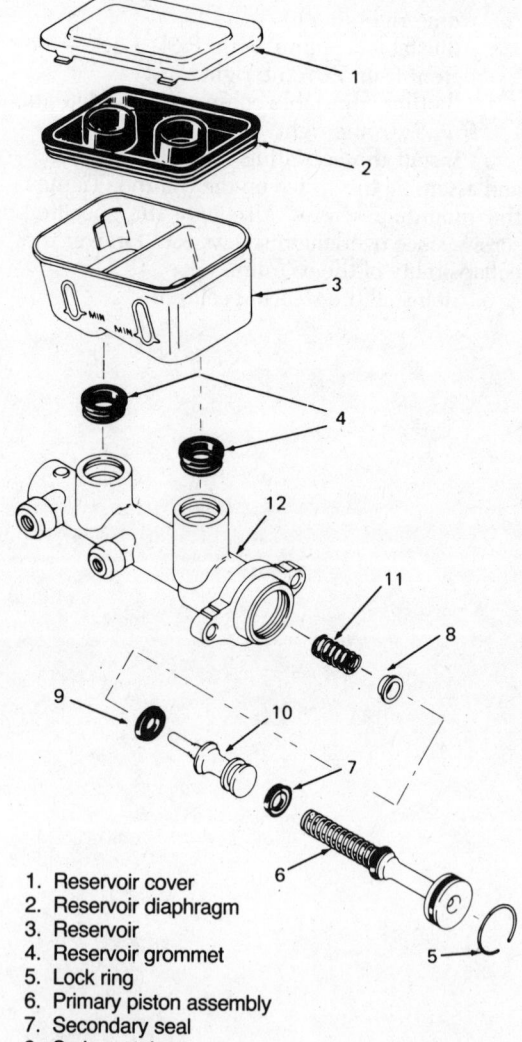

1. Reservoir cover
2. Reservoir diaphragm
3. Reservoir
4. Reservoir grommet
5. Lock ring
6. Primary piston assembly
7. Secondary seal
8. Spring retainer
9. Primary seal
10. Secondary piston
11. Spring
12. Cylinder body

Master cylinder, exploded view

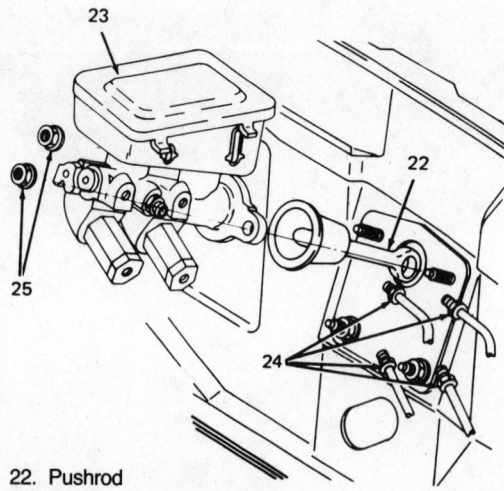

22. Pushrod
23. Master cylinder assembly
24. Tube nut
25. Nut

Master cylinder removal

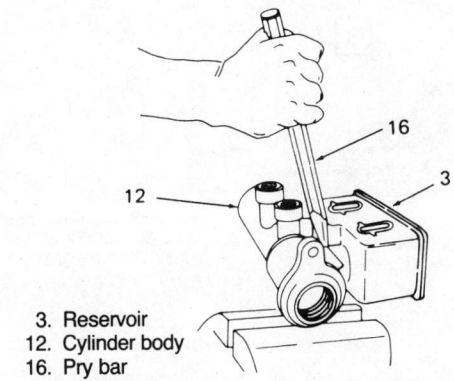

3. Reservoir
12. Cylinder body
16. Pry bar

Removing the reservoir

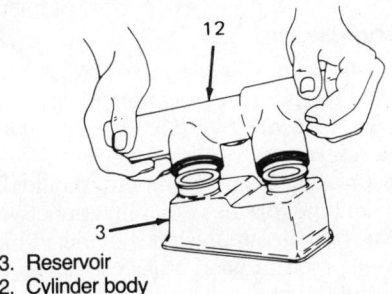

3. Reservoir
12. Cylinder body

Installing the reservoir

6. Using compressed air in the rear outlet port, force out the pistons, retainer and spring.

7. Wash all parts in denatured alcohol. Inspect all parts and replace any that are suspect. If any defect of any kind is found in the master cylinder bore, the master cylinder must be replaced. **THE MASTER CYLINDER CANNOT BE HONED. NO ABRASIVES ARE TO BE USED IN THE BORE!**

8. Assembly is the reverse of disassembly. Lubricate all rubber parts with clean brake fluid prior to assembly. Install the reservoir by pushing in with a rocking motion. Bleed the system.

Brake Bleeding

The hydraulic brake system must be free of air to operate properly. Air can enter the system when hydraulic parts are disconnected for servicing or replacement, or when the fluid level in the master cylinder reservoirs is very low. Air in the system will give the brake pedal a spongy feeling upon application.

The quickest and easiest of the two ways for system bleeding is the pressure method, but special equipment is needed to externally pressurize the hydraulic system. The other, more commonly used method of brake bleeding is done manually.

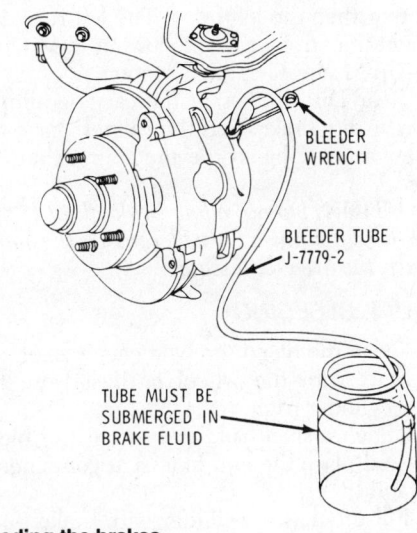

BLEEDER WRENCH

BLEEDER TUBE J-7779-2

TUBE MUST BE SUBMERGED IN BRAKE FLUID

Bleeding the brakes

BLEEDING SEQUENCE

Bleeding may be required at only one or two wheels or at the master cylinder, depending upon what point the system was opened to air. If after bleeding the cylinder caliper that was rebuilt or replaced and the pedal still has a spongy feeling upon application, it will be necessary to bleed the entire system. Bleed the system in the following order:

1. Master cylinder. If the cylinder is not equipped with bleeder screws, open the brake line(s) to the wheels slightly while pressure is applied to the brake pedal. Be sure to tighten the line before the brake pedal is released. The procedure for bench bleeding the master cylinder is covered below.

2. Power Brake Booster: If the unit is equipped with bleeder screws, it should be bled after the master cylinder. The car engine should be off and the brake pedal applied several times to exhaust any vacuum in the booster. If the unit is equipped with two bleeder screws, always bleed the higher one first.

3. Combination Valve: If equipped with a bleeder screw.

4. Front/Back Split Systems: Start with the wheel farthest away from the master cylinder, usually the right rear wheel. Bleed the other rear wheel then the right front and left front.

NOTE: *If you are unsuccessful in bleeding the front wheels, it me be necessary to deactivate the metering valve. This is accomplished by either pushing in, or pulling out a button or stem on the valve. The valve may be held by hand, with a special tool or taped, it should remain deactivated while the front brakes are bled.*

5. Diagonally Split System: Start with the

right rear then the left front. The left rear then the right front (refer to the following "GM Quick Take-Up Master Cylinder" section).

6. Rear Disc Brakes: If the car is equipped with rear disc brakes and the calipers have two bleeder screws, bleed the inner first then the outer.

CAUTION: *Do not allow brake fluid to spill on the car's finish, it will remove the paint. Flush the area with water.*

MANUAL BLEEDING

1. Clean the bleed screw at each wheel.
2. Start with the wheel farthest from the master cylinder (right rear).
3. Attach a small rubber hose to the bleed screw and place the end in a clear container of brake fluid.
4. Fill the master cylinder with brake fluid. (Check often during bleeding). Have an assistant slowly pump up the brake pedal and hold pressure.
5. Open the bleed screw about one-quarter turn, press the brake pedal to the floor, close the bleed screw and slowly release the pedal. Continue until no more air bubbles are forced from the cylinder on application of the brake pedal.
6. Repeat procedure on remaining wheel cylinders and calipers, still working from cylinder/caliper farthest from the master cylinder. Master cylinders equipped with bleed screws may be bled independently. When bleeding the Bendix-type dual master cylinder it is necessary to solidly cap one reservoir section while bleeding the other to prevent pressure loss through the cap vent hole.

CAUTION: *The bleeder valve at the wheel cylinder must be closed at the end of each stroke, and before the brake pedal is released, to insure that no air can enter the system. It is also important that the brake pedal be returned to the full up position so the piston in the master cylinder moves back enough to clear the bypass outlets.*

PRESSURE BLEEDING DISC BRAKES

Pressure bleeding disc brakes will close the metering valve and the front brakes will not bleed. For this reason it is necessary to manually hold the metering valve open during pressure bleeding. Never use a block or clamp to hold the valve open, and never force the valve stem beyond its normal position. Two different types of valves are used. The most common type requires the valve stem to be held in while bleeding the brakes, while the second type requires the valve stem to be held out (.060 in. minimum travel). Determine the type of visual inspection.

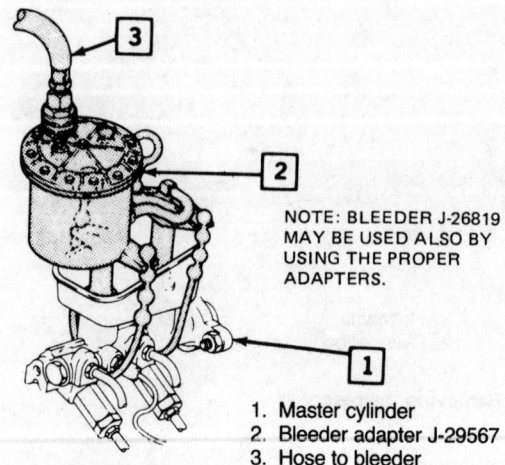

NOTE: BLEEDER J-26819 MAY BE USED ALSO BY USING THE PROPER ADAPTERS.

1. Master cylinder
2. Bleeder adapter J-29567
3. Hose to bleeder

Pressure bleeding

CAUTION: *Special adapters are required when pressure bleeding cylinders with plastic reservoirs.*

Pressure bleeding equipment should be diaphragm type; placing a diaphragm between the pressurized air supply and the brake fluid. This prevents moisture and other contaminants from entering the hydraulic system.

NOTE: *Front disc/rear drum equipped vehicles use a metering valve which closes off pressure to the front brakes under certain conditions. These systems contain manual release actuators, which must be engaged to pressure bleed the front brakes.*

1. Connect the tank hydraulic hose and adapter to the master cylinder.
2. Close hydraulic valve on the bleeder equipment.
3. Apply air pressure to the bleeder equipment

CAUTION: *Follow equipment manufacturer's recommendations for correct air pressure.*

4. Open the valve to bleed air out of the pressure hose to the master cylinder.

NOTE: *Never bleed this system using the secondary piston stopscrew on the bottom of many master cylinders.*

5. Open the hydraulic valve and bleed each wheel cylinder and caliper. Bleed rear brake system first when bleeding both front and rear systems.

FLUSHING HYDRAULIC BRAKE SYSTEMS

Hydraulic brake systems must be totally flushed if the fluid becomes contaminated with water, dirt or other corrosive chemicals. To flush, simply bleed the entire system until all fluid has been replaced with the correct type of new fluid.

BENCH BLEEDING MASTER CYLINDER

Bench bleeding the master cylinder before installing it on the car reduces the possibility of getting air into the lines.

1. Connect two short pieces of brake line to the outlet fittings, bend them until the free end is below the fluid level in the master cylinder reservoirs.

2. Fill the reservoirs with fresh brake fluid. Pump the piston until no more air bubbles appear in the reservoir(s).

3. Disconnect the two short lines, refill the master cylinder and securely install the cylinder cap(s).

4. Install the master cylinder on the car. Attach the lines but do not completely tighten them. Force any air that might have been trapped in the connection by slowly depressing the brake pedal. Tighten the lines before releasing the brake pedal.

GM QUICK TAKE-UP SYSTEM BLEEDING

Bleed the master cylinder as follows: disconnect the left front brake line at the master cylinder. Fill the cylinder with fluid until it flows from the opened port. Connect the line and tighten the fitting. Apply the brake pedal slowly one time and keep it applied. Loosen the same brake line fitting to allow any air to escape. Retighten the fitting and release the brake pedal slowly. Wait 15 seconds and repeat the procedure until all of the air is expelled. Bleed the right front connection in the same manner. Bleed the cylinders and calipers after you are sure all the air is out of the master cylinder.

CAUTION: *Rapid pumping will move the secondary piston down the bore and make it difficult to bleed the system. Always apply slow pedal pressure.*

Power Booster

REMOVAL AND INSTALLATION

1. From inside the car, detach the brake pushrod from the brake pedal.

2. Disconnect the hydraulic lines from the front of the master cylinder.

3. Remove the nuts from the mounting studs which hold the unit to the dash panel. Remove the unit and clean it prior to installation.

4. Install in reverse order of removal. Bleed system.

POWER BRAKE BOOSTER TROUBLESHOOTING

The following items are in addition to those listed in Chapter 9 "Troubleshooting". Check those items first.

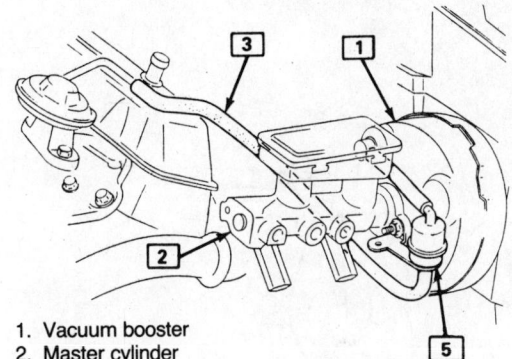

1. Vacuum booster
2. Master cylinder
3. Hose
5. Filter

Power booster and lines with a 3.0L engine

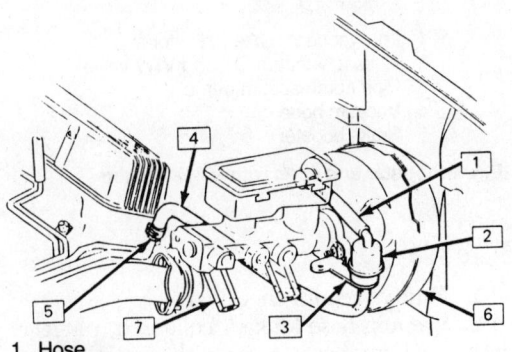

1. Hose
2. Filter asm
3. Support
4. Hose
5. Clamp
6. Booster asm
7 Master cylinder asm

Power booster and lines with a 3.8L engine

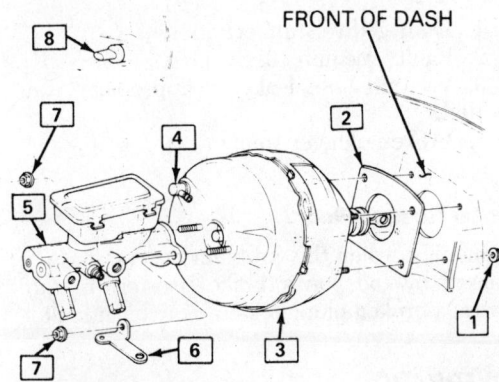

FRONT OF DASH

1. Nut (20 N·m/14 ft. lbs.)
2. Seal
3. Power booster
4. Check valve
5. Master cylinder
6. Vacuum switch bracket (diesel)
7. Nut (30 N·m/22 ft. lbs.)
8. Vacuum switch (gas)

Power booster and lines with a 4.1L engine

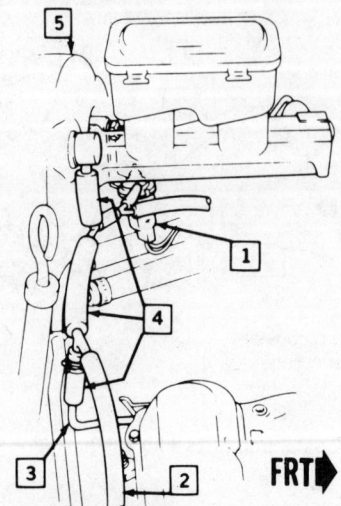

1. Low vacuum brake indicator switch
2. Hose to vacuum "T" and VRV valve
3. Pipe from vacuum pump
4. Vacuum hose
5. Brake booster

Power booster and lines with a 4.3L engine

Hard Pedal

1. Faulty vacuum check valve.
2. Vacuum hose kinked, collapsed, plugged leaky, or improperly connected.
3. Internal leak in unit.
4. Damaged vacuum cylinder.
5. Damaged valve plunger.
6. Broken or faulty springs.
7. Broken plunger stem.

Grabbing Brakes

1. Damaged vacuum cylinder.
2. Faulty vacuum check valve.
3. Vacuum hose leaky or improperly connected.
4. Broken plunger stem.

Pedal Goes to Floor

Generally, when this problem occurs, it is not caused by the power brake booster. In rare cases, a broken plunger stem may be at fault.

OVERHAUL

Most power brake boosters are serviced by replacement only. In many cases, repair parts are not available. A good many special tools are required for rebuilding these units. For these reasons, it would be most practical to replace a failed booster with a new or remanufactured unit.

FRONT DISC BRAKES

Brake Pads

REMOVAL AND INSTALLATION

1. Remove half of the brake fluid from the master cylinder.
2. Position a large C-clamp or pliers over the caliper with the screw end against the outboard brake pad. Tighten the clamp or pliers until the caliper is pushed out enough to bottom the piston.
3. Remove the C-clamp.
4. Remove the mounting pins, and lift off the caliper.
5. Unbend the outboard pad retaining tabs and lift out the pad.
6. Lift out the inboard pad.
7. Remove and discard the sleeves and bushings.
8. Lightly lubricate the new sleeves and bushings with silicone lubricant.
9. Installation is the reverse of removal. Make sure that the outboard pads are installed with the wear sensor at the leading edge of the pad. Bend the outboard pad tabs to retain the pad, using the method shown in the accompanying illustration. Torque the caliper mounting pins to 30-35 ft.lb.

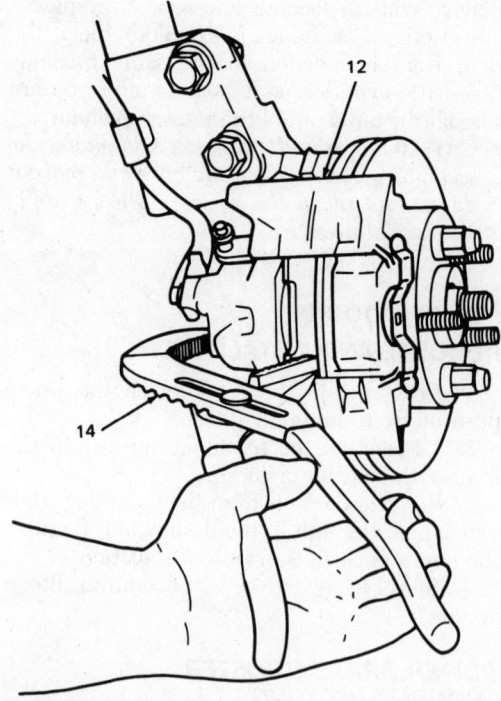

12. Caliper
14. Pliers

Compressing the piston

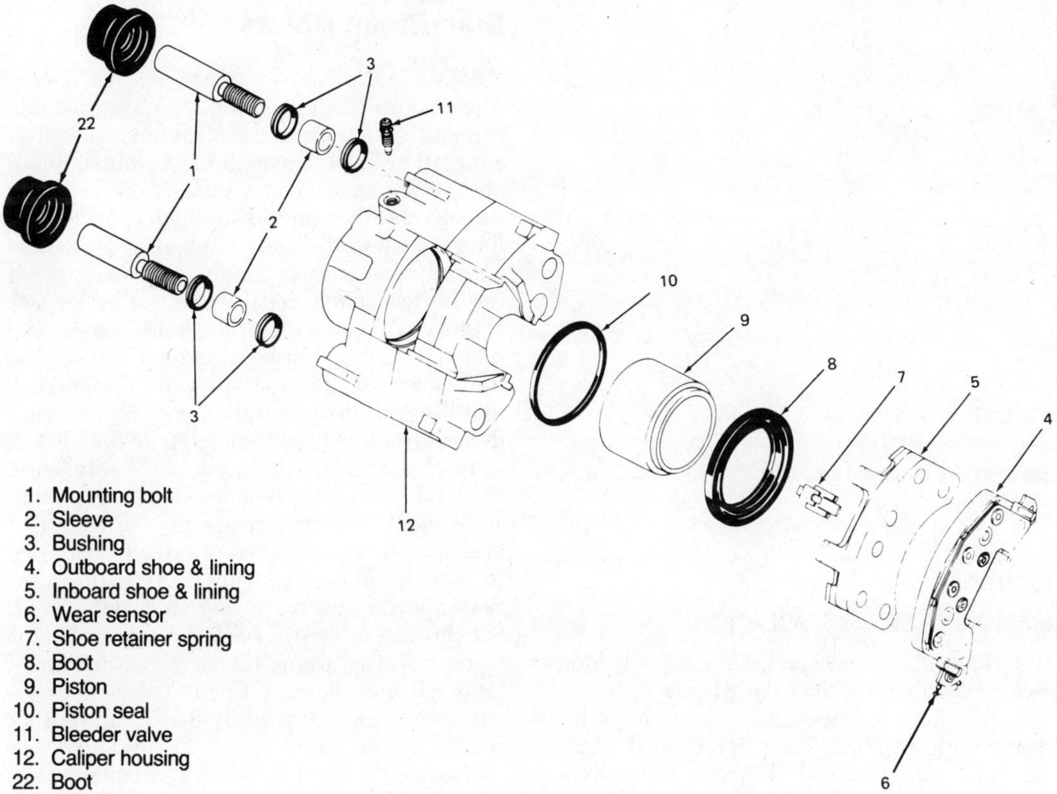

1. Mounting bolt
2. Sleeve
3. Bushing
4. Outboard shoe & lining
5. Inboard shoe & lining
6. Wear sensor
7. Shoe retainer spring
8. Boot
9. Piston
10. Piston seal
11. Bleeder valve
12. Caliper housing
22. Boot

Caliper exploded view

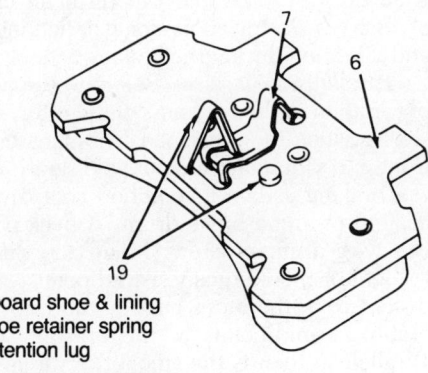

6. Inboard shoe & lining
7. Shoe retainer spring
19. Retention lug

Inboard pad and retainer

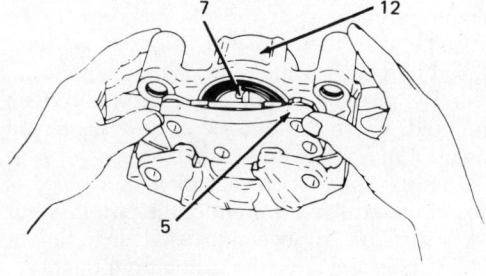

5. Inboard shoe & lining
7. Shoe retainer spring
12. Caliper housing

Installing the inboard pad

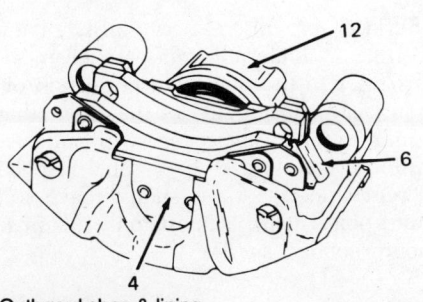

4. Outboard shoe & lining
6. Wear sensor
12. Caliper housing

Installing the outboard pad

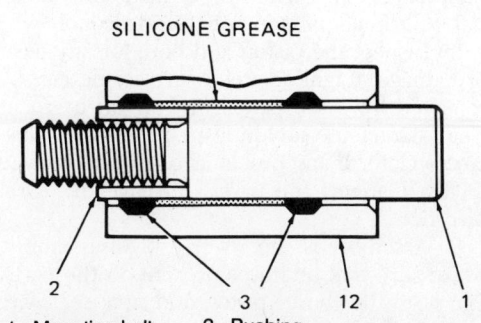

SILICONE GREASE

1. Mounting bolt   3. Bushing
2. Sleeve   12. Caliper housing

Lubrication points

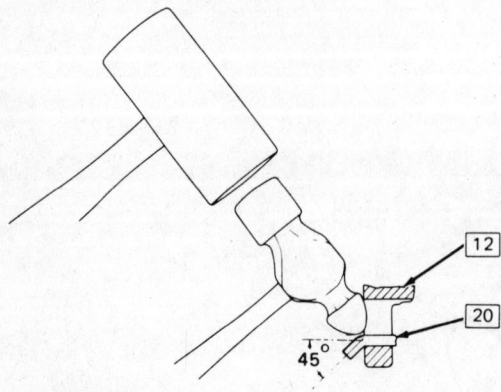

12. Caliper body
20. Outboard shoe tab

Bending the outboard pad tabs

Calipers

REMOVAL AND INSTALLATION

1. Raise the vehicle and safety support it on jackstands. Remove the front wheels.

2. Working on one side at a time only, disconnect the hydraulic inlet line from the caliper and plug the end. Remove the caliper mounting pins, and slide the caliper off the disc.

3. Remove the disc pads from the caliper or mounting adapter. If the old ones are to be reused, make them so that they can be reinstalled in their original positions.

4. Remove the caliper bleed screw and drain the fluid. Clean the outside of the caliper and mount it in a vise with padded jaws.

5. Place some shop rags or a thin piece of wood in the caliper, in front of the piston. Wear safety goggles. Apply compressed air to the inlet line hole and force the piston from the bore. **DON'T TRY TO CATCH THE PISTON WITH YOUR FINGERS! MASHED FINGERS COULD RESULT!**

6. Use a non-metallic device to pry the seal from the piston. Discard the seal and dust boot.

7. Clean all parts in denatured alcohol.

8. Inspect the piston and bore for any wear or damage. Light scratches or scoring can be removed from the piston or bore with crocus cloth. Don't use anything more abrasive than crocus cloth. If any problems cannot be reolved in this manner, the parts will have to be replaced.

9. Assembly is the reverse of disassembly. Make sure that no lint is present on the parts. Lubricate the bore, piston and new seal with clean brake fluid. Torque the bleeder screw to 110 in.lb. (9 ft.lb.). After caliper installation, bleed the brakes.

Disc Brake Rotors

RUNOUT

Manufacturers differ widely on permissible runout, but too much can sometimes be felt as a pulsation at the brake pedal. A wobble pump effect is created when a rotor is not perfectly smooth and the pad hits the high spots forcing fluid back into the master cylinder. This alternating pressure causes a pulsating feeling which can be felt at the pedal when the brakes are applied. This excessive runout also causes the brakes to be out of adjustment because disc brakes are self-adjusting; they are designed so that the pads drag on the rotor at all times and therefore automatically compensate for wear.

To check the actual runout of the rotor, first tighten the wheel spindle nut to a snug bearing adjustment, end-play removed. Fasten a dial indicator on the suspension at a convenient place so that the indicator stylus contacts the rotor face approximately one inch from its outer edge. Set the dial at zero. Check the total indicator reading while turning the rotor one full revolution. If the rotor is warped beyond the runout specification, it is likely that it can be successfully remachined.

"Lateral Runout": A wobbly movement of the rotor from side to side as it rotates. Excessive lateral runout causes the rotor faces to knock back the disc pads and can result in chatter, excessive pedal travel, pumping or fighting pedal and vibration during the braking action.

"Parallelism" (lack of): Refers to the amount of variation in the thickness of the rotor. Excessive variation can cause pedal vibration or fight, front end vibrations and possible "grab" during the braking action; a condition comparable to an "out-of-round brake drum." Check parallelism with a micrometer. "Mike" the thickness at eight or more equally spaced points, equally distant from the outer edge of the rotor, preferably at mid-points of the braking surface. Parallelism then is the amount of variation between maximum and minimum measurements.

"Surface or Micro-inch finish, flatness, smoothness": Different from parallelism, these terms refer to the degree of perfection of the flat surface on each side of the rotor; that is, the minute hills, valleys and swirls inherent in machining the surface. In a visual inspection, the remachined surface should have a find ground polish with, at most, only a faint trace of nondirectional swirls.

REMOVAL AND INSTALLATION

For rotor removal and installation, see Hub and Bearing Removal and Installation in Chapter 6.

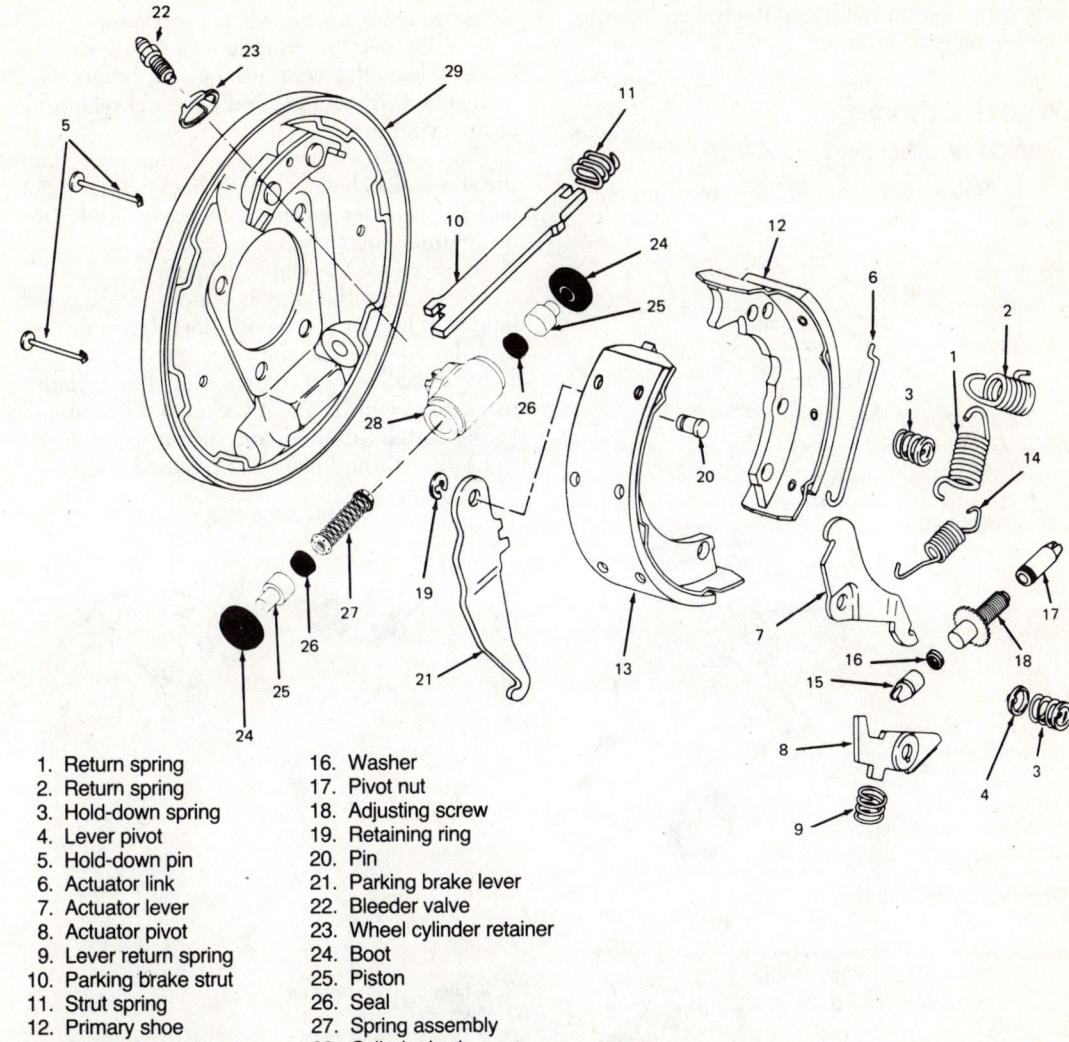

1. Return spring
2. Return spring
3. Hold-down spring
4. Lever pivot
5. Hold-down pin
6. Actuator link
7. Actuator lever
8. Actuator pivot
9. Lever return spring
10. Parking brake strut
11. Strut spring
12. Primary shoe
13. Secondary shoe
14. Adjusting screw spring
15. Socket
16. Washer
17. Pivot nut
18. Adjusting screw
19. Retaining ring
20. Pin
21. Parking brake lever
22. Bleeder valve
23. Wheel cylinder retainer
24. Boot
25. Piston
26. Seal
27. Spring assembly
28. Cylinder body
29. Backing plate

Drum brake components

DRUM BRAKES

Drum and Shoes

REMOVAL AND INSTALLATION

1. Raise and support the car safely on jack-stands.

2. Mark the relationship of the wheel-to-axle, and remove the wheels.

3. Mark the relationship of the drum-to-axle and remove the drum.

4. Using brake spring pliers, remove the return springs.

5. Using pliers, remove the holddown springs and lever pivot.

6. Remove the holddown pins.

7. Lift up on the actuating lever and remove the link.

8. Remove the actuator lever, pivot and return spring.

9. Remove the parking brake strut and spring.

10. Note the position of the adjusting spring, disconnect the parking brake cable and remove the brake shoes.

NOTE: *The left and right adjusters are not interchangeable.*

11. Inspect all brake parts. If any are questionable, replace them at this time.

12. Assembly is the reverse of disassembly. The threads on the adjusters must be cleaned and coated lightly and evenly with silicone brake

lubricant. Upon installation, the coil spring must not be over the adjuster.

Wheel Cylinders

REMOVAL AND INSTALLATION

1. Raise and support the rear on jack-stands.

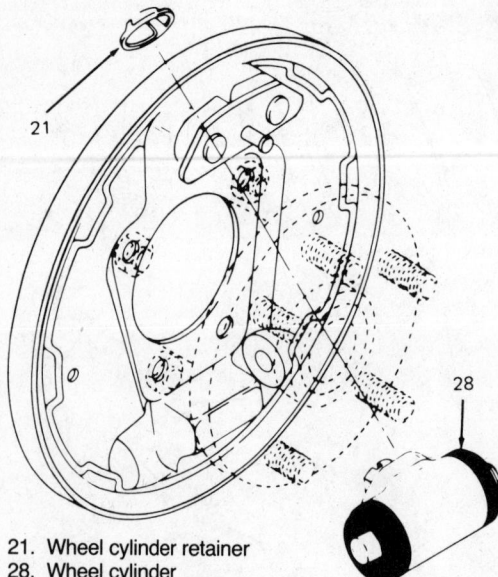

21. Wheel cylinder retainer
28. Wheel cylinder

Wheel cylinder mounting

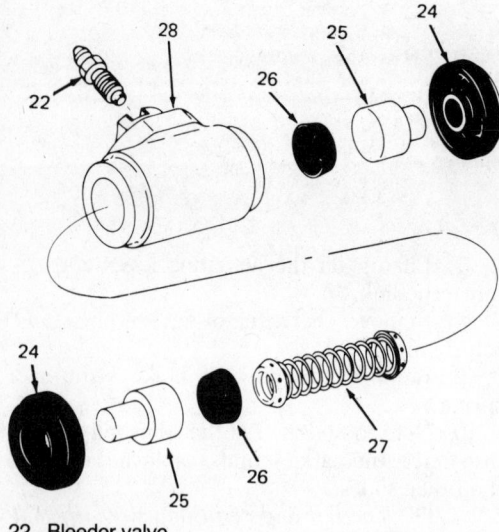

22. Bleeder valve
24. Boot
25. Piston
26. Seal
27. Spring assembly
28. Wheel cylinder body

Exploded view of a wheel cylinder

2. Matchmark the wheels and drum.
3. Remove the drum and brake shoes.
4. Clean the area around the brake tube connection, disconnect and cap the brake tube at the cylinder.
5. Insert two awls, 1/8 inch diameter, into the access slots bewteen the wheel cylinder pilot and the retainer locking tabs. Bend both tabs away simultaneously.
6. Remove the wheel cylinder.
7. Position the wheel cylinder in place and hold it there with a wood block between the cylinder and the flange.
8. Install a new retainer on the cylinder using a 1 1/8 inch 12 point socket and extension.
9. Install the shoes and drum.
10. Bleed the brakes as explained above.

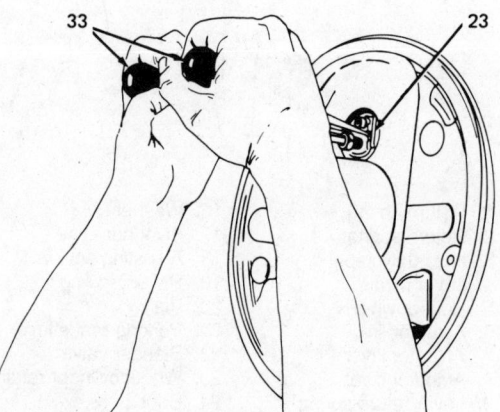

23. Wheel cylinder retainer
33. Awls

Removing wheel cylinder retainer

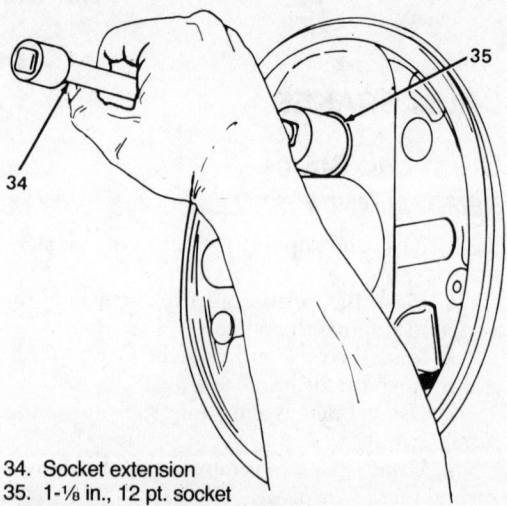

34. Socket extension
35. 1-1/8 in., 12 pt. socket

Installing wheel cylinder retainer

1. Right rear cable
2. Underbody
3. Guide
4. Bolt/screw 38 N·m (28 ft. lbs.)
5. Left rear cable
6. Cable asm–front
7. Cable asm–intermediate
8. Equalizer asm
9. Nut

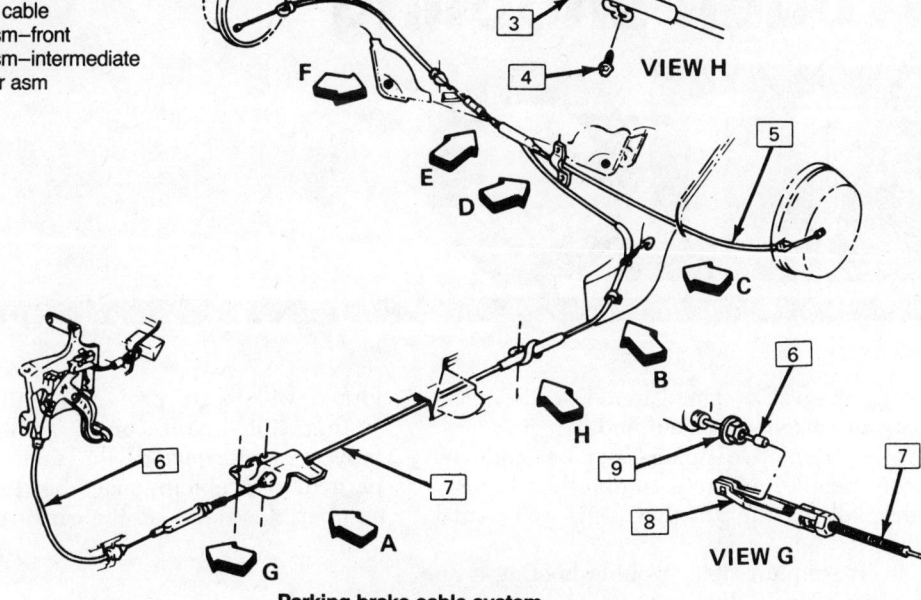

VIEW H

VIEW G

Parking brake cable system

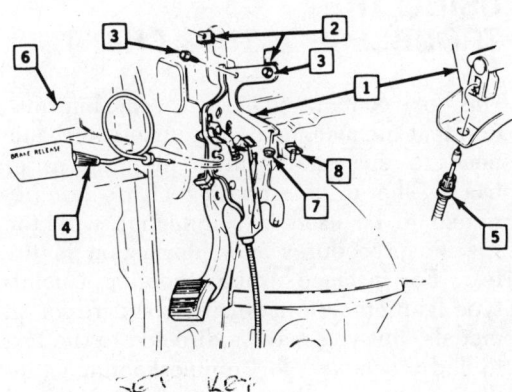

Parking brake lever assembly

Parking Brake

ADJUSTMENT

1. Depress the parking brake pedal exactly 3 ratchet clicks.

2. Raise the vehicle and support it with jackstands.

3. Check that the equalizer nut groove is liberally lubricated with chassis lube. Tighten the adjusting nut until the right rear wheel can just be turned to the rear with both hands, but is locked when forward rotation is attempted.

4. With the mechanism totally disengaged, both rear wheels should turn freely in either direction with no brake drag.

CAUTION: *Do not adjust the parking brake cable so tightly as to cause brake drag.*

5. Lower the vehicle.

Brake Specifications

Model	Master Cyl. Bore	Brake Disc			Brake Drum			Wheel Cyl. or Caliper Bore	
		Original Thickness	Minimum Thickness	Maximum Run-out	Orig. Inside Dia.	Max. Wear Limit	Maximum Machine O/S	Front	Rear
All	0.938	1.043	0.972	0.004	8.860	8.92	8.95	N.A.	①

① Except gasoline-engined station wagon: 0.74803 in. (19.0mm)
　Gasoline-engined station wagon: 0.811022 in. (20.6mm)

Troubleshooting

This section is designed to aid in the quick, accurate diagnosis of automotive problems. While automotive repairs can be made by many people, accurate troubleshooting is a rare skill for the amateur and professional alike.

In its simplest state, troubleshooting is an exercise in logic. It is essential to realize that an automobile is really composed of a series of systems. Some of these systems are interrelated; others are not. Automobiles operate within a framework of logical rules and physical laws, and the key to troubleshooting is a good understanding of all the automotive systems.

This section breaks the car or truck down into its component systems, allowing the problem to be isolated. The charts and diagnostic road maps list the most common problems and the most probable causes of trouble. Obviously it would be impossible to list every possible problem that could happen along with every possible cause, but it will locate MOST problems and eliminate a lot of unnecessary guesswork. The systematic format will locate problems within a given system, but, because many automotive systems are interrelated, the solution to your particular problem may be found in a number of systems on the car or truck.

USING THE TROUBLESHOOTING CHARTS

This book contains all of the specific information that the average do-it-yourself mechanic needs to repair and maintain his or her car or truck. The troubleshooting charts are designed to be used in conjunction with the specific procedures and information in the text. For instance, troubleshooting a point-type ignition system is fairly standard for all models, but you may be directed to the text to find procedures for troubleshooting an individual type of electronic ignition. You will also have to refer to the specification charts throughout the book for specifications applicable to your car or truck.

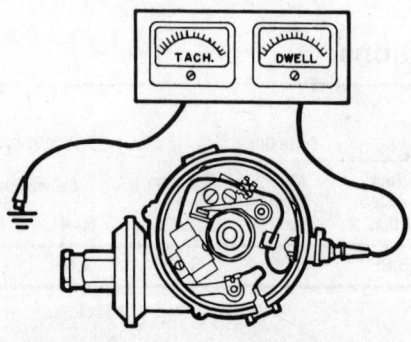

Tach-dwell hooked-up to distributor

TOOLS AND EQUIPMENT

The tools illustrated in Chapter 1 (plus two more diagnostic pieces) will be adequate to troubleshoot most problems. The two other tools needed are a voltmeter and an ohmmeter. These can be purchased separately or in combination, known as a VOM meter.

In the event that other tools are required, they will be noted in the procedures.

Troubleshooting Engine Problems

See Chapters 2, 3, 4 for more information and service procedures.

Index to Systems

System	To Test	Group
Battery	Engine need not be running	1
Starting system	Engine need not be running	2
Primary electrical system	Engine need not be running	3
Secondary electrical system	Engine need not be running	4
Fuel system	Engine need not be running	5
Engine compression	Engine need not be running	6
Engine vacuum	Engine must be running	7
Secondary electrical system	Engine must be running	8
Valve train	Engine must be running	9
Exhaust system	Engine must be running	10
Cooling system	Engine must be running	11
Engine lubrication	Engine must be running	12

Index to Problems

Problem: Symptom	Begin at Specific Diagnosis, Number
Engine Won't Start:	
Starter doesn't turn	1.1, 2.1
Starter turns, engine doesn't	2.1
Starter turns engine very slowly	1.1, 2.4
Starter turns engine normally	3.1, 4.1
Starter turns engine very quickly	6.1
Engine fires intermittently	4.1
Engine fires consistently	5.1, 6.1
Engine Runs Poorly:	
Hard starting	3.1, 4.1, 5.1, 8.1
Rough idle	4.1, 5.1, 8.1
Stalling	3.1, 4.1, 5.1, 8.1
Engine dies at high speeds	4.1, 5.1
Hesitation (on acceleration from standing stop)	5.1, 8.1
Poor pickup	4.1, 5.1, 8.1
Lack of power	3.1, 4.1, 5.1, 8.1
Backfire through the carburetor	4.1, 8.1, 9.1
Backfire through the exhaust	4.1, 8.1, 9.1
Blue exhaust gases	6.1, 7.1
Black exhaust gases	5.1
Running on (after the ignition is shut off)	3.1, 8.1
Susceptible to moisture	4.1
Engine misfires under load	4.1, 7.1, 8.4, 9.1
Engine misfires at speed	4.1, 8.4
Engine misfires at idle	3.1, 4.1, 5.1, 7.1, 8.4

Sample Section

Test and Procedure	Results and Indications	Proceed to
4.1—Check for spark: Hold each spark plug wire approximately ¼" from ground with gloves or a heavy, dry rag. Crank the engine and observe the spark.	→ If no spark is evident:	→4.2
	→ If spark is good in some cases:	→4.3
	→ If spark is good in all cases:	→4.6

Specific Diagnosis

This section is arranged so that following each test, instructions are given to proceed to another, until a problem is diagnosed.

Section 1—Battery

Test and Procedure	Results and Indications	Proceed to
1.1—Inspect the battery visually for case condition (corrosion, cracks) and water level.	If case is cracked, replace battery:	**1.4**
	If the case is intact, remove corrosion with a solution of baking soda and water (**CAUTION**: *do not get the solution into the battery*), and fill with water:	**1.2**

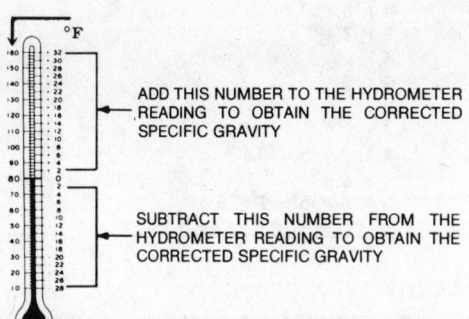
DIRT ON TOP OF BATTERY — PLUGGED VENT — CORROSION — LOOSE CABLE OR POSTS — CRACKS — LOW WATER LEVEL

Inspect the battery case

Test and Procedure	Results and Indications	Proceed to
1.2—Check the battery cable connections: Insert a screwdriver between the battery post and the cable clamp. Turn the headlights on high beam, and observe them as the screwdriver is gently twisted to ensure good metal to metal contact.	If the lights brighten, remove and clean the clamp and post; coat the post with petroleum jelly, install and tighten the clamp:	**1.4**
	If no improvement is noted:	**1.3**

TESTING BATTERY CABLE CONNECTIONS USING A SCREWDRIVER

| **1.3**—Test the state of charge of the battery using an individual cell tester or hydrometer. | If indicated, charge the battery. **NOTE:** *If no obvious reason exists for the low state of charge (i.e., battery age, prolonged storage), proceed to:* | **1.4** |

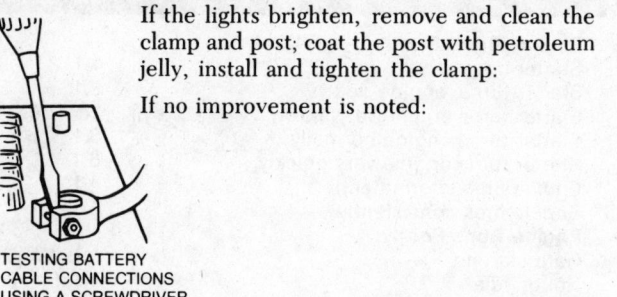

Specific Gravity (@ 80° F.)

Minimum	Battery Charge
1.260	100% Charged
1.230	75% Charged
1.200	50% Charged
1.170	25% Charged
1.140	Very Little Power Left
1.110	Completely Discharged

ADD THIS NUMBER TO THE HYDROMETER READING TO OBTAIN THE CORRECTED SPECIFIC GRAVITY

SUBTRACT THIS NUMBER FROM THE HYDROMETER READING TO OBTAIN THE CORRECTED SPECIFIC GRAVITY

The effects of temperature on battery specific gravity (left) and amount of battery charge in relation to specific gravity (right)

| **1.4**—Visually inspect battery cables for cracking, bad connection to ground, or bad connection to starter. | If necessary, tighten connections or replace the cables: | **2.1** |

Section 2—Starting System

See Chapter 3 for service procedures

Test and Procedure	Results and Indications	Proceed to
Note: Tests in Group 2 are performed with coil high tension lead disconnected to prevent accidental starting.		
2.1—Test the starter motor and solenoid: Connect a jumper from the battery post of the solenoid (or relay) to the starter post of the solenoid (or relay).	If starter turns the engine normally:	2.2
	If the starter buzzes, or turns the engine very slowly:	2.4
	If no response, replace the solenoid (or relay).	3.1
	If the starter turns, but the engine doesn't, ensure that the flywheel ring gear is intact. If the gear is undamaged, replace the starter drive.	3.1
2.2—Determine whether ignition override switches are functioning properly (clutch start switch, neutral safety switch), by connecting a jumper across the switch(es), and turning the ignition switch to "start".	If starter operates, adjust or replace switch:	3.1
	If the starter doesn't operate:	2.3
2.3—Check the ignition switch "start" position: Connect a 12V test lamp or voltmeter between the starter post of the solenoid (or relay) and ground. Turn the ignition switch to the "start" position, and jiggle the key.	If the lamp doesn't light or the meter needle doesn't move when the switch is turned, check the ignition switch for loose connections, cracked insulation, or broken wires. Repair or replace as necessary:	3.1
	If the lamp flickers or needle moves when the key is jiggled, replace the ignition switch.	3.3

Checking the ignition switch "start" position

STARTER RELAY
(IF EQUIPPED)

2.4—Remove and bench test the starter, according to specifications in the engine electrical section.	If the starter does not meet specifications, repair or replace as needed:	3.1
	If the starter is operating properly:	2.5
2.5—Determine whether the engine can turn freely: Remove the spark plugs, and check for water in the cylinders. Check for water on the dipstick, or oil in the radiator. Attempt to turn the engine using an 18″ flex drive and socket on the crankshaft pulley nut or bolt.	If the engine will turn freely only with the spark plugs out, and hydrostatic lock (water in the cylinders) is ruled out, check valve timing:	9.2
	If engine will not turn freely, and it is known that the clutch and transmission are free, the engine must be disassembled for further evaluation:	Chapter 3

Section 3—Primary Electrical System

Test and Procedure	Results and Indications	Proceed to
3.1—Check the ignition switch "on" position: Connect a jumper wire between the distributor side of the coil and ground, and a 12V test lamp between the switch side of the coil and ground. Remove the high tension lead from the coil. Turn the ignition switch on and jiggle the key.	If the lamp lights:	**3.2**
	If the lamp flickers when the key is jiggled, replace the ignition switch:	**3.3**
	If the lamp doesn't light, check for loose or open connections. If none are found, remove the ignition switch and check for continuity. If the switch is faulty, replace it:	**3.3**

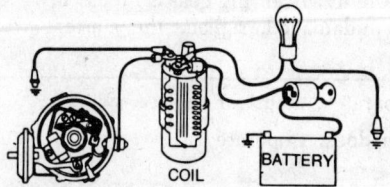

Checking the ignition switch "on" position

Test and Procedure	Results and Indications	Proceed to
3.2—Check the ballast resistor or resistance wire for an open circuit, using an ohmmeter. See Chapter 3 for specific tests.	Replace the resistor or resistance wire if the resistance is zero. **NOTE:** *Some ignition systems have no ballast resistor.*	**3.3**

 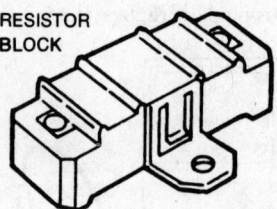

RESISTOR BLOCK

CALIBRATED RESISTANCE LEAD

Two types of resistors

Test and Procedure	Results and Indications	Proceed to
3.3—On point-type ignition systems, visually inspect the breaker points for burning, pitting or excessive wear. Gray coloring of the point contact surfaces is normal. Rotate the crankshaft until the contact heel rests on a high point of the distributor cam and adjust the point gap to specifications. On electronic ignition models, remove the distributor cap and visually inspect the armature. Ensure that the armature pin is in place, and that the armature is on tight and rotates when the engine is cranked. Make sure there are no cracks, chips or rounded edges on the armature.	If the breaker points are intact, clean the contact surfaces with fine emery cloth, and adjust the point gap to specifications. If the points are worn, replace them. On electronic systems, replace any parts which appear defective. If condition persists:	**3.4**

Test and Procedure	Results and Indications	Proceed to
3.4—On point-type ignition systems, connect a dwell-meter between the distributor primary lead and ground. Crank the engine and observe the point dwell angle. On electronic ignition systems, conduct a stator (magnetic pickup assembly) test. See Chapter 3.	On point-type systems, adjust the dwell angle if necessary. **NOTE:** *Increasing the point gap decreases the dwell angle and vice-versa.*	**3.6**
	If the dwell meter shows little or no reading;	**3.5**
	On electronic ignition systems, if the stator is bad, replace the stator. If the stator is good, proceed to the other tests in Chapter 3.	

WIDE GAP   NARROW GAP

CLOSE   OPEN

SMALL DWELL   LARGE DWELL

NORMAL DWELL   INSUFFICIENT DWELL   EXCESSIVE DWELL

Dwell is a function of point gap

3.5—On the point-type ignition systems, check the condenser for short: connect an ohmeter across the condenser body and the pigtail lead.	If any reading other than infinite is noted, replace the condenser	**3.6**

OHMMETER

Checking the condenser for short

3.6—Test the coil primary resistance: On point-type ignition systems, connect an ohmeter across the coil primary terminals, and read the resistance on the low scale. Note whether an external ballast resistor or resistance wire is used. On electronic ignition systems, test the coil primary resistance as in Chapter 3.	Point-type ignition coils utilizing ballast resistors or resistance wires should have approximately 1.0 ohms resistance. Coils with internal resistors should have approximately 4.0 ohms resistance. If values far from the above are noted, replace the coil.	**4.1**

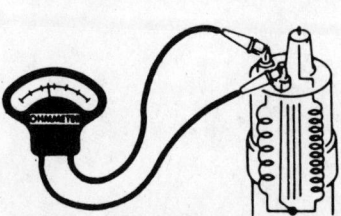

Check the coil primary resistance

Section 4—Secondary Electrical System

See Chapters 2–3 for service procedures

Test and Procedure	Results and Indications	Proceed to
4.1—Check for spark: Hold each spark plug wire approximately ¼″ from ground with gloves or a heavy, dry rag. Crank the engine, and observe the spark.	If no spark is evident:	4.2
	If spark is good in some cylinders:	4.3
	If spark is good in all cylinders:	4.6

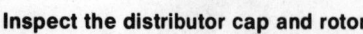

Check for spark at the plugs

Test and Procedure	Results and Indications	Proceed to
4.2—Check for spark at the coil high tension lead: Remove the coil high tension lead from the distributor and position it approximately ¼″ from ground. Crank the engine and observe spark. **CAUTION:** *This test should not be performed on engines equipped with electronic ignition.*	If the spark is good and consistent:	4.3
	If the spark is good but intermittent, test the primary electrical system starting at 3.3:	3.3
	If the spark is weak or non-existent, replace the coil high tension lead, clean and tighten all connections and retest. If no improvement is noted:	4.4
4.3—Visually inspect the distributor cap and rotor for burned or corroded contacts, cracks, carbon tracks, or moisture. Also check the fit of the rotor on the distributor shaft (where applicable).	If moisture is present, dry thoroughly, and retest per 4.1:	4.1
	If burned or excessively corroded contacts, cracks, or carbon tracks are noted, replace the defective part(s) and retest per 4.1:	4.1
	If the rotor and cap appear intact, or are only slightly corroded, clean the contacts thoroughly (including the cap towers and spark plug wire ends) and retest per 4.1:	
	If the spark is good in all cases:	4.6
	If the spark is poor in all cases:	4.5

CORRODED OR
LOOSE WIRE

EXCESSIVE WEAR
OF BUTTON

HIGH RESISTANCE
CARBON

ROTOR TIP
BURNED AWAY

Inspect the distributor cap and rotor

Test and Procedure	Results and Indications	Proceed to
4.4—Check the coil secondary resistance: On point-type systems connect an ohmmeter across the distributor side of the coil and the coil tower. Read the resistance on the high scale of the ohmmeter. On electronic ignition systems, see Chapter 3 for specific tests.	The resistance of a satisfactory coil should be between 4,000 and 10,000 ohms. If resistance is considerably higher (i.e., 40,000 ohms) replace the coil and retest per 4.1. **NOTE:** *This does not apply to high performance coils.*	

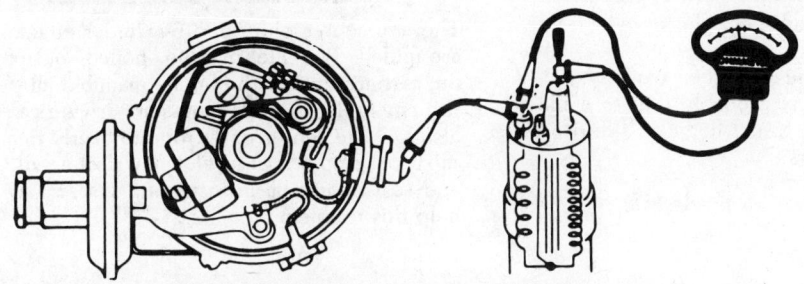

Testing the coil secondary resistance

4.5—Visually inspect the spark plug wires for cracking or brittleness. Ensure that no two wires are positioned so as to cause induction firing (adjacent and parallel). Remove each wire, one by one, and check resistance with an ohmmeter.	Replace any cracked or brittle wires. If any of the wires are defective, replace the entire set. Replace any wires with excessive resistance (over $8000\,\Omega$ per foot for suppression wire), and separate any wires that might cause induction firing.	**4.6**

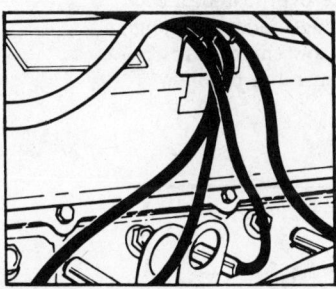

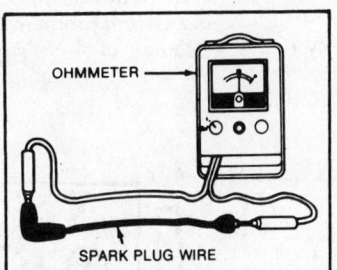

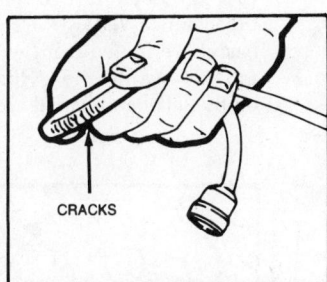

Misfiring can be the result of spark plug leads to adjacent, consecutively firing cylinders running parallel and too close together	On point-type ignition systems, check the spark plug wires as shown. On electronic ignitions, do not remove the wire from the distributor cap terminal; instead, test through the cap	Spark plug wires can be checked visually by bending them in a loop over your finger. This will reveal any cracks, burned or broken insulation. Any wire with cracked insulation should be replaced

4.6—Remove the spark plugs, noting the cylinders from which they were removed, and evaluate according to the color photos in the middle of this book.	See following.	**See following.**

Test and Procedure	Results and Indications	Proceed to
4.7—Examine the location of all the plugs.	The following diagrams illustrate some of the conditions that the location of plugs will reveal.	**4.8**

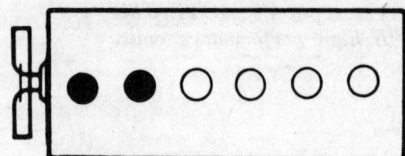

Two adjacent plugs are fouled in a 6-cylinder engine, 4-cylinder engine or either bank of a V-8. This is probably due to a blown head gasket between the two cylinders

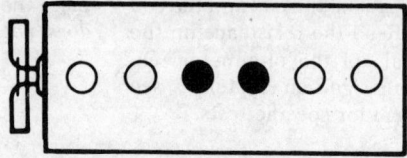

The two center plugs in a 6-cylinder engine are fouled. Raw fuel may be "boiled" out of the carburetor into the intake manifold after the engine is shut-off. Stop-start driving can also foul the center plugs, due to overly rich mixture. Proper float level, a new float needle and seat or use of an insulating spacer may help this problem

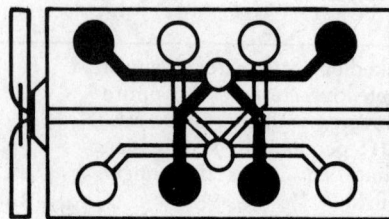

An unbalanced carburetor is indicated. Following the fuel flow on this particular design shows that the cylinders fed by the right-hand barrel are fouled from overly rich mixture, while the cylinders fed by the left-hand barrel are normal

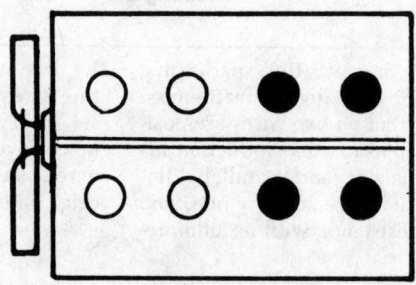

If the four rear plugs are overheated, a cooling system problem is suggested. A thorough cleaning of the cooling system may restore coolant circulation and cure the problem

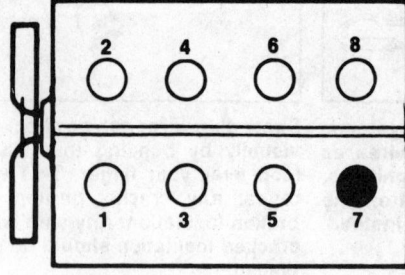

Finding one plug overheated may indicate an intake manifold leak near the affected cylinder. If the overheated plug is the second of two adjacent, consecutively firing plugs, it could be the result of ignition cross-firing. Separating the leads to these two plugs will eliminate cross-fire

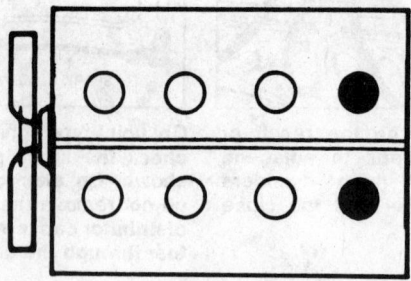

Occasionally, the two rear plugs in large, lightly used V-8's will become oil fouled. High oil consumption and smoky exhaust may also be noticed. It is probably due to plugged oil drain holes in the rear of the cylinder head, causing oil to be sucked in around the valve stems. This usually occurs in the rear cylinders first, because the engine slants that way

Test and Procedure	Results and Indications	Proceed to
4.8—Determine the static ignition timing. Using the crankshaft pulley timing marks as a guide, locate top dead center on the compression stroke of the number one cylinder.	The rotor should be pointing toward the No. 1 tower in the distributor cap, and, on electronic ignitions, the armature spoke for that cylinder should be lined up with the stator.	4.8
4.9—Check coil polarity: Connect a voltmeter negative lead to the coil high tension lead, and the positive lead to ground (**NOTE:** *Reverse the hook-up for positive ground systems*). Crank the engine momentarily. **Checking coil polarity**	If the voltmeter reads up-scale, the polarity is correct: If the voltmeter reads down-scale, reverse the coil polarity (switch the primary leads):	5.1 5.1

Section 5—Fuel System
See Chapter 4 for service procedures

Test and Procedure	Results and Indications	Proceed to
5.1—Determine that the air filter is functioning efficiently: Hold paper elements up to a strong light, and attempt to see light through the filter.	Clean permanent air filters in solvent (or manufacturer's recommendation), and allow to dry. Replace paper elements through which light cannot be seen:	5.2
5.2—Determine whether a flooding condition exists: Flooding is identified by a strong gasoline odor, and excessive gasoline present in the throttle bore(s) of the carburetor. **If the engine floods repeatedly, check the choke butterfly flap**	If flooding is not evident: If flooding is evident, permit the gasoline to dry for a few moments and restart. If flooding doesn't recur: If flooding is persistent:	5.3 5.7 5.5
5.3—Check that fuel is reaching the carburetor: Detach the fuel line at the carburetor inlet. Hold the end of the line in a cup (not styrofoam), and crank the engine.  **Check the fuel pump by disconnecting the output line (fuel pump-to-carburetor) at the carburetor and operating the starter briefly**	If fuel flows smoothly: If fuel doesn't flow (**NOTE:** *Make sure that there is fuel in the tank*), or flows erratically:	5.7 5.4

Test and Procedure	Results and Indications	Proceed to
5.4—Test the fuel pump: Disconnect all fuel lines from the fuel pump. Hold a finger over the input fitting, crank the engine (with electric pump, turn the ignition or pump on); and feel for suction.	If suction is evident, blow out the fuel line to the tank with low pressure compressed air until bubbling is heard from the fuel filler neck. Also blow out the carburetor fuel line (both ends disconnected):	5.7
	If no suction is evident, replace or repair the fuel pump: **NOTE:** *Repeated oil fouling of the spark plugs, or a no-start condition, could be the result of a ruptured vacuum booster pump diaphragm, through which oil or gasoline is being drawn into the intake manifold (where applicable).*	5.7
5.5—Occasionally, small specks of dirt will clog the small jets and orifices in the carburetor. With the engine cold, hold a flat piece of wood or similar material over the carburetor, where possible, and crank the engine.	If the engine starts, but runs roughly the engine is probably not run enough. If the engine won't start:	5.9
5.6—Check the needle and seat: Tap the carburetor in the area of the needle and seat.	If flooding stops, a gasoline additive (e.g., Gumout) will often cure the problem:	5.7
	If flooding continues, check the fuel pump for excessive pressure at the carburetor (according to specifications). If the pressure is normal, the needle and seat must be removed and checked, and/or the float level adjusted:	5.7
5.7—Test the accelerator pump by looking into the throttle bores while operating the throttle.	If the accelerator pump appears to be operating normally:	5.8
	If the accelerator pump is not operating, the pump must be reconditioned. Where possible, service the pump with the carburetor(s) installed on the engine. If necessary, remove the carburetor. Prior to removal:	5.8

Check for gas at the carburetor by looking down the carburetor throat while someone moves the accelerator

5.8—Determine whether the carburetor main fuel system is functioning: Spray a commercial starting fluid into the carburetor while attempting to start the engine.	If the engine starts, runs for a few seconds, and dies:	5.9
	If the engine doesn't start:	6.1

CHILTON'S
AUTO BODY REPAIR TIPS

**Tools and Materials • Step-by-Step Illustrated Procedures
How To Repair Dents, Scratches and Rust Holes
Spray Painting and Refinishing Tips**

With a little practice, basic body repair procedures can be mastered by any do-it-yourself mechanic. The step-by-step repairs shown here can be applied to almost any type of auto body repair.

TOOLS & MATERIALS

You may already have basic tools, such as hammers and electric drills. Other tools unique to body repair — body hammers, grinding attachments, sanding blocks, dent puller, half-round plastic file and plastic spreaders — are relatively inexpensive and can be obtained wherever auto parts or auto body repair parts are sold. Portable air compressors and paint spray guns can be purchased or rented.

Auto Body Repair Kits

The best and most often used products are available to the do-it-yourselfer in kit form, from major manufacturers of auto body repair products. The same manufacturers also merchandise the individual products for use by pros.

Kits are available to make a wide variety of repairs, including holes, dents and scratches and fiberglass, and offer the advantage of buying the materials you'll need for the job. There is little waste or chance of materials going bad from not being used. Many kits may also contain basic body-working tools such as body files, sanding blocks and spreaders. Check the contents of the kit before buying your tools.

BODY REPAIR TIPS

Safety

Many of the products associated with auto body repair and refinishing contain toxic chemicals. Read all labels before opening containers and store them in a safe place and manner.
• Wear eye protection (safety goggles) when using power tools or when performing any operation that involves the removal of any type of material.
• Wear lung protection (disposable mask or respirator) when grinding, sanding or painting.

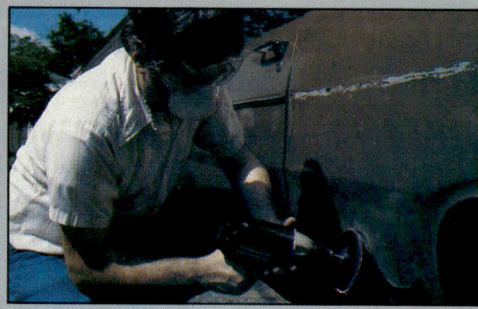

Sanding

1 Sand off paint before using a dent puller. When using a non-adhesive sanding disc, cover the back of the disc with an overlapping layer or two of masking tape and trim the edges. The disc will last considerably longer.

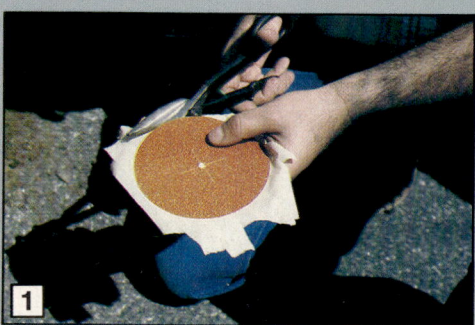

2 Use the circular motion of the sanding disc to grind *into* the edge of the repair. Grinding or sanding away from the jagged edge will only tear the sandpaper.

3 Use the palm of your hand flat on the panel to detect high and low spots. Do not use your fingertips. Slide your hand slowly back and forth.

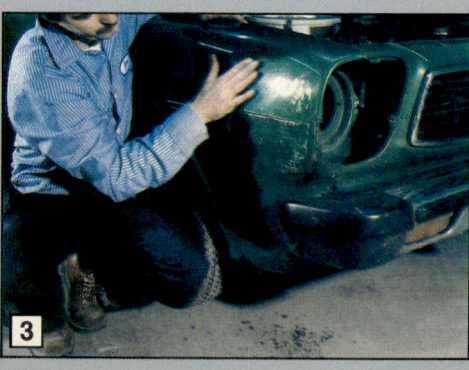

WORKING WITH BODY FILLER

Mixing The Filler

Cleanliness and proper mixing and application are extremely important. Use a clean piece of plastic or glass or a disposable artist's palette to mix body filler.

1 Allow plenty of time and follow directions. No useful purpose will be served by adding more hardener to make it cure (set-up) faster. Less hardener means more curing time, but the mixture dries harder; more hardener means less curing time but a softer mixture.

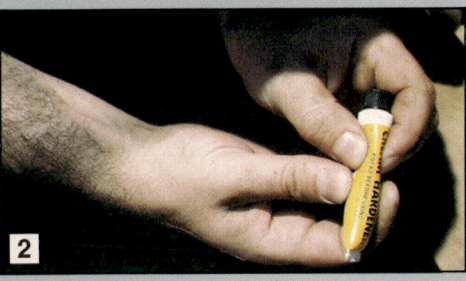

2 Both the hardener and the filler should be thoroughly kneaded or stirred before mixing. Hardener should be a solid paste and dispense like thin toothpaste. Body filler should be smooth, and free of lumps or thick spots.

Getting the proper amount of hardener in the filler is the trickiest part of preparing the filler. Use the same amount of hardener in cold or warm weather. For contour filler (thick coats), a bead of hardener twice the diameter of the filler is about right. There's about a 15% margin on either side, but, if in doubt use less hardener.

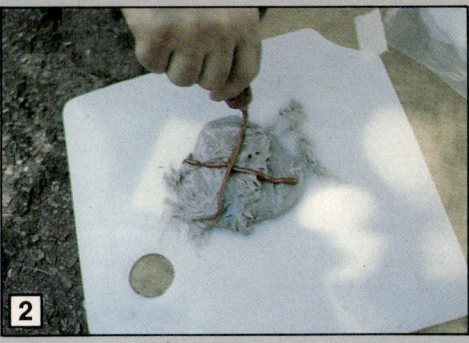

3 Mix the body filler and hardener by wiping across the mixing surface, picking the mixture up and wiping it again. Colder weather requires longer mixing times. Do not mix in a circular motion; this will trap air bubbles which will become holes in the cured filler.

Applying The Filler

1 For best results, filler should not be applied over 1/4″ thick.

Apply the filler in several coats. Build it up to above the level of the repair surface so that it can be sanded or grated down.

The first coat of filler must be pressed on with a firm wiping motion.

Apply the filler in one direction only. Working the filler back and forth will either pull it off the metal or trap air bubbles.

REPAIRING DENTS

Before you start, take a few minutes to study the damaged area. Try to visualize the shape of the panel before it was damaged. If the damage is on the left fender, look at the right fender and use it as a guide. If there is access to the panel from behind, you can reshape it with a body hammer. If not, you'll have to use a dent puller. Go slowly and work the metal a little at a time. Get the panel as straight as possible before applying filler.

1 This dent is typical of one that can be pulled out or hammered out from behind. Remove the headlight cover, headlight assembly and turn signal housing.

2 Drill a series of holes ½ the size of the end of the dent puller along the stress line. Make some trial pulls and assess the results. If necessary, drill more holes and try again. Do not hurry.

3 If possible, use a body hammer and block to shape the metal back to its original contours. Get the metal back as close to its original shape as possible. Don't depend on body filler to fill dents.

4 Using an 80-grit grinding disc on an electric drill, grind the paint from the surrounding area down to bare metal. Use a new grinding pad to prevent heat buildup that will warp metal.

5 The area should look like this when you're finished grinding. Knock the drill holes in and tape over small openings to keep plastic filler out.

6 Mix the body filler (see Body Repair Tips). Spread the body filler evenly over the entire area (see Body Repair Tips). Be sure to cover the area completely.

7 Let the body filler dry until the surface can just be scratched with your fingernail. Knock the high spots from the body filler with a body file ("Cheesegrater"). Check frequently with the palm of your hand for high and low spots.

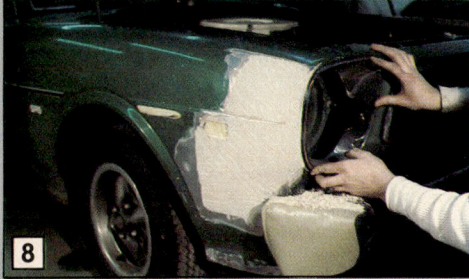

8 Check to be sure that trim pieces that will be installed later will fit exactly. Sand the area with 40-grit paper.

9 If you wind up with low spots, you may have to apply another layer of filler.

10 Knock the high spots off with 40-grit paper. When you are satisfied with the contours of the repair, apply a thin coat of filler to cover pin holes and scratches.

11 Block sand the area with 40-grit paper to a smooth finish. Pay particular attention to body lines and ridges that must be well-defined.

12 Sand the area with 400 paper and then finish with a scuff pad. The finished repair is ready for priming and painting (see Painting Tips).

Materials and photos courtesy of Ritt Jones Auto Body, Prospect Park, PA.

REPAIRING RUST HOLES

There are many ways to repair rust holes. The fiberglass cloth kit shown here is one of the most cost efficient for the owner because it provides a strong repair that resists cracking and moisture and is relatively easy to use. It can be used on large and small holes (with or without backing) and can be applied over contoured areas. Remember, however, that short of replacing an entire panel, no repair is a guarantee that the rust will not return.

1 Remove any trim that will be in the way. Clean away all loose debris. Cut away all the rusted metal. But be sure to leave enough metal to retain the contour or body shape.

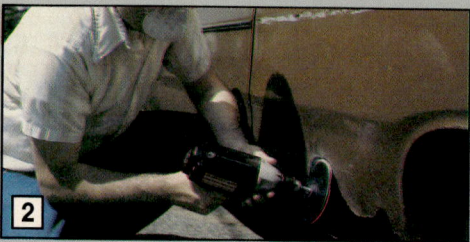

2 Grind away all traces of rust with a 24-grit grinding disc. Be sure to grind back 3-4 inches from the edge of the hole down to bare metal and be sure all traces of paint, primer and rust are removed.

3 Block sand the area with 80 or 100 grit sandpaper to get a clear, shiny surface and feathered paint edge. Tap the edges of the hole inward with a ball peen hammer.

4 If you are going to use release film, cut a piece about 2-3″ larger than the area you have sanded. Place the film over the repair and mark the sanded area on the film. Avoid any unnecessary wrinkling of the film.

5 Cut 2 pieces of fiberglass matte to match the shape of the repair. One piece should be about 1″ smaller than the sanded area and the second piece should be 1″ smaller than the first. Mix enough filler and hardener to saturate the fiberglass material (see Body Repair Tips).

6 Lay the release sheet on a flat surface and spread an even layer of filler, large enough to cover the repair. Lay the smaller piece of fiberglass cloth in the center of the sheet and spread another layer of filler over the fiberglass cloth. Repeat the operation for the larger piece of cloth.

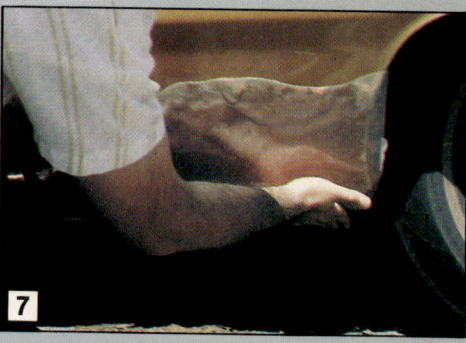

7 Place the repair material over the repair area, with the release film facing outward. Use a spreader and work from the center outward to smooth the material, following the body contours. Be sure to remove all air bubbles.

8 Wait until the repair has dried tack-free and peel off the release sheet. The ideal working temperature is 60°-90° F. Cooler or warmer temperatures or high humidity may require additional curing time. Wait longer, if in doubt.

9 Sand and feather-edge the entire area. The initial sanding can be done with a sanding disc on an electric drill if care is used. Finish the sanding with a block sander. Low spots can be filled with body filler; this may require several applications.

10 When the filler can just be scratched with a fingernail, knock the high spots down with a body file and smooth the entire area with 80-grit. Feather the filled areas into the surrounding areas.

11 When the area is sanded smooth, mix some topcoat and hardener and apply it directly with a spreader. This will give a smooth finish and prevent the glass matte from showing through the paint.

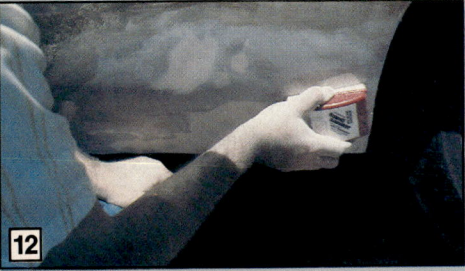

12 Block sand the topcoat smooth with finishing sandpaper (200 grit), and 400 grit. The repair is ready for masking, priming and painting (see Painting Tips).

Materials and photos courtesy Marson Corporation, Chelsea, Massachusetts

PAINTING TIPS

Preparation

1 SANDING — Use a 400 or 600 grit wet or dry sandpaper. Wet-sand the area with a 1/4 sheet of sandpaper soaked in clean water. Keep the paper wet while sanding. Sand the area until the repaired area tapers into the original finish.

2 CLEANING — Wash the area to be painted thoroughly with water and a clean rag. Rinse it thoroughly and wipe the surface dry until you're sure it's completely free of dirt, dust, fingerprints, wax, detergent or other foreign matter.

3 MASKING — Protect any areas you don't want to overspray by covering them with masking tape and newspaper. Be careful not get fingerprints on the area to be painted.

4 PRIMING — All exposed metal should be primed before painting. Primer protects the metal and provides an excellent surface for paint adhesion. When the primer is dry, wet-sand the area again with 600 grit wet-sandpaper. Clean the area again after sanding.

Painting Techniques

Paint applied from either a spray gun or a spray can (for small areas) will provide good results. Experiment on an old piece of metal to get the right combination before you begin painting.

SPRAYING VISCOSITY (SPRAY GUN ONLY)

Paint should be thinned to spraying viscosity according to the directions on the can. Use only the recommended thinner or reducer and the same amount of reduction regardless of temperature.

AIR PRESSURE (SPRAY GUN ONLY)

This is extremely important. Be sure you are using the proper recommended pressure.

TEMPERATURE

The surface to be painted should be approximately the same temperature as the surrounding air. Applying warm paint to a cold surface, or vice versa, will completely upset the paint characteristics.

THICKNESS

Spray with smooth strokes. In general, the thicker the coat of paint, the longer the drying time. Apply several thin coats about 30 seconds apart. The paint should remain wet long enough to flow out and no longer; heavier coats will only produce sags or wrinkles. Spray a light (fog) coat, followed by heavier color coats.

DISTANCE

The ideal spraying distance is 8″-12″ from the gun or can to the surface. Shorter distances will produce ripples, while greater distances will result in orange peel, dry film and poor color match and loss of material due to overspray.

OVERLAPPING

The gun or can should be kept at right angles to the surface at all times. Work to a wet edge at an even speed, using a 50% overlap and direct the center of the spray at the lower or nearest edge of the previous stroke.

RUBBING OUT (BLENDING) FRESH PAINT

Let the paint dry thoroughly. Runs or imperfections can be sanded out, primed and repainted.

Don't be in too big a hurry to remove the masking. This only produces paint ridges. When the finish has dried for at least a week, apply a small amount of fine grade rubbing compound with a clean, wet cloth. Use lots of water and blend the new paint with the surrounding area.

WRONG

Thin coat. Stroke too fast, not enough overlap, gun too far away.

CORRECT

Medium coat. Proper distance, good stroke, proper overlap.

WRONG

Heavy coat. Stroke too slow, too much overlap, gun too close.

Test and Procedure	Results and Indications	Proceed to
5.9—Uncommon fuel system malfunctions: See below:	If the problem is solved: If the problem remains, remove and recondition the carburetor.	6.1

Condition	Indication	Test	Prevailing Weather Conditions	Remedy
Vapor lock	Engine will not restart shortly after running.	Cool the components of the fuel system until the engine starts. Vapor lock can be cured faster by draping a wet cloth over a mechanical fuel pump.	Hot to very hot	Ensure that the exhaust manifold heat control valve is operating. Check with the vehicle manufacturer for the recommended solution to vapor lock on the model in question.
Carburetor icing	Engine will not idle, stalls at low speeds.	Visually inspect the throttle plate area of the throttle bores for frost.	High humidity, 32–40° F.	Ensure that the exhaust manifold heat control valve is operating, and that the intake manifold heat riser is not blocked.
Water in the fuel	Engine sputters and stalls; may not start.	Pump a small amount of fuel into a glass jar. Allow to stand, and inspect for droplets or a layer of water.	High humidity, extreme temperature changes.	For droplets, use one or two cans of commercial gas line anti-freeze. For a layer of water, the tank must be drained, and the fuel lines blown out with compressed air.

Section 6—Engine Compression
See Chapter 3 for service procedures

6.1—Test engine compression: Remove all spark plugs. Block the throttle wide open. Insert a compression gauge into a spark plug port, crank the engine to obtain the maximum reading, and record.	If compression is within limits on all cylinders:	7.1
	If gauge reading is extremely low on all cylinders:	6.2
	If gauge reading is low on one or two cylinders: (If gauge readings are identical and low on two or more adjacent cylinders, the head gasket must be replaced.)	6.2

Checking compression

6.2—Test engine compression (wet): Squirt approximately 30 cc. of engine oil into each cylinder, and retest per 6.1.	If the readings improve, worn or cracked rings or broken pistons are indicated:	See Chapter 3
	If the readings do not improve, burned or excessively carboned valves or a jumped timing chain are indicated: **NOTE:** *A jumped timing chain is often indicated by difficult cranking.*	7.1

Section 7—Engine Vacuum
See Chapter 3 for service procedures

Test and Procedure	Results and Indications	Proceed to
7.1—Attach a vacuum gauge to the intake manifold beyond the throttle plate. Start the engine, and observe the action of the needle over the range of engine speeds.	See below.	**See below**

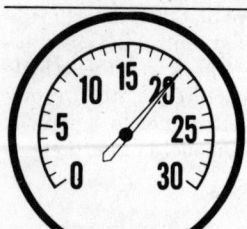

INDICATION: normal engine in good condition

Proceed to: 8.1

Normal engine
Gauge reading: steady, from 17–22 in./Hg.

INDICATION: sticking valves or ignition miss

Proceed to: 9.1, 8.3

Sticking valves
Gauge reading: intermittent fluctuation at idle

INDICATION: late ignition or valve timing, low compression, stuck throttle valve, leaking carburetor or manifold gasket

Proceed to: 6.1

Incorrect valve timing
Gauge reading: low (10–15 in./Hg) but steady

INDICATION: improper carburetor adjustment or minor intake leak.

Proceed to: 7.2

Carburetor requires adjustment
Gauge reading: drifting needle

INDICATION: ignition miss, blown cylinder head gasket, leaking valve or weak valve spring

Proceed to: 8.3, 6.1

Blown head gasket
Gauge reading: needle fluctuates as engine speed increases

INDICATION: burnt valve or faulty valve clearance. Needle will fall when defective valve operates

Proceed to: 9.1

Burnt or leaking valves
Gauge reading: steady needle, but drops regularly

INDICATION: choked muffler, excessive back pressure in system

Proceed to: 10.1

Clogged exhaust system
Gauge reading: gradual drop in reading at idle

INDICATION: worn valve guides

Proceed to: 9.1

Worn valve guides
Gauge reading: needle vibrates excessively at idle, but steadies as engine speed increases

White pointer = steady gauge hand　　　Black pointer = fluctuating gauge hand

Test and Procedure	Results and Indications	Proceed to
7.2—Attach a vacuum gauge per 7.1, and test for an intake manifold leak. Squirt a small amount of oil around the intake manifold gaskets, carburetor gaskets, plugs and fittings. Observe the action of the vacuum gauge.	If the reading improves, replace the indicated gasket, or seal the indicated fitting or plug: If the reading remains low:	8.1 7.3
7.3—Test all vacuum hoses and accessories for leaks as described in 7.2. Also check the carburetor body (dashpots, automatic choke mechanism, throttle shafts) for leaks in the same manner.	If the reading improves, service or replace the offending part(s): If the reading remains low:	8.1 6.1

Section 8—Secondary Electrical System
See Chapter 2 for service procedures

Test and Procedure	Results and Indications	Proceed to
8.1—Remove the distributor cap and check to make sure that the rotor turns when the engine is cranked. Visually inspect the distributor components.	Clean, tighten or replace any components which appear defective.	8.2
8.2—Connect a timing light (per manufacturer's recommendation) and check the dynamic ignition timing. Disconnect and plug the vacuum hose(s) to the distributor if specified, start the engine, and observe the timing marks at the specified engine speed.	If the timing is not correct, adjust to specifications by rotating the distributor in the engine: (Advance timing by rotating distributor opposite normal direction of rotor rotation, retard timing by rotating distributor in same direction as rotor rotation.)	8.3
8.3—Check the operation of the distributor advance mechanism(s): To test the mechanical advance, disconnect the vacuum lines from the distributor advance unit and observe the timing marks with a timing light as the engine speed is increased from idle. If the mark moves smoothly, without hesitation, it may be assumed that the mechanical advance is functioning properly. To test vacuum advance and/or retard systems, alternately crimp and release the vacuum line, and observe the timing mark for movement. If movement is noted, the system is operating.	If the systems are functioning: If the systems are not functioning, remove the distributor, and test on a distributor tester:	8.4 8.4
8.4—Locate an ignition miss: With the engine running, remove each spark plug wire, one at a time, until one is found that doesn't cause the engine to roughen and slow down.	When the missing cylinder is identified:	4.1

Section 9—Valve Train
See Chapter 3 for service procedures

Test and Procedure	Results and Indications	Proceed to
9.1—Evaluate the valve train: Remove the valve cover, and ensure that the valves are adjusted to specifications. A mechanic's stethoscope may be used to aid in the diagnosis of the valve train. By pushing the probe on or near push rods or rockers, valve noise often can be isolated. A timing light also may be used to diagnose valve problems. Connect the light according to manufacturer's recommendations, and start the engine. Vary the firing moment of the light by increasing the engine speed (and therefore the ignition advance), and moving the trigger from cylinder to cylinder. Observe the movement of each valve.	Sticking valves or erratic valve train motion can be observed with the timing light. The cylinder head must be disassembled for repairs.	**See Chapter 3**
9.2—Check the valve timing: Locate top dead center of the No. 1 piston, and install a degree wheel or tape on the crankshaft pulley or damper with zero corresponding to an index mark on the engine. Rotate the crankshaft in its direction of rotation, and observe the opening of the No. 1 cylinder intake valve. The opening should correspond with the correct mark on the degree wheel according to specifications.	If the timing is not correct, the timing cover must be removed for further investigation.	**See Chapter 3**

Section 10—Exhaust System

Test and Procedure	Results and Indications	Proceed to
10.1—Determine whether the exhaust manifold heat control valve is operating: Operate the valve by hand to determine whether it is free to move. If the valve is free, run the engine to operating temperature and observe the action of the valve, to ensure that it is opening.	If the valve sticks, spray it with a suitable solvent, open and close the valve to free it, and retest. If the valve functions properly: If the valve does not free, or does not operate, replace the valve:	**10.2** **10.2**
10.2—Ensure that there are no exhaust restrictions: Visually inspect the exhaust system for kinks, dents, or crushing. Also note that gases are flowing freely from the tailpipe at all engine speeds, indicating no restriction in the muffler or resonator.	Replace any damaged portion of the system:	**11.1**

Section 11—Cooling System
See Chapter 3 for service procedures

Test and Procedure	Results and Indications	Proceed to
11.1—Visually inspect the fan belt for glazing, cracks, and fraying, and replace if necessary. Tighten the belt so that the longest span has approximately ½″ play at its midpoint under thumb pressure (see Chapter 1).	Replace or tighten the fan belt as necessary:	**11.2**

Checking belt tension

11.2—Check the fluid level of the cooling system.	If full or slightly low, fill as necessary:	**11.5**
	If extremely low:	**11.3**
11.3—Visually inspect the external portions of the cooling system (radiator, radiator hoses, thermostat elbow, water pump seals, heater hoses, etc.) for leaks. If none are found, pressurize the cooling system to 14–15 psi.	If cooling system holds the pressure:	**11.5**
	If cooling system loses pressure rapidly, reinspect external parts of the system for leaks under pressure. If none are found, check dipstick for coolant in crankcase. If no coolant is present, but pressure loss continues:	**11.4**
	If coolant is evident in crankcase, remove cylinder head(s), and check gasket(s). If gaskets are intact, block and cylinder head(s) should be checked for cracks or holes. If the gasket(s) is blown, replace, and purge the crankcase of coolant:	**12.6**
	NOTE: *Occasionally, due to atmospheric and driving conditions, condensation of water can occur in the crankcase. This causes the oil to appear milky white. To remedy, run the engine until hot, and change the oil and oil filter.*	
11.4—Check for combustion leaks into the cooling system: Pressurize the cooling system as above. Start the engine, and observe the pressure gauge. If the needle fluctuates, remove each spark plug wire, one at a time, noting which cylinder(s) reduce or eliminate the fluctuation.	Cylinders which reduce or eliminate the fluctuation, when the spark plug wire is removed, are leaking into the cooling system. Replace the head gasket on the affected cylinder bank(s).	

Pressurizing the cooling system

Test and Procedure	Results and Indications	Proceed to
11.5—Check the radiator pressure cap: Attach a radiator pressure tester to the radiator cap (wet the seal prior to installation). Quickly pump up the pressure, noting the point at which the cap releases.	If the cap releases within ± 1 psi of the specified rating, it is operating properly:	**11.6**
	If the cap releases at more than ± 1 psi of the specified rating, it should be replaced:	**11.6**

Checking radiator pressure cap

Test and Procedure	Results and Indications	Proceed to
11.6—Test the thermostat: Start the engine cold, remove the radiator cap, and insert a thermometer into the radiator. Allow the engine to idle. After a short while, there will be a sudden, rapid increase in coolant temperature. The temperature at which this sharp rise stops is the thermostat opening temperature.	If the thermostat opens at or about the specified temperature:	**11.7**
	If the temperature doesn't increase: (If the temperature increases slowly and gradually, replace the thermostat.)	**11.7**
11.7—Check the water pump: Remove the thermostat elbow and the thermostat, disconnect the coil high tension lead (to prevent starting), and crank the engine momentarily.	If coolant flows, replace the thermostat and retest per 11.6:	**11.6**
	If coolant doesn't flow, reverse flush the cooling system to alleviate any blockage that might exist. If system is not blocked, and coolant will not flow, replace the water pump.	

Section 12—Lubrication
See Chapter 3 for service procedures

Test and Procedure	Results and Indications	Proceed to
12.1—Check the oil pressure gauge or warning light: If the gauge shows low pressure, or the light is on for no obvious reason, remove the oil pressure sender. Install an accurate oil pressure gauge and run the engine momentarily.	If oil pressure builds normally, run engine for a few moments to determine that it is functioning normally, and replace the sender.	—
	If the pressure remains low:	**12.2**
	If the pressure surges:	**12.3**
	If the oil pressure is zero:	**12.3**
12.2—Visually inspect the oil: If the oil is watery or very thin, milky, or foamy, replace the oil and oil filter.	If the oil is normal:	**12.3**
	If after replacing oil the pressure remains low:	**12.3**
	If after replacing oil the pressure becomes normal:	—

Test and Procedure	Results and Indications	Proceed to
12.3—Inspect the oil pressure relief valve and spring, to ensure that it is not sticking or stuck. Remove and thoroughly clean the valve, spring, and the valve body.	If the oil pressure improves: If no improvement is noted:	— **12.4**
12.4—Check to ensure that the oil pump is not cavitating (sucking air instead of oil): See that the crankcase is neither over nor underfull, and that the pickup in the sump is in the proper position and free from sludge.	Fill or drain the crankcase to the proper capacity, and clean the pickup screen in solvent if necessary. If no improvement is noted:	**12.5**
12.5—Inspect the oil pump drive and the oil pump:	If the pump drive or the oil pump appear to be defective, service as necessary and retest per 12.1: If the pump drive and pump appear to be operating normally, the engine should be disassembled to determine where blockage exists:	**12.1** **See Chapter 3**
12.6—Purge the engine of ethylene glycol coolant: Completely drain the crankcase and the oil filter. Obtain a commercial butyl cellosolve base solvent, designated for this purpose, and follow the instructions precisely. Following this, install a new oil filter and refill the crankcase with the proper weight oil. The next oil and filter change should follow shortly thereafter (1000 miles).		

TROUBLESHOOTING EMISSION CONTROL SYSTEMS

See Chapter 4 for procedures applicable to individual emission control systems used on specific combinations of engine/transmission/model.

TROUBLESHOOTING THE CARBURETOR
See Chapter 4 for service procedures

Carburetor problems cannot be effectively isolated unless all other engine systems (particularly ignition and emission) are functioning properly and the engine is properly tuned.

Condition	Possible Cause
Engine cranks, but does not start	1. Improper starting procedure 2. No fuel in tank 3. Clogged fuel line or filter 4. Defective fuel pump 5. Choke valve not closing properly 6. Engine flooded 7. Choke valve not unloading 8. Throttle linkage not making full travel 9. Stuck needle or float 10. Leaking float needle or seat 11. Improper float adjustment
Engine stalls	1. Improperly adjusted idle speed or mixture **Engine hot** 2. Improperly adjusted dashpot 3. Defective or improperly adjusted solenoid 4. Incorrect fuel level in fuel bowl 5. Fuel pump pressure too high 6. Leaking float needle seat 7. Secondary throttle valve stuck open 8. Air or fuel leaks 9. Idle air bleeds plugged or missing 10. Idle passages plugged **Engine Cold** 11. Incorrectly adjusted choke 12. Improperly adjusted fast idle speed 13. Air leaks 14. Plugged idle or idle air passages 15. Stuck choke valve or binding linkage 16. Stuck secondary throttle valves 17. Engine flooding—high fuel level 18. Leaking or misaligned float
Engine hesitates on acceleration	1. Clogged fuel filter 2. Leaking fuel pump diaphragm 3. Low fuel pump pressure 4. Secondary throttle valves stuck, bent or misadjusted 5. Sticking or binding air valve 6. Defective accelerator pump 7. Vacuum leaks 8. Clogged air filter 9. Incorrect choke adjustment (engine cold)
Engine feels sluggish or flat on acceleration	1. Improperly adjusted idle speed or mixture 2. Clogged fuel filter 3. Defective accelerator pump 4. Dirty, plugged or incorrect main metering jets 5. Bent or sticking main metering rods 6. Sticking throttle valves 7. Stuck heat riser 8. Binding or stuck air valve 9. Dirty, plugged or incorrect secondary jets 10. Bent or sticking secondary metering rods. 11. Throttle body or manifold heat passages plugged 12. Improperly adjusted choke or choke vacuum break.
Carburetor floods	1. Defective fuel pump. Pressure too high. 2. Stuck choke valve 3. Dirty, worn or damaged float or needle valve/seat 4. Incorrect float/fuel level 5. Leaking float bowl

Condition	Possible Cause
Engine idles roughly and stalls	1. Incorrect idle speed 2. Clogged fuel filter 3. Dirt in fuel system or carburetor 4. Loose carburetor screws or attaching bolts 5. Broken carburetor gaskets 6. Air leaks 7. Dirty carburetor 8. Worn idle mixture needles 9. Throttle valves stuck open 10. Incorrectly adjusted float or fuel level 11. Clogged air filter
Engine runs unevenly or surges	1. Defective fuel pump 2. Dirty or clogged fuel filter 3. Plugged, loose or incorrect main metering jets or rods 4. Air leaks 5. Bent or sticking main metering rods 6. Stuck power piston 7. Incorrect float adjustment 8. Incorrect idle speed or mixture 9. Dirty or plugged idle system passages 10. Hard, brittle or broken gaskets 11. Loose attaching or mounting screws 12. Stuck or misaligned secondary throttle valves
Poor fuel economy	1. Poor driving habits 2. Stuck choke valve 3. Binding choke linkage 4. Stuck heat riser 5. Incorrect idle mixture 6. Defective accelerator pump 7. Air leaks 8. Plugged, loose or incorrect main metering jets 9. Improperly adjusted float or fuel level 10. Bent, misaligned or fuel-clogged float 11. Leaking float needle seat 12. Fuel leak 13. Accelerator pump discharge ball not seating properly 14. Incorrect main jets
Engine lacks high speed performance or power	1. Incorrect throttle linkage adjustment 2. Stuck or binding power piston 3. Defective accelerator pump 4. Air leaks 5. Incorrect float setting or fuel level 6. Dirty, plugged, worn or incorrect main metering jets or rods 7. Binding or sticking air valve 8. Brittle or cracked gaskets 9. Bent, incorrect or improperly adjusted secondary metering rods 10. Clogged fuel filter 11. Clogged air filter 12. Defective fuel pump

TROUBLESHOOTING FUEL INJECTION PROBLEMS

Each fuel injection system has its own unique components and test procedures, for which it is impossible to generalize. Refer to Chapter 4 of this Repair & Tune-Up Guide for specific test and repair procedures, if the vehicle is equipped with fuel injection.

TROUBLESHOOTING ELECTRICAL PROBLEMS

See Chapter 5 for service procedures

For any electrical system to operate, it must make a complete circuit. This simply means that the power flow from the battery must make a complete circle. When an electrical component is operating, power flows from the battery to the component, passes through the component causing it to perform its function (lighting a light bulb), and then returns to the battery through the ground of the circuit. This ground is usually (but not always) the metal part of the car or truck on which the electrical component is mounted.

Perhaps the easiest way to visualize this is to think of connecting a light bulb with two wires attached to it to the battery. If one of the two wires attached to the light bulb were attached to the negative post of the battery and the other were attached to the positive post of the battery, you would have a complete circuit. Current from the battery would flow to the light bulb, causing it to light, and return to the negative post of the battery.

The normal automotive circuit differs from this simple example in two ways. First, instead of having a return wire from the bulb to the battery, the light bulb returns the current to the battery through the chassis of the vehicle. Since the negative battery cable is attached to the chassis and the chassis is made of electrically conductive metal, the chassis of the vehicle can serve as a ground wire to complete the circuit. Secondly, most automotive circuits contain switches to turn components on and off as required.

Every complete circuit from a power source must include a component which is using the power from the power source. If you were to disconnect the light bulb from the wires and touch the two wires together (don't do this) the power supply wire to the component would be grounded before the normal ground connection for the circuit.

Because grounding a wire from a power source makes a complete circuit—less the required component to use the power—this phenomenon is called a short circuit. Common causes are: broken insulation (exposing the metal wire to a metal part of the car or truck), or a shorted switch.

Some electrical components which require a large amount of current to operate also have a relay in their circuit. Since these circuits carry a large amount of current, the thickness of the wire in the circuit (gauge size) is also greater. If this large wire were connected from the component to the control switch on the instrument panel, and then back to the component, a voltage drop would occur in the circuit. To prevent this potential drop in voltage, an electromagnetic switch (relay) is used. The large wires in the circuit are connected from the battery to one side of the relay, and from the opposite side of the relay to the component. The relay is normally open, preventing current from passing through the circuit. An additional, smaller, wire is connected from the relay to the control switch for the circuit. When the control switch is turned on, it grounds the smaller wire from the relay and completes the circuit. This closes the relay and allows current to flow from the battery to the component. The horn, headlight, and starter circuits are three which use relays.

It is possible for larger surges of current to pass through the electrical system of your car or truck. If this surge of current were to reach an electrical component, it could burn it out. To prevent this, fuses, circuit breakers or fusible links are connected into the current supply wires of most of the major electrical systems. When an electrical current of excessive power passes through the component's fuse, the fuse blows out and breaks the circuit, saving the component from destruction.

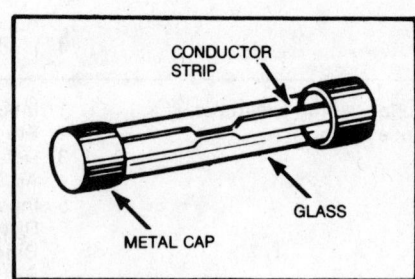

Typical automotive fuse

A circuit breaker is basically a self-repairing fuse. The circuit breaker opens the circuit the same way a fuse does. However, when either the short is removed from the circuit or the surge subsides, the circuit breaker resets itself and does not have to be replaced as a fuse does.

A fuse link is a wire that acts as a fuse. It is normally connected between the starter relay and the main wiring harness. This connection is usually under the hood. The fuse link (if installed) protects all the

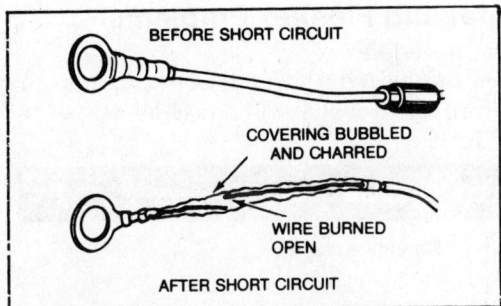

Most fusible links show a charred, melted insulation when they burn out

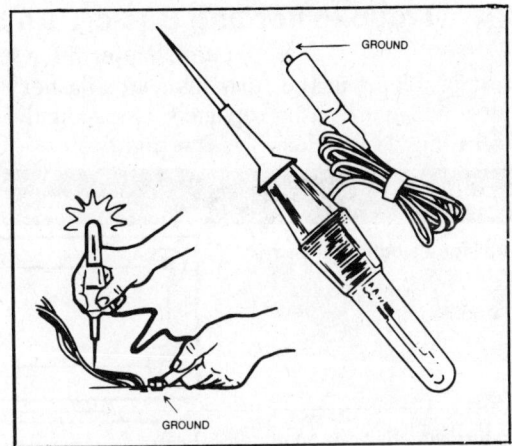

The test light will show the presence of current when touched to a hot wire and grounded at the other end

chassis electrical components, and is the probable cause of trouble when none of the electrical components function, unless the battery is disconnected or dead.

Electrical problems generally fall into one of three areas:

1. The component that is not functioning is not receiving current.

2. The component itself is not functioning.

3. The component is not properly grounded.

The electrical system can be checked with a test light and a jumper wire. A test light is a device that looks like a pointed screwdriver with a wire attached to it and has a light bulb in its handle. A jumper wire is a piece of insulated wire with an alligator clip attached to each end.

If a component is not working, you must follow a systematic plan to determine which of the three causes is the villain.

1. Turn on the switch that controls the inoperable component.

2. Disconnect the power supply wire from the component.

3. Attach the ground wire on the test light to a good metal ground.

4. Touch the probe end of the test light to the end of the power supply wire that was disconnected from the component. If the component is receiving current, the test light will go on.

NOTE: *Some components work only when the ignition switch is turned on.*

If the test light does not go on, then the problem is in the circuit between the battery and the component. This includes all the switches, fuses, and relays in the system. Follow the wire that runs back to the battery. The problem is an open circuit between the

battery and the component. If the fuse is blown and, when replaced, immediately blows again, there is a short circuit in the system which must be located and repaired. If there is a switch in the system, bypass it with a jumper wire. This is done by connecting one end of the jumper wire to the power supply wire into the switch and the other end of the jumper wire to the wire coming out of the switch. If the test light lights with the jumper wire installed, the switch or whatever was bypassed is defective.

NOTE: *Never substitute the jumper wire for the component, since it is required to use the power from the power source.*

5. If the bulb in the test light goes on, then the current is getting to the component that is not working. This eliminates the first of the three possible causes. Connect the power supply wire and connect a jumper wire from the component to a good metal ground. Do this with the switch which controls the component turned on, and also the ignition switch turned on if it is required for the component to work. If the component works with the jumper wire installed, then it has a bad ground. This is usually caused by the metal area on which the component mounts to the chassis being coated with some type of foreign matter.

6. If neither test located the source of the trouble, then the component itself is defective. Remember that for any electrical system to work, all connections must be clean and tight.

Troubleshooting Basic Turn Signal and Flasher Problems
See Chapter 5 for service procedures

Most problems in the turn signals or flasher system can be reduced to defective flashers or bulbs, which are easily replaced. Occasionally, the turn signal switch will prove defective.

F = Front   R = Rear   ● = Lights off   ○ = Lights on

Condition		Possible Cause
Turn signals light, but do not flash	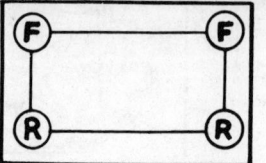	Defective flasher
No turn signals light on either side	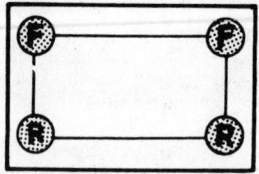	Blown fuse. Replace if defective. Defective flasher. Check by substitution. Open circuit, short circuit or poor ground.
Both turn signals on one side don't work	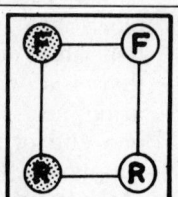	Bad bulbs. Bad ground in both (or either) housings.
One turn signal light on one side doesn't work	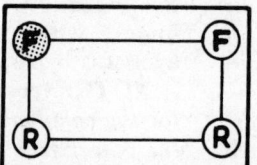	Defective bulb. Corrosion in socket. Clean contacts. Poor ground at socket.
Turn signal flashes too fast or too slowly	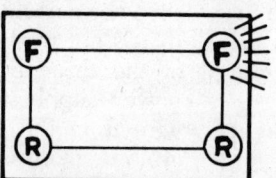	Check any bulb on the side flashing too fast. A heavy-duty bulb is probably installed in place of a regular bulb. Check the bulb flashing too slowly. A standard bulb was probably installed in place of a heavy-duty bulb. Loose connections or corrosion at the bulb socket.
Indicator lights don't work in either direction		Check if the turn signals are working. Check the dash indicator lights. Check the flasher by substitution.
One indicator light doesn't light		On systems with one dash indicator: See if the lights work on the same side. Often the filaments have been reversed in systems combining stoplights with taillights and turn signals. Check the flasher by substitution. On systems with two indicators: Check the bulbs on the same side. Check the indicator light bulb. Check the flasher by substitution.

Troubleshooting Lighting Problems
See Chapter 5 for service procedures

Condition	Possible Cause
One or more lights don't work, but others do	1. Defective bulb(s) 2. Blown fuse(s) 3. Dirty fuse clips or light sockets 4. Poor ground circuit
Lights burn out quickly	1. Incorrect voltage regulator setting or defective regulator 2. Poor battery/alternator connections
Lights go dim	1. Low/discharged battery 2. Alternator not charging 3. Corroded sockets or connections 4. Low voltage output
Lights flicker	1. Loose connection 2. Poor ground. (Run ground wire from light housing to frame) 3. Circuit breaker operating (short circuit)
Lights "flare"—Some flare is normal on acceleration—If excessive, see "Lights Burn Out Quickly"	High voltage setting
Lights glare—approaching drivers are blinded	1. Lights adjusted too high 2. Rear springs or shocks sagging 3. Rear tires soft

Troubleshooting Dash Gauge Problems

Most problems can be traced to a defective sending unit or faulty wiring. Occasionally, the gauge itself is at fault. See Chapter 5 for service procedures.

Condition	Possible Cause
COOLANT TEMPERATURE GAUGE	
Gauge reads erratically or not at all	1. Loose or dirty connections 2. Defective sending unit. 3. Defective gauge. To test a bi-metal gauge, remove the wire from the sending unit. Ground the wire for an instant. If the gauge registers, replace the sending unit. To test a magnetic gauge, disconnect the wire at the sending unit. With ignition ON gauge should register COLD. Ground the wire; gauge should register HOT.
AMMETER GAUGE—TURN HEADLIGHTS ON (DO NOT START ENGINE). NOTE REACTION	
Ammeter shows charge Ammeter shows discharge Ammeter does not move	1. Connections reversed on gauge 2. Ammeter is OK 3. Loose connections or faulty wiring 4. Defective gauge

Condition	Possible Cause

OIL PRESSURE GAUGE

Gauge does not register or is inaccurate	1. On mechanical gauge, Bourdon tube may be bent or kinked. 2. Low oil pressure. Remove sending unit. Idle the engine briefly. If no oil flows from sending unit hole, problem is in engine. 3. Defective gauge. Remove the wire from the sending unit and ground it for an instant with the ignition ON. A good gauge will go to the top of the scale. 4. Defective wiring. Check the wiring to the gauge. If it's OK and the gauge doesn't register when grounded, replace the gauge. 5. Defective sending unit.

ALL GAUGES

All gauges do not operate All gauges read low or erratically All gauges pegged	1. Blown fuse 2. Defective instrument regulator 3. Defective or dirty instrument voltage regulator 4. Loss of ground between instrument voltage regulator and frame 5. Defective instrument regulator

WARNING LIGHTS

Light(s) do not come on when ignition is ON, but engine is not started Light comes on with engine running	1. Defective bulb 2. Defective wire 3. Defective sending unit. Disconnect the wire from the sending unit and ground it. Replace the sending unit if the light comes on with the ignition ON. 4. Problem in individual system 5. Defective sending unit

Troubleshooting Clutch Problems

It is false economy to replace individual clutch components. The pressure plate, clutch plate and throwout bearing should be replaced as a set, and the flywheel face inspected, whenever the clutch is overhauled. See Chapter 6 for service procedures.

Condition	Possible Cause
Clutch chatter	1. Grease on driven plate (disc) facing 2. Binding clutch linkage or cable 3. Loose, damaged facings on driven plate (disc) 4. Engine mounts loose 5. Incorrect height adjustment of pressure plate release levers 6. Clutch housing or housing to transmission adapter misalignment 7. Loose driven plate hub
Clutch grabbing	1. Oil, grease on driven plate (disc) facing 2. Broken pressure plate 3. Warped or binding driven plate. Driven plate binding on clutch shaft
Clutch slips	1. Lack of lubrication in clutch linkage or cable (linkage or cable binds, causes incomplete engagement) 2. Incorrect pedal, or linkage adjustment 3. Broken pressure plate springs 4. Weak pressure plate springs 5. Grease on driven plate facings (disc)

Troubleshooting Clutch Problems (cont.)

Condition	Possible Cause
Incomplete clutch release	1. Incorrect pedal or linkage adjustment or linkage or cable binding 2. Incorrect height adjustment on pressure plate release levers 3. Loose, broken facings on driven plate (disc) 4. Bent, dished, warped driven plate caused by overheating
Grinding, whirring grating noise when pedal is depressed	1. Worn or defective throwout bearing 2. Starter drive teeth contacting flywheel ring gear teeth. Look for milled or polished teeth on ring gear.
Squeal, howl, trumpeting noise when pedal is being released (occurs during first inch to inch and one-half of pedal travel)	Pilot bushing worn or lack of lubricant. If bushing appears OK, polish bushing with emery cloth, soak lube wick in oil, lube bushing with oil, apply film of chassis grease to clutch shaft pilot hub, reassemble. NOTE: Bushing wear may be due to misalignment of clutch housing or housing to transmission adapter
Vibration or clutch pedal pulsation with clutch disengaged (pedal fully depressed)	1. Worn or defective engine transmission mounts 2. Flywheel run out. (Flywheel run out at face not to exceed 0.005") 3. Damaged or defective clutch components

Troubleshooting Manual Transmission Problems
See Chapter 6 for service procedures

Condition	Possible Cause
Transmission jumps out of gear	1. Misalignment of transmission case or clutch housing. 2. Worn pilot bearing in crankshaft. 3. Bent transmission shaft. 4. Worn high speed sliding gear. 5. Worn teeth or end-play in clutch shaft. 6. Insufficient spring tension on shifter rail plunger. 7. Bent or loose shifter fork. 8. Gears not engaging completely. 9. Loose or worn bearings on clutch shaft or mainshaft. 10. Worn gear teeth. 11. Worn or damaged detent balls.
Transmission sticks in gear	1. Clutch not releasing fully. 2. Burred or battered teeth on clutch shaft, or sliding sleeve. 3. Burred or battered transmission mainshaft. 4. Frozen synchronizing clutch. 5. Stuck shifter rail plunger. 6. Gearshift lever twisting and binding shifter rail. 7. Battered teeth on high speed sliding gear or on sleeve. 8. Improper lubrication, or lack of lubrication. 9. Corroded transmission parts. 10. Defective mainshaft pilot bearing. 11. Locked gear bearings will give same effect as stuck in gear.
Transmission gears will not synchronize	1. Binding pilot bearing on mainshaft, will synchronize in high gear only. 2. Clutch not releasing fully. 3. Detent spring weak or broken. 4. Weak or broken springs under balls in sliding gear sleeve. 5. Binding bearing on clutch shaft, or binding countershaft. 6. Binding pilot bearing in crankshaft. 7. Badly worn gear teeth. 8. Improper lubrication. 9. Constant mesh gear not turning freely on transmission mainshaft. Will synchronize in that gear only.

Condition	Possible Cause
Gears spinning when shifting into gear from neutral	1. Clutch not releasing fully. 2. In some cases an extremely light lubricant in transmission will cause gears to continue to spin for a short time after clutch is released. 3. Binding pilot bearing in crankshaft.
Transmission noisy in all gears	1. Insufficient lubricant, or improper lubricant. 2. Worn countergear bearings. 3. Worn or damaged main drive gear or countergear. 4. Damaged main drive gear or mainshaft bearings. 5. Worn or damaged countergear anti-lash plate.
Transmission noisy in neutral only	1. Damaged main drive gear bearing. 2. Damaged or loose mainshaft pilot bearing. 3. Worn or damaged countergear anti-lash plate. 4. Worn countergear bearings.
Transmission noisy in one gear only	1. Damaged or worn constant mesh gears. 2. Worn or damaged countergear bearings. 3. Damaged or worn synchronizer.
Transmission noisy in reverse only	1. Worn or damaged reverse idler gear or idler bushing. 2. Worn or damaged mainshaft reverse gear. 3. Worn or damaged reverse countergear. 4. Damaged shift mechanism.

TROUBLESHOOTING AUTOMATIC TRANSMISSION PROBLEMS

Keeping alert to changes in the operating characteristics of the transmission (changing shift points, noises, etc.) can prevent small problems from becoming large ones. If the problem cannot be traced to loose bolts, fluid level, misadjusted linkage, clogged filters or similar problems, you should probably seek professional service.

Transmission Fluid Indications

The appearance and odor of the transmission fluid can give valuable clues to the overall condition of the transmission. Always note the appearance of the fluid when you check the fluid level or change the fluid. Rub a small amount of fluid between your fingers to feel for grit and smell the fluid on the dipstick.

If the fluid appears:	It indicates:
Clear and red colored	Normal operation
Discolored (extremely dark red or brownish) or smells burned	Band or clutch pack failure, usually caused by an overheated transmission. Hauling very heavy loads with insufficient power or failure to change the fluid often result in overheating. Do not confuse this appearance with newer fluids that have a darker red color and a strong odor (though not a burned odor).
Foamy or aerated (light in color and full of bubbles)	1. The level is too high (gear train is churning oil) 2. An internal air leak (air is mixing with the fluid). Have the transmission checked professionally.
Solid residue in the fluid	Defective bands, clutch pack or bearings. Bits of band material or metal abrasives are clinging to the dipstick. Have the transmission checked professionally.
Varnish coating on the dipstick	The transmission fluid is overheating

TROUBLESHOOTING DRIVE AXLE PROBLEMS

First, determine when the noise is most noticeable.

Drive Noise: Produced under vehicle acceleration.

Coast Noise: Produced while coasting with a closed throttle.

Float Noise: Occurs while maintaining constant speed (just enough to keep speed constant) on a level road.

External Noise Elimination

It is advisable to make a thorough road test to determine whether the noise originates in the rear axle or whether it originates from the tires, engine, transmission, wheel bearings or road surface. Noise originating from other places cannot be corrected by servicing the rear axle.

ROAD NOISE

Brick or rough surfaced concrete roads produce noises that seem to come from the rear axle. Road noise is usually identical in Drive or Coast and driving on a different type of road will tell whether the road is the problem.

TIRE NOISE

Tire noise can be mistaken as rear axle noise, even though the tires on the front are at fault. Snow tread and mud tread tires or tires worn unevenly will frequently cause vibrations which seem to originate elsewhere; *temporarily, and for test purposes only,* inflate the tires to 40–50 lbs. This will significantly alter the noise produced by the tires, but will not alter noise from the rear axle. Noises from the rear axle will normally cease at speeds below 30 mph on coast, while tire noise will continue at lower tone as speed is decreased. The rear axle noise will usually change from drive conditions to coast conditions, while tire noise will not. Do not forget to lower the tire pressure to normal after the test is complete.

ENGINE/TRANSMISSION NOISE

Determine at what speed the noise is most pronounced, then stop in a quiet place. With the transmission in Neutral, run the engine through speeds corresponding to road speeds where the noise was noticed. Noises produced with the vehicle standing still are coming from the engine or transmission.

FRONT WHEEL BEARINGS

Front wheel bearing noises, sometimes confused with rear axle noises, will not change when comparing drive and coast conditions. While holding the speed steady, lightly apply the footbrake. This will often cause wheel bearing noise to lessen, as some of the weight is taken off the bearing. Front wheel bearings are easily checked by jacking up the wheels and spinning the wheels. Shaking the wheels will also determine if the wheel bearings are excessively loose.

REAR AXLE NOISES

Eliminating other possible sources can narrow the cause to the rear axle, which normally produces noise from worn gears or bearings. Gear noises tend to peak in a narrow speed range, while bearing noises will usually vary in pitch with engine speeds.

Noise Diagnosis

The Noise Is:	Most Probably Produced By:
1. Identical under Drive or Coast	Road surface, tires or front wheel bearings
2. Different depending on road surface	Road surface or tires
3. Lower as speed is lowered	Tires
4. Similar when standing or moving	Engine or transmission
5. A vibration	Unbalanced tires, rear wheel bearing, unbalanced driveshaft or worn U-joint
6. A knock or click about every two tire revolutions	Rear wheel bearing
7. Most pronounced on turns	Damaged differential gears
8. A steady low-pitched whirring or scraping, starting at low speeds	Damaged or worn pinion bearing
9. A chattering vibration on turns	Wrong differential lubricant or worn clutch plates (limited slip rear axle)
10. Noticed only in Drive, Coast or Float conditions	Worn ring gear and/or pinion gear

Troubleshooting Steering & Suspension Problems

Condition	Possible Cause
Hard steering (wheel is hard to turn)	1. Improper tire pressure 2. Loose or glazed pump drive belt 3. Low or incorrect fluid 4. Loose, bent or poorly lubricated front end parts 5. Improper front end alignment (excessive caster) 6. Bind in steering column or linkage 7. Kinked hydraulic hose 8. Air in hydraulic system 9. Low pump output or leaks in system 10. Obstruction in lines 11. Pump valves sticking or out of adjustment 12. Incorrect wheel alignment
Loose steering (too much play in steering wheel)	1. Loose wheel bearings 2. Faulty shocks 3. Worn linkage or suspension components 4. Loose steering gear mounting or linkage points 5. Steering mechanism worn or improperly adjusted 6. Valve spool improperly adjusted 7. Worn ball joints, tie-rod ends, etc.
Veers or wanders (pulls to one side with hands off steering wheel)	1. Improper tire pressure 2. Improper front end alignment 3. Dragging or improperly adjusted brakes 4. Bent frame 5. Improper rear end alignment 6. Faulty shocks or springs 7. Loose or bent front end components 8. Play in Pitman arm 9. Steering gear mountings loose 10. Loose wheel bearings 11. Binding Pitman arm 12. Spool valve sticking or improperly adjusted 13. Worn ball joints
Wheel oscillation or vibration transmitted through steering wheel	1. Low or uneven tire pressure 2. Loose wheel bearings 3. Improper front end alignment 4. Bent spindle 5. Worn, bent or broken front end components 6. Tires out of round or out of balance 7. Excessive lateral runout in disc brake rotor 8. Loose or bent shock absorber or strut
Noises (see also "Troubleshooting Drive Axle Problems")	1. Loose belts 2. Low fluid, air in system 3. Foreign matter in system 4. Improper lubrication 5. Interference or chafing in linkage 6. Steering gear mountings loose 7. Incorrect adjustment or wear in gear box 8. Faulty valves or wear in pump 9. Kinked hydraulic lines 10. Worn wheel bearings
Poor return of steering	1. Over-inflated tires 2. Improperly aligned front end (excessive caster) 3. Binding in steering column 4. No lubrication in front end 5. Steering gear adjusted too tight
Uneven tire wear (see "How To Read Tire Wear")	1. Incorrect tire pressure 2. Improperly aligned front end 3. Tires out-of-balance 4. Bent or worn suspension parts

HOW TO READ TIRE WEAR

The way your tires wear is a good indicator of other parts of the suspension. Abnormal wear patterns are often caused by the need for simple tire maintenance, or for front end alignment.

Excessive wear at the center of the tread indicates that the air pressure in the tire is consistently too high. The tire is riding on the center of the tread and wearing it prematurely. Occasionally, this wear pattern can result from outrageously wide tires on narrow rims. The cure for this is to replace either the tires or the wheels.

This type of wear usually results from consistent under-inflation. When a tire is under-inflated, there is too much contact with the road by the outer treads, which wear prematurely. When this type of wear occurs, and the tire pressure is known to be consistently correct, a bent or worn steering component or the need for wheel alignment could be indicated.

Feathering is a condition when the edge of each tread rib develops a slightly rounded edge on one side and a sharp edge on the other. By running your hand over the tire, you can usually feel the sharper edges before you'll be able to see them. The most common causes of feathering are incorrect toe-in setting or deteriorated bushings in the front suspension.

When an inner or outer rib wears faster than the rest of the tire, the need for wheel alignment is indicated. There is excessive camber in the front suspension, causing the wheel to lean too much putting excessive load on one side of the tire. Misalignment could also be due to sagging springs, worn ball joints, or worn control arm bushings. Be sure the vehicle is loaded the way it's normally driven when you have the wheels aligned.

Cups or scalloped dips appearing around the edge of the tread almost always indicate worn (sometimes bent) suspension parts. Adjustment of wheel alignment alone will seldom cure the problem. Any worn component that connects the wheel to the suspension can cause this type of wear. Occasionally, wheels that are out of balance will wear like this, but wheel imbalance usually shows up as bald spots between the outside edges and center of the tread.

Second-rib wear is usually found only in radial tires, and appears where the steel belts end in relation to the tread. It can be kept to a minimum by paying careful attention to tire pressure and frequently rotating the tires. This is often considered normal wear but excessive amounts indicate that the tires are too wide for the wheels.

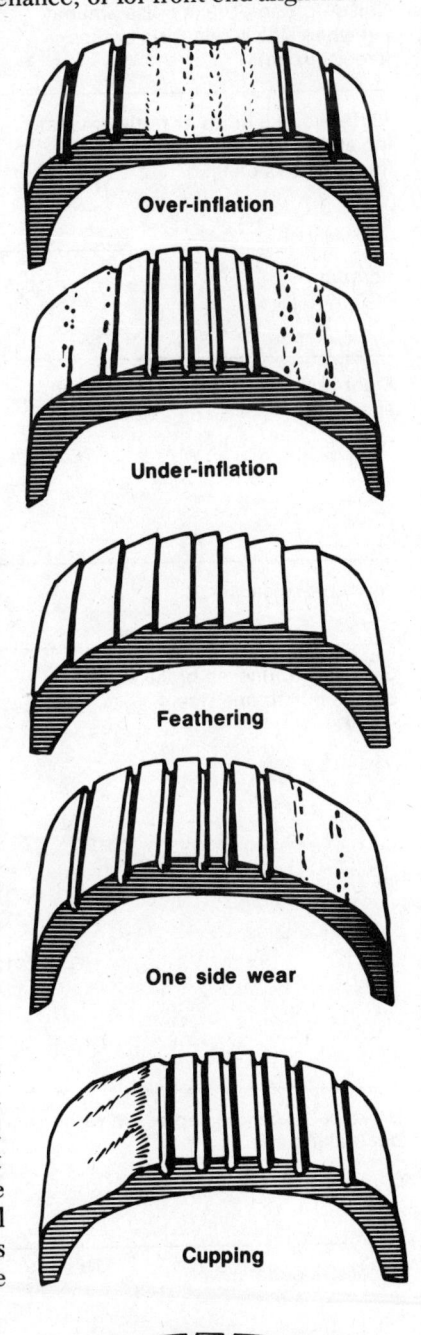

Over-inflation

Under-inflation

Feathering

One side wear

Cupping

Second-rib wear

Troubleshooting Disc Brake Problems

Condition	Possible Cause
Noise—groan—brake noise emanating when slowly releasing brakes (creep-groan)	Not detrimental to function of disc brakes—no corrective action required. (This noise may be eliminated by slightly increasing or decreasing brake pedal efforts.)
Rattle—brake noise or rattle emanating at low speeds on rough roads, (front wheels only).	1. Shoe anti-rattle spring missing or not properly positioned. 2. Excessive clearance between shoe and caliper. 3. Soft or broken caliper seals. 4. Deformed or misaligned disc. 5. Loose caliper.
Scraping	1. Mounting bolts too long. 2. Loose wheel bearings. 3. Bent, loose, or misaligned splash shield.
Front brakes heat up during driving and fail to release	1. Operator riding brake pedal. 2. Stop light switch improperly adjusted. 3. Sticking pedal linkage. 4. Frozen or seized piston. 5. Residual pressure valve in master cylinder. 6. Power brake malfunction. 7. Proportioning valve malfunction.
Leaky brake caliper	1. Damaged or worn caliper piston seal. 2. Scores or corrosion on surface of cylinder bore.
Grabbing or uneven brake action—Brakes pull to one side	1. Causes listed under "Brakes Pull". 2. Power brake malfunction. 3. Low fluid level in master cylinder. 4. Air in hydraulic system. 5. Brake fluid, oil or grease on linings. 6. Unmatched linings. 7. Distorted brake pads. 8. Frozen or seized pistons. 9. Incorrect tire pressure. 10. Front end out of alignment. 11. Broken rear spring. 12. Brake caliper pistons sticking. 13. Restricted hose or line. 14. Caliper not in proper alignment to braking disc. 15. Stuck or malfunctioning metering valve. 16. Soft or broken caliper seals. 17. Loose caliper.
Brake pedal can be depressed without braking effect	1. Air in hydraulic system or improper bleeding procedure. 2. Leak past primary cup in master cylinder. 3. Leak in system. 4. Rear brakes out of adjustment. 5. Bleeder screw open.
Excessive pedal travel	1. Air, leak, or insufficient fluid in system or caliper. 2. Warped or excessively tapered shoe and lining assembly. 3. Excessive disc runout. 4. Rear brake adjustment required. 5. Loose wheel bearing adjustment. 6. Damaged caliper piston seal. 7. Improper brake fluid (boil). 8. Power brake malfunction. 9. Weak or soft hoses.

Troubleshooting Disc Brake Problems (cont.)

Condition	Possible Cause
Brake roughness or chatter (pedal pumping)	1. Excessive thickness variation of braking disc. 2. Excessive lateral runout of braking disc. 3. Rear brake drums out-of-round. 4. Excessive front bearing clearance.
Excessive pedal effort	1. Brake fluid, oil or grease on linings. 2. Incorrect lining. 3. Frozen or seized pistons. 4. Power brake malfunction. 5. Kinked or collapsed hose or line. 6. Stuck metering valve. 7. Scored caliper or master cylinder bore. 8. Seized caliper pistons.
Brake pedal fades (pedal travel increases with foot on brake)	1. Rough master cylinder or caliper bore. 2. Loose or broken hydraulic lines/connections. 3. Air in hydraulic system. 4. Fluid level low. 5. Weak or soft hoses. 6. Inferior quality brake shoes or fluid. 7. Worn master cylinder piston cups or seals.

Troubleshooting Drum Brakes

Condition	Possible Cause
Pedal goes to floor	1. Fluid low in reservoir. 2. Air in hydraulic system. 3. Improperly adjusted brake. 4. Leaking wheel cylinders. 5. Loose or broken brake lines. 6. Leaking or worn master cylinder. 7. Excessively worn brake lining.
Spongy brake pedal	1. Air in hydraulic system. 2. Improper brake fluid (low boiling point). 3. Excessively worn or cracked brake drums. 4. Broken pedal pivot bushing.
Brakes pulling	1. Contaminated lining. 2. Front end out of alignment. 3. Incorrect brake adjustment. 4. Unmatched brake lining. 5. Brake drums out of round. 6. Brake shoes distorted. 7. Restricted brake hose or line. 8. Broken rear spring. 9. Worn brake linings. 10. Uneven lining wear. 11. Glazed brake lining. 12. Excessive brake lining dust. 13. Heat spotted brake drums. 14. Weak brake return springs. 15. Faulty automatic adjusters. 16. Low or incorrect tire pressure.

Condition	Possible Cause
Squealing brakes	1. Glazed brake lining. 2. Saturated brake lining. 3. Weak or broken brake shoe retaining spring. 4. Broken or weak brake shoe return spring. 5. Incorrect brake lining. 6. Distorted brake shoes. 7. Bent support plate. 8. Dust in brakes or scored brake drums. 9. Linings worn below limit. 10. Uneven brake lining wear. 11. Heat spotted brake drums.
Chirping brakes	1. Out of round drum or eccentric axle flange pilot.
Dragging brakes	1. Incorrect wheel or parking brake adjustment. 2. Parking brakes engaged or improperly adjusted. 3. Weak or broken brake shoe return spring. 4. Brake pedal binding. 5. Master cylinder cup sticking. 6. Obstructed master cylinder relief port. 7. Saturated brake lining. 8. Bent or out of round brake drum. 9. Contaminated or improper brake fluid. 10. Sticking wheel cylinder pistons. 11. Driver riding brake pedal. 12. Defective proportioning valve. 13. Insufficient brake shoe lubricant.
Hard pedal	1. Brake booster inoperative. 2. Incorrect brake lining. 3. Restricted brake line or hose. 4. Frozen brake pedal linkage. 5. Stuck wheel cylinder. 6. Binding pedal linkage. 7. Faulty proportioning valve.
Wheel locks	1. Contaminated brake lining. 2. Loose or torn brake lining. 3. Wheel cylinder cups sticking. 4. Incorrect wheel bearing adjustment. 5. Faulty proportioning valve.
Brakes fade (high speed)	1. Incorrect lining. 2. Overheated brake drums. 3. Incorrect brake fluid (low boiling temperature). 4. Saturated brake lining. 5. Leak in hydraulic system. 6. Faulty automatic adjusters.
Pedal pulsates	1. Bent or out of round brake drum.
Brake chatter and shoe knock	1. Out of round brake drum. 2. Loose support plate. 3. Bent support plate. 4. Distorted brake shoes. 5. Machine grooves in contact face of brake drum (Shoe Knock). 6. Contaminated brake lining. 7. Missing or loose components. 8. Incorrect lining material. 9. Out-of-round brake drums. 10. Heat spotted or scored brake drums. 11. Out-of-balance wheels.

Troubleshooting Drum Brakes (cont.)

Condition	Possible Cause
Brakes do not self adjust	1. Adjuster screw frozen in thread. 2. Adjuster screw corroded at thrust washer. 3. Adjuster lever does not engage star wheel. 4. Adjuster installed on wrong wheel.
Brake light glows	1. Leak in the hydraulic system. 2. Air in the system. 3. Improperly adjusted master cylinder pushrod. 4. Uneven lining wear. 5. Failure to center combination valve or proportioning valve.

Mechanic's Data

General Conversion Table

Multiply By	To Convert	To	
		LENGTH	
2.54	Inches	Centimeters	.3937
25.4	Inches	Millimeters	.03937
30.48	Feet	Centimeters	.0328
.304	Feet	Meters	3.28
.914	Yards	Meters	1.094
1.609	Miles	Kilometers	.621
		VOLUME	
.473	Pints	Liters	2.11
.946	Quarts	Liters	1.06
3.785	Gallons	Liters	.264
.016	Cubic inches	Liters	61.02
16.39	Cubic inches	Cubic cms.	.061
28.3	Cubic feet	Liters	.0353
		MASS (Weight)	
28.35	Ounces	Grams	.035
.4536	Pounds	Kilograms	2.20
—	To obtain	From	Multiply by

Multiply By	To Convert	To	
		AREA	
.645	Square inches	Square cms.	.155
.836	Square yds.	Square meters	1.196
		FORCE	
4.448	Pounds	Newtons	.225
.138	Ft./lbs.	Kilogram/meters	7.23
1.36	Ft./lbs.	Newton-meters	.737
.112	In./lbs.	Newton-meters	8.844
		PRESSURE	
.068	Psi	Atmospheres	14.7
6.89	Psi	Kilopascals	.145
		OTHER	
1.104	Horsepower (DIN)	Horsepower (SAE)	.9861
.746	Horsepower (SAE)	Kilowatts (KW)	1.34
1.60	Mph	Km/h	.625
.425	Mpg	Km/1	2.35
—	To obtain	From	Multiply by

Tap Drill Sizes

National Coarse or U.S.S.

Screw & Tap Size	Threads Per Inch	Use Drill Number
No. 5	.40	.39
No. 6	.32	.36
No. 8	.32	.29
No. 10	.24	.25
No. 12	.24	.17
1/4	.20	8
5/16	.18	F
3/8	.16	5/16
7/16	.14	U
1/2	.13	27/64
9/16	.12	31/64
5/8	.11	17/32
3/4	.10	21/32
7/8	9	49/64

National Coarse or U.S.S.

Screw & Tap Size	Threads Per Inch	Use Drill Number
1	8	7/8
1 1/8	7	63/64
1 1/4	7	1 7/64
1 1/2	6	1 11/32

National Fine or S.A.E.

Screw & Tap Size	Threads Per Inch	Use Drill Number
No. 5	.44	.37
No. 6	.40	.33
No. 8	.36	.29
No. 10	.32	.21

National Fine or S.A.E.

Screw & Tap Size	Threads Per Inch	Use Drill Number
No. 12	.28	.15
1/4	.28	3
5/16	.24	1
3/8	.24	Q
7/16	.20	W
1/2	.20	29/64
9/16	.18	33/64
5/8	.18	37/64
3/4	.16	11/16
7/8	.14	13/16
1 1/8	.12	1 3/64
1 1/4	.12	1 11/64
1 1/2	.12	1 27/64

Drill Sizes In Decimal Equivalents

Inch	Decimal	Wire	mm
1/64	.0156		.39
	.0157		.4
	.0160	78	
	.0165		.42
	.0173		.44
	.0177		.45
	.0180	77	
	.0181		.46
	.0189		.48
	.0197		.5
	.0200	76	
	.0210	75	
	.0217		.55
	.0225	74	
	.0236		.6
	.0240	73	
	.0250	72	
	.0256		.65
	.0260	71	
	.0276		.7
	.0280	70	
	.0292	69	
	.0295		.75
	.0310	68	
1/32	.0312		.79
	.0315		.8
	.0320	67	
	.0330	66	
	.0335		.85
	.0350	65	
	.0354		.9
	.0360	64	
	.0370	63	
	.0374		.95
	.0380	62	
	.0390	61	
	.0394		1.0
	.0400	60	
	.0410	59	
	.0413		1.05
	.0420	58	
	.0430	57	
	.0433		1.1
	.0453		1.15
	.0465	56	
3/64	.0469		1.19
	.0472		1.2
	.0492		1.25
	.0512		1.3
	.0520	55	
	.0531		1.35
	.0550	54	
	.0551		1.4
	.0571		1.45
	.0591		1.5
	.0595	53	
	.0610		1.55
1/16	.0625		1.59
	.0630		1.6
	.0635	52	
	.0650		1.65
	.0669		1.7
	.0670	51	
	.0689		1.75
	.0700	50	
	.0709		1.8
	.0728		1.85

Inch	Decimal	Wire	mm
	.0730	49	
	.0748		1.9
	.0760	48	
	.0768		1.95
5/64	.0781		1.98
	.0785	47	
	.0787		2.0
	.0807		2.05
	.0810	46	
	.0820	45	
	.0827		2.1
	.0846		2.15
	.0860	44	
	.0866		2.2
	.0886		2.25
	.0890	43	
	.0906		2.3
	.0925		2.35
	.0935	42	
3/32	.0938		2.38
	.0945		2.4
	.0960	41	
	.0965		2.45
	.0980	40	
	.0981		2.5
	.0995	39	
	.1015	38	
	.1024		2.6
	.1040	37	
	.1063		2.7
	.1065	36	
	.1083		2.75
7/64	.1094		2.77
	.1100	35	
	.1102		2.8
	.1110	34	
	.1130	33	
	.1142		2.9
	.1160	32	
	.1181		3.0
	.1200	31	
	.1220		3.1
1/8	.1250		3.17
	.1260		3.2
	.1280		3.25
	.1285	30	
	.1299		3.3
	.1339		3.4
	.1360	29	
	.1378		3.5
	.1405	28	
9/64	.1406		3.57
	.1417		3.6
	.1440	27	
	.1457		3.7
	.1470	26	
	.1476		3.75
	.1495	25	
	.1496		3.8
	.1520	24	
	.1535		3.9
	.1540	23	
5/32	.1562		3.96
	.1570	22	
	.1575		4.0
	.1590	21	
	.1610	20	

Inch	Decimal	Wire & Letter	mm
	.1614		4.1
	.1654		4.2
	.1660	19	
	.1673		4.25
	.1693		4.3
	.1695	18	
11/64	.1719		4.36
	.1730	17	
	.1732		4.4
	.1770	16	
	.1772		4.5
	.1800	15	
	.1811		4.6
	.1820	14	
	.1850	13	
	.1850		4.7
	.1870		4.75
3/16	.1875		4.76
	.1890		4.8
	.1890	12	
	.1910	11	
	.1929		4.9
	.1935	10	
	.1960	9	
	.1969		5.0
	.1990	8	
	.2008		5.1
	.2010	7	
13/64	.2031		5.16
	.2040	6	
	.2047		5.2
	.2055	5	
	.2067		5.25
	.2087		5.3
	.2090	4	
	.2126		5.4
	.2130	3	
	.2165		5.5
7/32	.2188		5.55
	.2205		5.6
	.2210	2	
	.2244		5.7
	.2264		5.75
	.2280	1	
	.2283		5.8
	.2323		5.9
	.2340	A	
15/64	.2344		5.95
	.2362		6.0
	.2380	B	
	.2402		6.1
	.2420	C	
	.2441		6.2
	.2460	D	
	.2461		6.25
	.2480		6.3
1/4	.2500	E	6.35
	.2520		6.
	.2559		6.5
	.2570	F	
	.2598		6.6
	.2610	G	
	.2638		6.7
17/64	.2656		6.74
	.2657		6.75
	.2660	H	
	.2677		6.8

Inch	Decimal	Letter	mm
	.2717		6.9
	.2720	I	
	.2756		7.0
	.2770	J	
	.2795		7.1
	.2810	K	
9/32	.2812		7.14
	.2835		7.2
	.2854		7.25
	.2874		7.3
	.2900	L	
	.2913		7.4
	.2950	M	
	.2953		7.5
19/64	.2969		7.54
	.2992		7.6
	.3020	N	
	.3031		7.7
	.3051		7.75
	.3071		7.8
	.3110		7.9
5/16	.3125		7.93
	.3150		8.0
	.3160	O	
	.3189		8.1
	.3228		8.2
	.3230	P	
	.3248		8.25
	.3268		8.3
21/64	.3281		8.33
	.3307		8.4
	.3320	Q	
	.3346		8.5
	.3386		8.6
	.3390	R	
	.3425		8.7
11/32	.3438		8.73
	.3445		8.75
	.3465		8.8
	.3480	S	
	.3504		8.9
	.3543		9.0
	.3580	T	
	.3583		9.1
23/64	.3594		9.12
	.3622		9.2
	.3642		9.25
	.3661		9.3
	.3680	U	
	.3701		9.4
	.3740		9.5
3/8	.3750		9.52
	.3770	V	
	.3780		9.6
	.3819		9.7
	.3839		9.75
	.3858		9.8
	.3860	W	
	.3898		9.9
25/64	.3906		9.92
	.3937		10.0
	.3970	X	
	.4040	Y	
	.4062		10.31
13/32	.4130	Z	
	.4134		10.5
27/64	.4219		10.71

Inch	Decimal	mm
	.4331	11.0
7/16	.4375	11.11
	.4528	11.5
29/64	.4531	11.51
15/32	.4688	11.90
	.4724	12.0
31/64	.4844	12.30
	.4921	12.5
1/2	.5000	12.70
33/64	.5156	13.09
17/32	.5312	13.49
	.5315	13.5
35/64	.5469	13.89
	.5512	14.0
9/16	.5625	14.28
	.5709	14.5
37/64	.5781	14.68
	.5906	15.0
19/32	.5938	15.08
39/64	.6094	15.47
	.6102	15.5
5/8	.6250	15.87
	.6299	16.0
41/64	.6406	16.27
	.6496	16.5
21/32	.6562	16.66
	.6693	17.0
43/64	.6719	17.06
11/16	.6875	17.46
	.6890	17.5
45/64	.7031	17.85
	.7087	18.0
23/32	.7188	18.25
	.7283	18.5
47/64	.7344	18.65
	.7480	19.0
3/4	.7500	19.05
49/64	.7656	19.44
	.7677	19.5
25/32	.7812	19.84
	.7874	20.0
51/64	.7969	20.24
	.8071	20.5
13/16	.8125	20.63
	.8268	21.0
53/64	.8281	21.03
27/32	.8438	21.43
	.8465	21.5
55/64	.8594	21.82
	.8661	22.0
7/8	.8750	22.22
	.8858	22.5
57/64	.8906	22.62
	.9055	23.0
29/32	.9062	23.01
59/64	.9219	23.41
	.9252	23.5
15/16	.9375	23.81
	.9449	24.0
61/64	.9531	24.2
	.9646	24.5
31/32	.9688	24.6
	.9843	25.0
63/64	.9844	25.0
1	1.0000	25.4

Index

Chilton's Repair & Tune-Up Guides

The Complete line covers domestic cars, imports, trucks, vans, RV's and 4-wheel drive vehicles.

RTUG Title	Part No.	RTUG Title	Part No.
AMC 1975-82	7199	**Corvair 1960-69**	6691
Covers all U.S. and Canadian models		Covers all U.S. and Canadian models	
Aspen/Volare 1976-80	6637	**Corvette 1953-62**	6576
Covers all U.S. and Canadian models		Covers all U.S. and Canadian models	
Audi 1970-73	5902	**Corvette 1963-84**	6843
Covers all U.S. and Canadian models.		Covers all U.S. and Canadian models	
Audi 4000/5000 1978-81	7028	**Cutlass 1970-85**	6933
Covers all U.S. and Canadian models including turbocharged and diesel engines		Covers all U.S. and Canadian models	
Barracuda/Challenger 1965-72	5807	**Dart/Demon 1968-76**	6324
Covers all U.S. and Canadian models		Covers all U.S. and Canadian models	
Blazer/Jimmy 1969-82	6931	**Datsun 1961-72**	5790
Covers all U.S. and Canadian 2- and 4-wheel drive models, including diesel engines		Covers all U.S. and Canadian models of Nissan Patrol; 1500, 1600 and 2000 sports cars; Pick-Ups; 410, 411, 510, 1200 and 240Z	
BMW 1970-82	6844	**Datsun 1973-80 Spanish**	7083
Covers U.S. and Canadian models		**Datsun/Nissan F-10, 310, Stanza, Pulsar 1977-86**	7196
Buick/Olds/Pontiac 1975-85	7308	Covers all U.S. and Canadian models	
Covers all U.S. and Canadian full size rear wheel drive models		**Datsun/Nissan Pick-Ups 1970-84**	6816
Cadillac 1967-84	7462	Covers all U.S and Canadian models	
Covers all U.S. and Canadian rear wheel drive models		**Datsun/Nissan Z & ZX 1970-86**	6932
Camaro 1967-81	6735	Covers all U.S. and Canadian models	
Covers all U.S. and Canadian models		**Datsun/Nissan 1200, 210, Sentra 1973-86**	7197
Camaro 1982-85	7317	Covers all U.S. and Canadian models	
Covers all U.S. and Canadian models		**Datsun/Nissan 200SX, 510, 610, 710, 810, Maxima 1973-84**	7170
Capri 1970-77	6695	Covers all U.S. and Canadian models	
Covers all U.S. and Canadian models		**Dodge 1968-77**	6554
Caravan/Voyager 1984-85	7482	Covers all U.S. and Canadian models	
Covers all U.S. and Canadian models		**Dodge Charger 1967-70**	6486
Century/Regal 1975-85	7307	Covers all U.S. and Canadian models	
Covers all U.S. and Canadian rear wheel drive models, including turbocharged engines		**Dodge/Plymouth Trucks 1967-84**	7459
Champ/Arrow/Sapporo 1978-83	7041	Covers all $1/_2$, $3/_4$, and 1 ton 2- and 4-wheel drive U.S. and Canadian models, including diesel engines	
Covers all U.S. and Canadian models		**Dodge/Plymouth Vans 1967-84**	6934
Chevette/1000 1976-86	6836	Covers all $1/_2$, $3/_4$, and 1 ton U.S. and Canadian models of vans, cutaways and motor home chassis	
Covers all U.S. and Canadian models		**D-50/Arrow Pick-Up 1979-81**	7032
Chevrolet 1968-85	7135	Covers all U.S. and Canadian models	
Covers all U.S. and Canadian models		**Fairlane/Torino 1962-75**	6320
Chevrolet 1968-79 Spanish	7082	Covers all U.S. and Canadian models	
Chevrolet/GMC Pick-Ups 1970-82 Spanish	7468	**Fairmont/Zephyr 1978-83**	6965
Chevrolet/GMC Pick-Ups and Suburban 1970-86	6936	Covers all U.S. and Canadian models	
Covers all U.S. and Canadian $1/_2$, $3/_4$ and 1 ton models, including 4-wheel drive and diesel engines		**Fiat 1969-81**	7042
		Covers all U.S. and Canadian models	
Chevrolet LUV 1972-81	6815	**Fiesta 1978-80**	6846
Covers all U.S. and Canadian models		Covers all U.S. and Canadian models	
Chevrolet Mid-Size 1964-86	6840	**Firebird 1967-81**	5996
Covers all U.S. and Canadian models of 1964-77 Chevelle, Malibu and Malibu SS; 1974-77 Laguna; 1978-85 Malibu; 1970-86 Monte Carlo; 1964-84 El Camino, including diesel engines		Covers all U.S. and Canadian models	
		Firebird 1982-85	7345
		Covers all U.S. and Canadian models	
		Ford 1968-79 Spanish	7084
Chevrolet Nova 1986	7658	**Ford Bronco 1966-83**	7140
Covers all U.S. and Canadian models		Covers all U.S. and Canadian models	
Chevy/GMC Vans 1967-84	6930	**Ford Bronco II 1984**	7408
Covers all U.S. and Canadian models of $1/_2$, $3/_4$, and 1 ton vans, cutaways, and motor home chassis, including diesel engines		Covers all U.S. and Canadian models	
		Ford Courier 1972-82	6983
		Covers all U.S. and Canadian models	
Chevy S-10 Blazer/GMC S-15 Jimmy 1982-85	7383	**Ford/Mercury Front Wheel Drive 1981-85**	7055
Covers all U.S. and Canadian models		Covers all U.S. and Canadian models Escort, EXP, Tempo, Lynx, LN-7 and Topaz	
Chevy S-10/GMC S-15 Pick-Ups 1982-85	7310	**Ford/Mercury/Lincoln 1968-85**	6842
Covers all U.S. and Canadian models		Covers all U.S. and Canadian models of FORD Country Sedan, Country Squire, Crown Victoria, Custom, Custom 500, Galaxie 500, LTD through 1982, Ranch Wagon, and XL; MERCURY Colony Park, Commuter, Marquis through 1982, Gran Marquis, Monterey and Park Lane; LINCOLN Continental and Towne Car	
Chevy II/Nova 1962-79	6841		
Covers all U.S. and Canadian models			
Chrysler K- and E-Car 1981-85	7163		
Covers all U.S. and Canadian front wheel drive models			
Colt/Challenger/Vista/Conquest 1971-85	7037		
Covers all U.S. and Canadian models			
Corolla/Carina/Tercel/Starlet 1970-85	7036	**Ford/Mercury/Lincoln Mid-Size 1971-85**	6696
Corona/Cressida/Crown/Mk.II/Camry/Van 1970-84	7044	Covers all U.S. and Canadian models of FORD Elite, 1983-85 LTD, 1977-79 LTD II, Ranchero, Torino, Gran Torino, 1977-85 Thunderbird; MERCURY 1972-85 Cougar,	
Covers all U.S. and Canadian models			

continued on next page

RTUG Title	Part No.	RTUG Title	Part No.
1983-85 Marquis, Montego, 1980-85 XR-7; LINCOLN 1982-85 Continental, 1984-85 Mark VII, 1978-80 Versailles		**Mercedes-Benz 1974-84** Covers all U.S. and Canadian models	6809
Ford Pick-Ups 1965-86 Covers all $\frac{1}{2}$, $\frac{3}{4}$ and 1 ton, 2- and 4-wheel drive U.S. and Canadian pick-up, chassis cab and camper models, including diesel engines	6913	**Mitsubishi, Cordia, Tredia, Starion, Galant 1983-85** Covers all U.S. and Canadian models	7583
		MG 1961-81 Covers all U.S. and Canadian models	6780
Ford Pick-Ups 1965-82 Spanish	7469	**Mustang/Capri/Merkur 1979-85** Covers all U.S. and Canadian models	6963
Ford Ranger 1983-84 Covers all U.S. and Canadian models	7338	**Mustang/Cougar 1965-73** Covers all U.S. and Canadian models	6542
Ford Vans 1961-86 Covers all U.S. and Canadian $\frac{1}{2}$, $\frac{3}{4}$ and 1 ton van and cutaway chassis models, including diesel engines	6849	**Mustang II 1974-78** Covers all U.S. and Canadian models	6812
		Omni/Horizon/Rampage 1978-84 Covers all U.S. and Canadian models of DODGE omni, Miser, 024, Charger 2.2; PLYMOUTH Horizon, Miser, TC3, TC3 Tourismo; Rampage	6845
GM A-Body 1982-85 Covers all front wheel drive U.S. and Canadian models of BUICK Century, CHEVROLET Celebrity, OLDSMOBILE Cutlass Ciera and PONTIAC 6000	7309		
		Opel 1971-75 Covers all U.S. and Canadian models	6575
GM C-Body 1985 Covers all front wheel drive U.S. and Canadian models of BUICK Electra Park Avenue and Electra T-Type, CADILLAC Fleetwood and deVille, OLDSMOBILE 98 Regency and Regency Brougham	7587	**Peugeot 1970-74** Covers all U.S. and Canadian models	5982
		Pinto/Bobcat 1971-80 Covers all U.S. and Canadian models	7027
		Plymouth 1968-76 Covers all U.S. and Canadian models	6552
		Pontiac Fiero 1984-85 Covers all U.S. and Canadian models	7571
GM J-Car 1982-85 Covers all U.S. and Canadian models of BUICK Skyhawk, CHEVROLET Cavalier, CADILLAC Cimarron, OLDSMOBILE Firenza and PONTIAC 2000 and Sunbird	7059	**Pontiac Mid-Size 1974-83** Covers all U.S. and Canadian models of Ventura, Grand Am, LeMans, Grand LeMans, GTO, Phoenix, and Grand Prix	7346
		Porsche 924/928 1976-81 Covers all U.S. and Canadian models	7048
GM N-Body 1985-86 Covers all U.S. and Canadian models of front wheel drive BUICK Somerset and Skylark, OLDSMOBILE Calais, and PONTIAC Grand Am	7657	**Renault 1975-85** Covers all U.S. and Canadian models	7165
		Roadrunner/Satellite/Belvedere/GTX 1968-73 Covers all U.S. and Canadian models	5821
GM X-Body 1980-85 Covers all U.S. and Canadian models of BUICK Skylark, CHEVROLET Citation, OLDSMOBILE Omega and PONTIAC Phoenix	7049	**RX-7 1979-81** Covers all U.S. and Canadian models	7031
		SAAB 99 1969-75 Covers all U.S. and Canadian models	5988
GM Subcompact 1971-80 Covers all U.S. and Canadian models of BUICK Skyhawk (1975-80), CHEVROLET Vega and Monza, OLDSMOBILE Starfire, and PONTIAC Astre and 1975-80 Sunbird	6935	**SAAB 900 1979-85** Covers all U.S. and Canadian models	7572
		Snowmobiles 1976-80 Covers Arctic Cat, John Deere, Kawasaki, Polaris, Ski-Doo and Yamaha	6978
		Subaru 1970-84 Covers all U.S. and Canadian models	6982
Granada/Monarch 1975-82 Covers all U.S. and Canadian models	6937	**Tempest/GTO/LeMans 1968-73** Covers all U.S. and Canadian models	5905
Honda 1973-84 Covers all U.S. and Canadian models	6980	**Toyota 1966-70** Covers all U.S. and Canadian models of Corona, MkII, Corolla, Crown, Land Cruiser, Stout and Hi-Lux	5795
International Scout 1967-73 Covers all U.S. and Canadian models	5912		
Jeep 1945-87 Covers all U.S. and Canadian CJ-2A, CJ-3A, CJ-3B, CJ-5, CJ-6, CJ-7, Scrambler and Wrangler models	6817	**Toyota 1970-79 Spanish**	7467
		Toyota Celica/Supra 1971-85 Covers all U.S. and Canadian models	7043
Jeep Wagoneer, Commando, Cherokee, Truck 1957-86 Covers all U.S. and Canadian models of Wagoneer, Cherokee, Grand Wagoneer, Jeepster, Jeepster Commando, J-100, J-200, J-300, J-10, J20, FC-150 and FC-170	6739	**Toyota Trucks 1970-85** Covers all U.S. and Canadian models of pickups, Land Cruiser and 4Runner	7035
		Valiant/Duster 1968-76 Covers all U.S. and Canadian models	6326
Laser/Daytona 1984-85 Covers all U.S. and Canadian models	7563	**Volvo 1956-69** Covers all U.S. and Canadian models	6529
Maverick/Comet 1970-77 Covers all U.S. and Canadian models	6634	**Volvo 1970-83** Covers all U.S. and Canadian models	7040
Mazda 1971-84 Covers all U.S. and Canadian models of RX-2, RX-3, RX-4, 808, 1300, 1600, Cosmo, GLC and 626	6981	**VW Front Wheel Drive 1974-85** Covers all U.S. and Canadian models	6962
		VW 1949-71 Covers all U.S. and Canadian models	5796
Mazda Pick-Ups 1972-86 Covers all U.S. and Canadian models	7659	**VW 1970-79 Spanish**	7081
Mercedes-Benz 1959-70 Covers all U.S. and Canadian models	6065	**VW 1970-81** Covers all U.S. and Canadian Beetles, Karmann Ghia, Fastback, Squareback, Vans, 411 and 412	6837
Mereceds-Benz 1968-73 Covers all U.S. and Canadian models	5907		

Chilton's Repair & Tune-Up Guides are available at your local retailer or by mailing a check or money order for **$12.50** plus **$2.25** to cover postage and handling to:

Chilton Book Company
Dept. DM
Radnor, PA 19089

NOTE: When ordering be sure to include your name & address, book part No. & title.